CASSAVA

Edited by **Viduranga Waisundara**

Cassava
http://dx.doi.org/10.5772/intechopen.69424
Edited by Viduranga Waisundara

Contributors

Andri Frediansyah, Adriano Simões, Daniel Gomes Coelho, Moab Andrade, Domingos Mélo Neto, Aline Sousa, Kelem Silva Fonseca, Fred Brito, Rainiério Silva, Vincent Fondong, M. E. Chrissie Rey, Shadrack Kwadwo Amponsah, Ahmad Addo, Gangadharan Byju, Martin Anikwe, Ejike Ikenganyia, Paul Iji, Apeh A Omede, Emmanuel Ahiwe, Daniel Schwantes, César Ricardo Teixeira Tarley, Affonso Celso Gonçalves Jr., Marcelo Angelo Campagnolo, Andreia Da Paz Schiller, Jéssica Manfrin, Douglas Cardoso Dragunski, Andre Antoine Fanou, Amégnikin Valerien Zinsou, Kerstin Wydra, Robert Ndjouenkeu, Santhosh Mithra Velayudhan Santhakumari, Seena Radhakrishnan, Divya Lekshmanan, Luiz Jcb Carvalho, James Anderson, Songbi Chen, Chikelu Mba, Münevver Doğramaci, F Filho Josefino, Priscila G. Figueiredo, Suarau Oshunsanya, Viduranga Waisundara, Angela Alleyne, Sammy Aso, Simeon Achinewhu, Arthur Teixeira

Published by InTech
Janeza Trdine 9, 51000 Rijeka, Croatia

Publishing Process Manager Martina Usljebrka
Technical Editor SPi Global
Cover InTech Design team

First published January, 2018
Printed in Croatia
Legal deposit, Croatia: National and University Library in Zagreb

Additional hard copies can be obtained from orders@intechopen.com

Cassava, Edited by Viduranga Waisundara
p. cm.
Print ISBN 978-953-51-3740-5
Online ISBN 978-953-51-3741-2

For EU product safety concerns:
IN TECH d.o.o., Prolaz Marije Krucifikse Kozulić 3, 51000 Rijeka, Croatia,
info@intechopen.com or visit our website at intechopen.com.

Meet the editor

Dr. Viduranga Waisundara obtained her PhD degree from the Department of Chemistry, National University of Singapore in Food Science and Technology, in 2010. She was a lecturer at Temasek Polytechnic, Singapore, from July 2009 to March 2013. Following this, she relocated to her motherland Sri Lanka and spearheaded the Functional Food Product Development Project at the National Institute of Fundamental Studies from April 2013 to October 2016. She is currently a senior lecturer on a temporary basis, at the Department of Food Technology, Faculty of Technology, Rajarata University of Sri Lanka. Dr. Waisundara is a prolific writer with many research publications and articles in newspapers and magazines. She is also the current Global Harmonization Initiative (GHI) Ambassador for Sri Lanka.

Contents

Preface

Cassava (*Manihot esculenta* Crantz) has been used as a staple food by many nations. It is also known as manioc, yucca and tapioca. Cassava has the greatest conversion in terms of transforming solar energy into soluble carbohydrates per unit of area. Among the starchy staples, cassava gives a carbohydrate production that is about 40% higher than rice and 25% more than maize. In addition to the tremendous potentials of cassava to be improved agronomically and genetically, the crop is noted to be the highest yield crop under marginal conditions. It is grown widely in several countries in sub-Sahara Africa and Madagascar. Given these facts, it is an important crop when it comes to providing energy and nutrition, as well as a means of ensuring food security for many of the countries where it has been grown in abundance.

This book primarily focuses on three aspects: (1) cassava diseases, (2) improving cassava production and (3) postharvest processing of cassava. These are vital factors that determine the future of cassava production and its value as an agricultural crop. Biotechnological applications have been the primary focus when it comes to these three aspects. This is so because such interventions are necessary to assure sufficient production of cassava for the demand from the consumers. Harvesting plays a critical role in the cassava production value chain. Development of labour-saving technology for cassava harvesting has become the most critical challenge in cassava transformation worldwide.

I would like to extend my most sincere gratitude to the authors who have generously contributed chapters to this book, without whom this project would not have been a success. Also, I would like to give my heartfelt thanks to InTech Publishers with whom I have been working in quite a number of book projects of similar nature; the experiences have always been enjoyable, and I look forward to working with them in many more projects in the future as well. Last but not least, my appreciation goes to Ms. Martina Usljebrka, the Publishing Process Manager assigned to this book, who has rendered her utmost support in putting the materials together.

In conclusion, it is hoped that this book will be of value to both scientific and non-scientific communities to make informed choices about cassava, as well as to see it as a crop that would curb the incidence of food insecurity.

Dr. Viduranga Waisundara
Department of Food Technology,
Faculty of Technology,
Rajarata University of Sri Lanka,
Mihintale, Sri Lanka

Section 1

Introduction

Chapter 1

Introductory Chapter: Cassava as a Staple Food

Viduranga Y. Waisundara

Additional information is available at the end of the chapter

http://dx.doi.org/10.5772/intechopen.70324

1. Introduction

Cassava (*Manihot esculenta* Crantz) has been used as a staple food of many nations. It is also known as manioc, yucca and tapioca. Its origins lie in Latin America, where it was discovered by the indigenous Indian population more than 4000 years ago [1]. After its discovery by the European traders who came to Central America, the crop was taken to Africa as well as Asia for food security purposes and for the extraction of starch [1]. The plant can be grown throughout the year and is known to exist under severe climates, being particularly suited to conditions of low nutrient availability and able to survive drought [2]. Its tuber—the swollen root of the plant—is the most popular form of consumption, although the leaves are also consumed at times for medicinal purposes.

Cassava has the greatest conversion in terms of transforming solar energy into soluble carbohydrates per unit of area [3]. Among the starchy staples, cassava gives a carbohydrate production that is about 40% higher than rice and 25% more than maize [3]. Cassava also consists of essential micronutrients, such as vitamins A, B and C, iron and Zinc, even though it is considered not having a limited nutritional value [4]. It is a major source of carbohydrate for many populations, and it is the third largest source of carbohydrate in the world with Africa being the largest centre of production [4]. Cassava is increasingly popular with African farmers because of its agricultural advantages and potential to feed rapidly increasing populations. Nigeria is the largest producer of cassava out of all the African countries [5]. It is the third largest producer of cassava in the world after Brazil and almost double the production of Indonesia and Thailand [6]. It is noteworthy in this aspect that households under stress from HIV/AIDS are switching from high-input to low-input farming systems that involve cassava [7]. With these developments, cassava has undoubtedly been touted as one of the major crops around the world as a source of income as well as for food security purposes.

Bearing these in mind, this book primarily focuses on the contemporary importance of cassava as a crop that requires biotechnological interventions, improvements of postharvest management and farming practices. The subsequent subtitles in this introductory chapter provide brief overviews of aspects which may or may not be covered in the content chapters, so that the voids and gaps are filled.

2. Cassava processing

Freshly harvested cassava roots start deteriorating almost immediately after harvest. This is due to its high moisture content. Thus, the best form of preservation of cassava is drying into pellets or chips or processing into flour. The traditional methods of processing cassava roots into various types of food have been adapted to suit the many attributes of the plant such as root yield, spoilage, cyanide content, nutrient content and process ability [8]. Nevertheless, with increasing populations, indigenous methods of cultivation and processing of cassava have been transformed by modern scientific knowledge for use in industrial operations [8]. In this aspect, mechanization of cassava processing plays a pivotal role in removing the negative attributes of the traditional processing techniques and promoting timely large-scale processing of the tubers in hygienic environments [6].

Cassava processing operations are often preceded by peeling, and for this task, many models and mechanisms have been developed throughout the years [9, 10]. Apart from the peelers, various types of cassava-grating machines have also been developed [11, 12]. Other unit operations involved in cassava processing include dewatering of cassava pulp (mash), drying and frying, which are still majorly carried out manually [13, 14]. There are several factors nevertheless, to be considered before the usage and implication of technical devices for these unit operations. The survey by Quaye et al. [15] in Ghana revealed the following major aspects and considerations for adopting a new cassava processing technology: (1) affordability of the technologies in term of cost implication and profit margin, (2) efficiency of the machine, (3) labour required to operate the machine and (4) simplicity.

3. Cassava products

A wide range of products can be made from cassava, although it is commonly used as raw material for the food industry. The freshly peeled tubers are eaten as a vegetable after boiling or roasting. When boiled and pounded into a paste, the tubers are often added to soups and stews—which is called 'Fufu' in Nigeria [1]. It can also be consumed as sundried chips, which is known as 'Kokonte' in West Africa, and consumed after cooking or being ground into flour. Cassava flour is used in the preparation of bread, biscuits, confectionary, pasta and couscous-like products and in the production of adhesives [1]. The fermentation of cassava brings a new line of food products altogether.

Fermentation, either naturally or with selected microbial inoculums, has been extensively used to enhance the nutrient potentials of cassava and its by-products both for human and livestock consumption [16]. For the fermentation of cassava, two popular fermentation techniques, namely, the liquid substrate or submerged fermentation technique and the solid substrate fermentation are used [16]. The cassava roots, peels, leaves and pomace are the typically used parts of the plant, which are subjected to fermentation. The fermentation process has also played a significant role in the nutritional enhancement of the agro-industrial by-products generated through the harvesting and processing of cassava roots.

Apart from the food industry, cassava starch is used for textiles and the paper industry, and in the manufacture of plywood and veneer adhesives and glucose and dextrin syrups. Through fermentation, it can also be used for alcohol production, and as a waste material, it can be processed to biogas [17].

4. Nutritive and anti-nutritive properties of cassava

The composition of cassava, and thus its nutritional properties, depends on the specific tissue (root or leaf) being consumed. These aspects in turn, depend on several factors, such as geographic location, variety, age of the plant and environmental conditions [18]. Cassava roots are typically known to be the primary source of energy [11]. The leaves on the other hand provide protein, vitamins and minerals [19]. However, cassava roots and leaves are deficient in the sulphur-containing amino acids, methionine and cysteine, and some nutrients are not optimally distributed within the rest of the plant's physiology [20].

Cassava also contains its own share of anti-nutrients, which have either positive or negative effects on the health, depending upon the amount of the component being ingested [21]. They basically interfere with the digestibility and uptake of some nutrients. Nevertheless, depending on the amount consumed, these substances can also bring benefits to humans. Cyanide is the most toxic factor restricting the consumption of cassava roots and leaves. Several health disorders and diseases have been reported in cassava-eating populations, owing to the presence of improperly processed cyanide [18]. The consumption of lower cyanide amounts is not lethal but long-term intake could cause severe health problems such as tropical neuropathy [22]. The nitrate content in cassava leaves ranges from 43 to 310 mg/100 g DM (dry matter) [21]. Cassava-eating populations are naturally exposed to high amounts of cyanide, nitrates and nitrites—chemical compounds which are known to contribute to the risk of developing stomach cancer [22]. Cassava-eating individuals tend to have a high amount of thiocyanate in the stomach due to cyanide detoxification by the body, which may catalyse the formation of carcinogenic nitrosamines [18, 21, 22].

5. Cassava for ensuring food security

Food security has become a growing concern around the world. Coupled with inadequate caloric intake, food insecurity is a major cause of death and morbidity in the world, particularly in developing countries [23]. The major staples of rice, wheat, maize and soybean are now recognized as not being the complete solution to world food security [24]. Diversification of farming of agricultural crops and food production has been recognized as a need, extending towards coarse grains, roots and tubers, pulses and oilseeds [24]. In this aspect, cassava has been recognized as a crop that is able to address the global food security needs around the world. It has been biotechnologically manipulated for better growth and higher crop production for this purpose.

Good yield progress has been achieved for cassava crops after relatively few decades of genetic improvement compared with other staples, which are being bred and harvested for food security purposes [24]. Adoption of new varieties of cassava has been strong in Thailand, Vietnam and Nigeria [25–27]. Given the current practice of minimal use of inputs, great scope also exists for closing the large yield gap of cassava production through better agronomy [24]. For this purpose, commercialisation of the cassava cultivation in Sub-Saharan Africa should help close the gap by providing stimulus for farmers to invest in more inputs [28].

6. Postharvest deterioration of cassava roots

Given the marginal environments where cassava is grown, its postharvest processing is frequently affected by large distances to the processing centres and deficient transport infrastructure, specifically roads [29]. Cassava roots are also bulky, containing approximately 65% water, which leads extensively to the postharvest physiological deterioration (PPD) [29]. The short shelf life of the roots hinders many of the marketing options by increasing the likelihood of losses and thereby increasing the overall marketing costs [29]. In addition, the access to urban markets and processing facilities is restricted to production sites that are relatively close to them [30, 31].

Research to date concerning the study of PPD has mostly focused on biochemical signalling events several hours after harvest [32]. Upon examination of physiological and biochemical changes occurring after cassava root detachment, changes in the nature and type of volatile compounds emitted, secondary metabolites accumulated, and changes in the expression of key genes in reactive oxygen species (ROS) turnover had been primarily observed [33, 34]. Nevertheless, based on combined proteomics data, enzymatic activities, and lipid peroxidation assays, Vanderschuren et al. for instance [35] have identified glutathione peroxidase as a candidate for reducing PPD. Further, in this study, transgenic cassava overexpressing a cytosolic glutathione peroxidase in storage roots showed delayed PPD and reduced lipid peroxidation as well as decreased hydrogen peroxide accumulation [35].

7. Conclusions

Cassava continues to grow as a crop of importance around the world for curbing food security issues as well as a means of income and livelihood. While its versatility as not only a food source for humans but also for animals, as well as a means of biofuel has been recognized, further research needs to be targeted towards the biofortification of cassava, so that the plant can be advocated as a contributor of essential vitamins and minerals.

As a crop which is heavily subjected to biotechnological interventions, transgenic interferences looking into the clonal propagation of crops have the potential to accelerate product development and address genetic constraints which may impede traditional breeding approaches. This could be considered as a vital approach for promoting cassava as a means of preventing food security issues. To be successful, however, as highlighted by Vanderschuren et al. [35], crop biofortification programs must develop integrated management practices by which molecular biologists, breeders, agronomists, nutritionists, educators, economists, farmers and consumers are all engaged in product development and delivery.

As an objective of this book, it is hoped that readers would see the importance of cassava, its research and cultivation aspects as a vital means of livelihood for feeding the global population, which is likely to grow in the subsequent years. As a crop, it has many applications, thus, having the ability to generate revenue and income for developing countries. It is hoped that cultivation of this crop will be seen as a positive means of agriculture, and its existing agricultural, postharvest and processing issues will gain rapid attention for remediation.

Author details

Viduranga Y. Waisundara

Address all correspondence to: viduranga@gmail.com

Technology Degree Programme, Faculty of Applied Sciences, Rajarata University of Sri Lanka, Mihintale, Sri Lanka

References

[1] Akinpelu AO, Amamgbo LEF, Olojede AO, Oyekale AS. Health implications of cassava production and consumption. Journal of Agriculture and Social Research. 2011;**11**(2):118-125

[2] Burrell MM. Starch: The need for improved quality or quantity – An overview. Journal of Experimental Botany. 2003;**54**(382):451-456

[3] Tonukari NJ. Cassava and the future of starch. Electronic Journal of Biotechnology. 2004;**7**(1). DOI: 10.4067/S0717-34582004000100003

[4] Adenle AA, Aworh OC, Akromah R, Parayilet G. Developing GM super cassava for improved health and food security: Future challenges in Africa. Agriculture and Food Security. 2012;**1**:1-15

[5] Egesi C, Mbanaso E, Ogbe F, Okogbenin E, Fregene M. Development of cassava varieties with high value root quality through induced mutations and marker-aided breeding. Umudike Annual Report. 2006;**2006**:2-6

[6] Oriola KO, Raji AO. Trends at mechanizing cassava postharvest processing operations. International Journal of Engineering and Technology. 2013;**3**(9):879-887

[7] FAO. The Impact of HIV/AIDS on the Agricultural Sector [Internet]. 2008. Available from: www.fao.org/DOCREP/005/Y4636E/y4636e05.htm [Accessed: 30 June 2017]

[8] Falade KO, Akingbala JO. Utilization of cassava for food. Food Reviews International. 2010;**27**(1):51-83

[9] Agbetoye LAS. Improving the technology of cassava harvesting and processing mechanization for food security in Nigeria. In: Proceedings of the International Conference on Science and Technology; August 14-19, 2005; Federal University of Technology; Akure, Ondo State, Nigeria. 2005. pp.196-206

[10] Akande FB, Adebayo AO, Busari RA. Design, fabrication and testing of manually operated cassava. Proceedings of the Nigerian Institution of Agricultural Engineers. 2008:1-14

[11] Kolawole OP, Agbetoye LAS. Engineering research to improve cassava processing technology. International Journal of Food Engineering. 2007;**3**(6):1-2

[12] Kolawole OP, Agbetoye LAS, Ogunlowo AS. Strength and elastic properties of cassava tuber. International Journal of Food Engineering. 2007;**3**(5):1-10

[13] Kolawole OP, Agbetoye LAS, Ogunlowo AS. Cassava mash dewatering parameters. International Journal of Food Engineering. 2007;**3**(1):4

[14] Kolawole OP, Agbetoye LAS, Ogunlowo AS. Sustaining world food security with improved cassava processing technologies: The Nigerian experience. Sustainability. 2010;**2**:3681-3694

[15] Quaye W, Gayin J, Yawson I, Plahar WA. Characteristics of various cassava processing methods and the adoption requirements in Ghana. Journal of Root Crops. 2009; **35**(1):59-68

[16] Aro SO. Improvement in the nutritive quality of cassava and its by-products through microbial fermentation. African Journal of Biotechnology. 2008;**7**(25):4789-4797

[17] Lawrence K, Anadajayasekeram P, Ochieng C. A synthesis/lesson-learning study of the research carried out on root and tuber crops commissioned through the DFID RNRRS research programmes between 1995 and 2005. 2006. Crop Protection Programme (CPP) of the UK Department for Internal Development (DFID), East Kilbride, UK; R1182

[18] Montagnac JA, Davis CR, Tanumihardjo SA. Nutritional value of cassava for use as a staple food and recent advances for improvement. Comprehensive Reviews in Food Science and Food Safety. 2009;**8**(3):181-194

[19] Tewe OO, Lutaladio N. Cassava for Livestock Feed in Sub-Saharan Africa. Rome, Italy: FAO; 2004

[20] Montagnac JA, Davis CR, Tanumihardjo SA. Processing techniques to reduce toxicity and antinutrients of cassava for use as a staple food. Comprehensive Reviews in Food Science and Food Safety. 2009;**8**:17-27

[21] Wobeto C, Corrêa AD, De Abreu CMP, Dos Santos CD, Pereira HV. Antinutrients in the cassava (Manihot esculenta Crantz) leaf powder at three ages of the plant. Science Technology Alimentaires. 2007;**27**:108-112

[22] Ernesto M, Cardoso AP, Nicala D, Mirione E, Massaza F, Cliff J, Haque MR, Bradbury JH. Persistent konzo and cyanide toxicity from cassava in northern Mozambique. Acta Tropica. 2002;**82**:357-362

[23] Sayre R, Beeching JR, Cahoon EB, Egesi C, Fauquet C, Fellman J, Fregene M, Gruissem W, Mallowa S, Manary M, Maziya-Dixon B, Mbanaso A, Schachtman DP, Siritunga D, Taylor N, Vanderschuren H, Zhang P. The BioCassava plus program: Biofortification of cassava for sub-Saharan Africa. Annual Reviews in Plant Biology. 2011;**62**:251-272

[24] Fischer T, Byerlee D, Edmeades G. Crop yields and Global Food Security: Will Yield Increase Continue to Feed the World? 1st ed. Australia: Grains Research & Development Corporation; 2012. p. 634

[25] Dixon AGO, Akorada MO, Okechukwu RU, Ogbe F, Ilona P, Sanni LO, et al. Fast track participatory approach to release of elite cassava genotypes for various uses in Nigeria's cassava economy. Euphytica. 2008;**160**:1-13

[26] El-Sharkawy MA. Physiological characteristics of cassava tolerance to prolonged drought in the tropics: Implications for breeding cultivars adapted to seasonally dry and semiarid environments. Brazilian Journal of Plant Physiology. 2007;**19**:257-286

[27] Fermont AM, van Asten PJA, Tittonell P, van Wijk MT, Giller KE. Closing the cassava yield gap: An analysis of smallholder farms in East Africa. Field Crops Research. 2009;**112**:24-36

[28] Okechukwu RU, Dixon AGO. Genetic gains from 30 years of cassava breeding in Nigeria for storage root yield and disease resistance in elite cassava genotypes. Journal of Crop Improvement. 2008;**22**:181-208

[29] Morante N, Sánchez T, Ceballos H, Calle F, Pérez JC, Egesi C, Cuambe CE, Escobar AF, Ortiz D, Chávez AL, Fregene M. Tolerance to postharvest physiological deterioration in cassava roots. Crop Science. 2010;**50**:1333-1338

[30] Reilly K, Han Y, Tohme J, Beeching JR. Isolation and characterisation of a cassava catalase expressed during post-harvest physiological deterioration. Biochimica et Biophysica Acta. 2001;**1518**:317-323

[31] Pfeiffer WH, McClaferty B. HarvestPlus: Breeding crops for better nutrition. Crop Science. 2007;**47**:S88-S105

[32] Iyer S, Mattinson S, Fellman JK. Study of the early events leading to cassava root postharvest deterioration. Tropical Plant Biology. 2010;**3**:151-165

[33] Reilly K, Gómez-Vásquez R, Buschmann H, Tohme J, Beeching JR. Oxidative stress responses during post-harvest physiological deterioration. Plant Molecular Biology. 2003;**53**:669-685

[34] Reilly K, Bernal D, Cortés DF, Gómez-Vásquez R, Tohme J, Beeching JR. Towards identifying the full set of genes expressed during cassava post-harvest physiological deterioration. Plant Molecular Biology. 2007;**64**:87-203

[35] Vanderschuren H, Nyaboga E, Poon JS, Baerenfaller K, Grossman J, Hirsch-Hoffman M, Kirchgessner N, Nanni P, Gruissem W. Large-scale proteomics of the cassava storage root and identification of a target gene to reduce postharvest deterioration. The Plant Cell. 2014;**26**:1939-1924

Section 2

Cassava Diseases

Chapter 2

Cassava Bacterial Blight: A Devastating Disease of Cassava

André Antoine Fanou,
Valerien Amégnikin Zinsou and Kerstin Wydra

Additional information is available at the end of the chapter

http://dx.doi.org/10.5772/intechopen.71527

Abstract

Cassava (*Manihot esculenta* Crantz) with its long life cycle is affected by several diseases of which cassava bacterial blight (CBB) is the major bacterial disease in the cassava belt worldwide. The epidemiological and ecological investigations undertaken on the disease showed that the causal agent, the bacterium *Xanthomonas axonopodis* pv. *manihotis* (Xam), possesses several means for survival and dissemination that may play an important role as inoculum sources for the infection when favorable conditions occur, and the subsequent damage of the plant causing severe yield losses. In fact, Xam survives epiphytically on some weeds occurring in and around cassava fields without developing blight symptoms. Investigating the survival period over the seasons, a longer survival exceeding 5 months has been observed in non-decayed cassava debris. Also, some insects in cassava field like the variegated grasshopper (*Zonocerus variegatus*) vehicles the pathogen for some time. Over seasons Xam also survives often latently, in cassava stems which are then used for establishing new plantations. In regional disease surveys across ecozones in West Africa, no zone of preference has been found. Though, comparing the development of the disease and the damages caused in yield loss trials in two agro-eco-zones over 2 years, CBB was more pronounced and caused higher yield and biomass losses in the forest savannah transition zone than in the dry savannah where symptom development was positively correlated with the rainfall patterns. The detailed knowledge of the epidemiology, disease development, survival and dissemination, of the reaction of cassava varieties towards CBB such as physiological resistance mechanisms, identification of genetic resistance (QTL) and the background of observed field resistance as well as of the influence of planting time and cropping pattern allows to recommend integrated management measures such as sanitation, intercropping, removal of diseased leaves, management of planting dates according to ecozone, soil amendments, use of resistant genotypes.

Keywords: cassava, cassava bacterial blight, *Xanthomonas axonopodis* pv. *manihotis*, disease

1. Introduction

Cassava bacterial blight (CBB) was first reported in Brazil [1] and later observed in several countries of the cassava production belt worldwide [2–8]. A diagnostic survey in Africa (Ghana, Benin, Nigeria and Cameroon) revealed that CBB is present in all ecozones, but with variable incidence and severity [9, 10]. It is the second most devastating disease after Cassava Mosaic Virus Disease Complex and may cause more damage to the crop than any other bacterial disease. The disease causes losses of fresh roots and also of planting material [11, 12]. Root yield reduction level may vary with the susceptibility of cassava cultivars, the climatic conditions and the inoculums pressure. The poor yield of storage roots due to severe outbreaks of the disease can affect the population as well as the livestock in areas where cassava is a major staple food. A low accumulation of starch in the roots due to CBB was observed [13]. Under favorable ecological conditions, wilting of leaves and leaf fall due to CBB can be high. This loss of leaves can affect the availability of leafy vegetables for humans and reduces cash income in communities where cassava leaves are sold. As CBB affects systemically cassava stems, this leads to shortages in the supply of healthy (bacteria-free) planting materials. The causal agent is a Gram-negative bacterium of the genus *Xanthomonas*. It was first named *Bacillus manihotis* Arthaud-Berthet, then *Phytomonas manihotis* (Arthaud-Berthet and Bondar) Viegas, later *Xanthomonas manihotis* (Arthaud-Berthet) Starr, and then *Xanthomonas campestris* pv. *manihotis* (Berthet and Bondar [14]). Two decades ago, on the basis of genotypic investigations, Vauterin et al. [15] proposed a reclassification of *Xanthomonas*, renaming the CBB pathogen *Xanthomonas axonopodis* pv. *manihotis* (Xam). The cells of Xam are motile and have polar, monotrichous flagellation. The strains do not produce a yellow pigment on sugar containing media, which is exceptional for the genus *Xanthomonas*, which normally grows in yellow colonies. The colonies of the strains on agar are mucoid, convex and round. Xam is an obligate aerobic bacterium and uses oxygen as a terminal electron acceptor [16]. It grows optimally between 25 and 30°C. Its development is favored between pH 6.5 and 7.2. The pathogen causes various symptoms. A study on microflora of cassava leaves revealed the presence of the pathogen on apparently healthy leaves collected from fields that in which some plants showed CBB symptoms (canker with exudates on stems). With the begin of the rainfall at the end of the dry season, this residual epiphytic population of Xam [17, 18] multiplies and penetrates the leaves' tissues through epidermal wounds and through natural openings like stomata. After few days to 1 week, the first symptoms are visible as translucent water-soaked spots when observed against the light. These translucent spots on the abaxial surface of the leaves become angular dark green spots limited by veins and are irregularly distributed on the lamina. Later, the spots enlarge, neighboring spots join together to form large brown patches. In the lesions, droplets of creamy white exudates that become yellow are observed. These exudates are also visible on stems, and often on leaf petioles under high air humidity. The following days, the affected parts of leaves coalesce and show, including also the leaf tips a superficially burnt appearance, the blight symptoms. The leaf blight is due to production of toxins by the bacterium, such as 3-(methylthio) propionic acid [19], tiglic acid, phenylacetic acid cyclopentanecarboxylic acid [20]. From the leaves, the bacteria move systemically into the petiole and stem and continue to multiply discontinuously throughout the plant, blocking the movement of water and nutrients in the vascular system of the woody stem and inducing leaf wilting. Petioles of wilted leaves typically remain attached horizontally to the main stem axis for a while, before the base of the petiole collapses. Symptomatically, this is a typical symptom, differentiating CBB from leaf wilt caused by anthracnose disease (*Colletotrichum gloeosporioides*), where wilted leaves and petioles hang downwards directly from the stem. Progressively, wilted leaves fall causing defoliation of the shoot tip. Finally, the non-lignified soft tissue at the top of the growing shoot dies giving plants a characteristic candle stick symptom or tip dieback. Newly growing shoots at the lower stem part or the stem basis also start wilting and soon show tip dieback. Due to the systemic nature of the disease, a characteristic brownish discoloration of the vascular system can easily be observed in stems. In newly planted fields, primary CBB symptoms are the wilting of the young germinating sprouts shortly followed by tip dieback right after infected cuttings have been planted. Field observations during surveys in cassava production areas in Africa revealed that the disease is more spread and more devastating in the savannah and the forest savannah transition zone than in the dense forest zone [9, 21], while later surveys showed an increased disease pressure also in rain forest zones [22]. The cycle of CBB is characterized by two phases, a parasitic phase during the rainy season and a survival phase during the dry season [23]. During the survival phase, the pathogen survives in apparently healthy stems and as epiphyte on leaves. During the parasitic phase, the bacteria multiply with the beginning of the rainy season on the leaves, and later symptoms occur on aerial parts of the cassava plant. The symptoms development is favored by rainfall, high temperature, high relative humidity, occurrence of insect vectors and wounds on the leaves, as well as high differences between day and night temperatures. The disease causes variable harvest losses depending on the cassava variety's susceptibility, the virulence of the strains of Xam and the environmental conditions. The CBB pathogen can be disseminated by several means which serve as inoculums sources.

2. Epidemiological investigations and disease management

2.1. Potential sources of inoculums and their implication in the epidemiology

Cassava bacterial blight appears suddenly in a newly planted cassava fields as well as in old, established plantations after the end of the dry season. This sudden apparition of the disease has led to numerous studies on the means and times of survival and dissemination of Xam within and between cassava plantations under variable environmental conditions. These studies are very important to understand the epidemiology of the disease.

The vegetative propagation using cuttings of mature stems is the almost exclusive method used for producing cassava. The cuttings used by farmers to establish a new plantation are habitually collected from fields of the previous season and are mostly not free of diseases. Cassava stems are often infected by Xam [24–30]. The pathogen has been detected in cassava plants using indirect immunofluorescent technique [28]. Following the distribution of the pathogen in the vascular system, these authors concluded that the distribution is discontinuous. In order to develop sanitation measures in areas where the disease is prevalent, the distribution of Xam in the cassava stems (upper part, middle part, basal part and lateral branches) of resistant, medium resistant and susceptible varieties in relation to the ecozone and the strain of the Xam has been investigated by selecting 24 cassava varieties classified in susceptible, semi-susceptible or resistant from previous screening trials [31]. Xam has been detected in the three categories of cassava varieties and the pathogen is present in the upper part, the middle part and the basal part of the varieties. It has also been observed that Xam colonized the whole stem, or that the colonization was discontinuous. No preferential zone of pathogen concentration in stems was found [31]. The high stem infection observed in the susceptible variety BEN 86052 (64%) in comparison to the resistant I30572 (36%) support the results of Kpémoua [32] reporting that the tissues of susceptible cultivars are more favorable to the systemic invasion. In these plants, the mechanisms of protection like deposit of tyloses developed tardily and also lytic pockets were formed around the protoxylem and extended to the phloem and cortex in the susceptible cultivar [33]. On the contrary, in the resistant cultivars, the tyloses appeared early and differentiated specifically with production of phenolic compounds which slowed down the multiplication and the evolution of the pathogen in the tissues [30]. Cicatrisation tissue forms around the lytic pockets in the cortical parenchyma and in the phloem [33]. However, our studies indicated an average stem infection with Xam of 33% in the xylem of cassava varieties which were presumed to be resistant to CBB, and derived from a high rainfall region, whereas, this stem infection was 53% for the semi-resistant varieties and 57% for the susceptible varieties. The infection of the xylem also of the resistant clones may be influenced by the climatic conditions and the high virulence of the inoculated strain, closely linked to the non-formation of cicatrisation tissue in the cortical parenchyma and in the phloem during the interaction of Xam with the plant. During the vegetation period, no entirely dry month was recorded, the average monthly temperature ranged between 25 and 29°C and the relative humidity between 59 and 85%. These conditions have certainly favored a rapid multiplication of the pathogen. Considering the aggressiveness of the strain GSPB 2511, the mechanisms of resistance of the plant may have been overcome by the pathogen and an accumulation of Xam cells in the basal part of the stem of resistant clones and distribution in the xylem of the whole plant occurred. The detection of Xam in stems of the variety BEN 86052 without dieback, demonstrates that apparently healthy plants can lodge the bacterium, which is a potential risk for the selection of "healthy" cuttings for the next plantation. On the other hand, although we also did not detect the pathogen in cuttings from plants without dieback of the variety I30572, it does not necessarily indicate that an apparently healthy resistant plant will be completely Xam-free. Considering the continuous or/and discontinuous distribution of Xam in cassava stems, all attempts to get pathogen-free cuttings by selecting some apparently healthy plants from a contaminated field will not be reliable. The systemic colonization permits a preservation of the pathogen through the unfavorable dry season to the next cropping season. The plant pathogenic bacteria can survive inside the host plant for over 1 year [34]. Thus, the survival of Xam in cassava tissues and especially stem cuttings used to establish a new cassava field plays an important role in the epidemiology of CBB. The use of infected cuttings is the most important means of "continual" survival and spread of the pathogen from one cropping season to the next one and from region to region [35, 36]. Eighty-six percent of young plants deriving from cuttings originating from infected cassava fields developed cassava bacterial blight symptoms [2] proving that the primary symptoms of CBB derive from infected cuttings. Using Xam-contaminating cuttings to establish cassava plantation affects seriously the root yield. Comparing the yields of cassava plots planted with Xam-free cuttings and infected cuttings, Otim-Nape [37] obtained a reduction in fresh root yield from 40.1 to 26.6 t/ha.

Several early investigations have shown that Xam survives on some weeds in cassava fields, while also seemingly contradictory reports that Xam does not have alternative hosts [38, 39], or on the possible existence of alternative hosts for Xam were published [40, 41]. To confirm or infirm one these reports, experiments under field and glasshouse conditions were conducted at the International Institute of Tropical Agriculture (IITA), Benin station. After spray-inoculation of cassava fields with an Xam strain marked by streptomycin and rifampicin for easier detection, the occurring weeds (*Brachiaria deflexa, Cassia mimosoides, Commelina benghalensis, Cyathula prostrata, Dactyloctenium aegyptium, Digitaria horizontalis, Euphorbia heterophylla, Mariscus alternifolius, Mucuna cochinchinensis, Physalis angulata, Pupalia lappacea, Solanum nigrum, Talinum triangulare, Tridax procumbens, Vernonia cinerea*) in this field were sampled weekly and tested for the survival of Xam [42]. The number of weeds harboring the pathogen increased gradually and reached 73% 2 weeks after artificial spray-inoculation of cassava plants. Some weeds lodged a high epiphytic population of the marked Xam strain, but the survival period from the spray-inoculation did not reach 37 days [42]. Typical CBB symptoms were never observed on any of the tested weed species. During the experiment, the marked Xam has been never detected on *V. cinerea, M. cochinchinensis, C. mimosoides* and *C. benghalensis*. In the glasshouse, 13 of these weed species (except *S. nigrum* and *C. prostrata*) transplanted in pots have been infiltration-inoculated with a highly virulent strain marked with resistance against streptomycin and rifampicin to determine whether the infiltrated leaves would develop similar CBB symptoms. The pathogen proved to be present on all tested weed species up to 25 days and multiplied on these weeds except *P. lappacea* [42]. Contrarily to field experiment during which Xam has not been detected on four weed species, three of these species (*C. mimosoides, M. cochinchinensis* and *V. cinerea*) harbored the infiltrated pathogen during at least 25 days post inoculation with a long survival period on or in *V. cinerea* that reached at least 60 days [42]. The survival of Xam did not reach up to 2.5 months in or on any of the weed species, and CBB symptoms were not observed.

Under field conditions as well as glasshouse conditions, Xam survived epiphytically on weeds without developing CBB symptoms. Also various other authors reported an epiphytic survival of Xam on cassava plants or on weeds [17, 18, 43–46]. The bacteria obviously survive and multiply without causing apparent damage to the weeds leading to the confirmation that Xam does not have alternative host as it had also been concluded by Ikotun [38] and Amusa et al. [39] during their studies hosts. In contrast, *Manihot glaziovii*, variegated ornamental cassava, *Euphorbia pulcherrima* and *Pedilanthus tithymaloides* [40], *Amaranthus* species, *Panicum fasciculatum, Sida* species, *Sorghum halepense* and several species belonging to the Euphorbiaceae in Venezuela [41] have been identified as possible alternative hosts for Xam. The duration of the survival varied greatly depending on the weed species and the bacterial strain. After spray-inoculation the pathogen survived only for 12 days in *Euphorbia repanda* (Euphorbiaceae), 7 days in *Ricinus communis* (Euphorbiaceae), 5 days in *Phaseolus vulgaris* (Leguminoseae), *Nicotiana tabacum* (Solanaceae), *Lycopersium esculentum* (Solanaceae) and *Physalis angulata*

(Solanaceae), and 3 days in *Amaranthus dubius* (Amaranthaceae) [38], while Fanou et al. [42] obtained survival up to 45 days on species of the family Euphorbiaceae and Solanaceae and up to 30days on species of the family Amaranthaceae. The maximal survival period of Xam corresponded to the vegetative cycle of the annual weeds studied. Thus, Xam could not be detected when the weeds reached the end of their growth cycle and dried, and therefore we conclude, that the survival of Xam on or in weeds may play an important role in the spread of CBB during the growing season. In the epidemiology of foliar pathogens, survival of cells on non-host plants, especially weeds, may have a far reaching significance. The role of weeds as inoculum sources for phytopathogens [47] and generally for xanthomonads [48] for disease development on susceptible hosts has been suggested in several cases. In cassava growing areas, weeds are most of the time found close to and between cassava fields. These weeds are the habitat for a wide range of insects (Orthoptera, Coleoptera, Diptera, etc.), for certain animals, and serve as niche for insect-feeding birds during the rainy season. The movement of men, insects, birds and animals through contaminated weeds and cassava plantations, especially during or after rain or in the early morning, may contribute to pathogen spread. Strong winds or wind-driven rains may transport the bacteria from weeds to cassava plants, within and among cassava plantations, additionally causing wounds on leaves and thereby increase the entrance points for the bacteria.

Cassava debris is another sources of inoculum. During the plant vegetation, cassava leaves fall and remain as debris on the soil for an extended period. Especially varieties highly susceptible to CBB shed their infected leaves. The survival time of Xam on and in these infected leaves and the role of infected debris on the perpetuation of the disease are questions of interest in the epidemiology of CBB. Survival experiments of Xam in debris under controlled conditions [49], under field conditions [28, 50] and when the debris are buried during the dry season [51] have been conducted, but details on the survival period of the pathogen and trials on infected buried debris during the rainy season lacked. Thus, studies have been undertaken to determine the survival of marked Xam strain with resistance against streptomycin and rifampicin under various ecological conditions in and on leaves on the soil surface and when leaves are buried [42]. Under field conditions, the survival period of Xam varies and is negatively correlated with the rainfall. With increasing rainfall, the survival period of Xam in debris laid on the soil surface, slightly covered and buried at 25cm to 30cm, reduces [42]. Also, the survival period depends on where the debris was located. The population of Xam in debris decreased more rapidly and reached zero when the debris were buried than when they were left on the soil surface [42]. Under glasshouse conditions, a long survival period up to 5 months was obtained when the debris have been kept in dry condition [42]. The short survival period of Xam in slightly covered or buried debris recorded by Fanou et al. [42] is similar to those obtained by Thaveechai et al. [52] who reported survival of Xam for 21–49 days in infected cassava tissues buried in the soil under field conditions of Thailand. However, a long survival period of 60days was observed when infected cassava debris were buried in 10 cm soil depth under field conditions during the dry season [51]. The survival of the pathogen in debris on the soil surface with high CFU counts compared to the covered or buried debris obtained by Fanou et al. [42] confirmed the findings of Ikotun [50] who observed that the survival of Xam is restricted to debris on the soil surface and the upper 5 cm of soil. It can be concluded that rainfall and soil humidity as well as the depth of leaves buried in the soil play a decisive role in the decay of the debris and high rainfall and soil humidity as well as leaves deeper buried in soil contribute to the short survival of Xam in debris. Under dry conditions in the glasshouse, Xam survived longer than 5 months. These findings corroborate the results recorded by Persley [49] that Xam survived for up to 180 days in debris in soils kept at 30°C in the laboratory under dry conditions. Other authors reported even longer survival times: the pathogen survived for up to 1 year without losing its pathogenicity in highly contaminated cassava debris kept in the laboratory at 25°C and at 70% relative humidity [23, 28], for more than 30 months in dried infected cassava stems [25] and for up to 22 months or even several years under dry conditions at room temperature [53] own observations.

It is concluded from these experiments, that in highly contaminated cassava plantations, infected cassava leaves falling at the end of the rainy season may conserve the pathogen during the 5 to 7 months of the dry period and constitute an inoculum source for the new cropping season, while infected leaves falling during the rainy season can contribute to the dissemination in the field. Wind-driven rain and water splash may transfer the bacteria from infected plant debris to new cassava plantations. Thus, in fields where successive cassava plantings are practised, the infected debris on the soil surface may favor the initial infection of lower leaves of newly grown plants in close proximity to the soil surface.

Most of the insects that are associated to cassava during its long growth cycle, are feeding on cassava leaves. Especially the leaves infected by CBB are preferred by the grasshopper *Zonocerus variegatus* [54, 55]. Several studies and field observations reported *Z. variegatus* as vector of plant diseases. According to Refs [56, 57] *Z. variegatus* transmits okra mosaic disease with an efficacy of 10% and cowpea mosaic disease with an efficacy of 19%, respectively. Terry [36] suggested a probable role of *Z. variegatus* in the transmission of CBB. Forty percent of the insects collected on infected cassava plants, lodged the pathogenic bacteria in their alimentary canal and faeces [58]. Studies on the transmission of Xam to cassava plants by *Z. variegatus* have been conducted under glasshouse conditions where healthy cassava plants on which Xam-contaminated *Z. variegatus* had fed developed CBB symptoms [54]. Therefore, *Z. variegatus* is supposed to be involved in the survival and transmission of Xam. Likewise, studies have been initiated to determine whether *Z. variegatus* may be involved in the dissemination of CBB during the rainy season [31]. After infecting in cage cassava plants with an Xam marked strain with resistance against streptomycin and rifampicin, *Z. variegatus* have been released on these plants for 1 week. Then, the insects were transferred on healthy cassava plants in another cage where someday later, CBB symptoms have been observed. Dissecting the insects after the digestion of infected leaves and analyzing the faeces, Xam was accumulated in the faeces which lodged more bacteria than the mandibles, legs and the alimentary canal [59]. The locomotion organs of the insect always carried the pathogen. When the insects were fed exclusively on infected leaves in a cage in the glasshouse, the pathogen was found in a greater number on all the organs and in the faeces than when the insects were fed on infected plants in the field. In both cases, the number of bacteria per organ varied according to the organ as follows: faeces > alimentary canal > legs > mandibles [31]. Also, Daniel et al. [58], Daniel and Boher [28], Bani [54] and Zandjanakou-Tachin et al. [59] detected Xam on the exoskeleton (wash water) in the digestive system and in faeces of *Z. variegatus* collected from infected cassava fields and in the insects gut using immunofluorescence microscopy [59]. When insects contaminated by Xam have been transferred onto healthy plants under glasshouse conditions in order to determine the survival time of the bacteria on or in the organs, all the organs and the faeces carried a high number of the pathogen on the transfer day. After 1 week, the bacteria were no longer detected on the mandibles, on the legs and in the peritrophic membrane, but some bacteria survived in the faeces, and few bacteria have been also found in the alimentary canal. Two weeks after transfer, living bacteria were no longer detectable on or in the insects or faeces [31, 59]. The limited survival time on mandibles, legs (less than 1 week), in the digestive system and faeces (less than 2 weeks) indicated that Xam did not multiply on or in these organs and in faeces. In contrast, Bani [54] detected Xam in the alimentary canal 2 months after feeding of *Z. variegatus* on infected plants, and Daniel and Boher [28] suggested that Xam could survive and multiply in the alimentary canal when Xam was detected on *Z. variegatus* during the dry period when no CBB lesions were observed on the leaves. When Xam-infested faeces were placed on scarified cassava leaves, on leaves wounded with holes, or on the adaxial and abaxial surface of intact leaves and the plants were kept in the glasshouse, angular leaf spots were observed on the scarified leaves and on the border of the holes of wounded leaves 5 days later. Symptoms appeared after 7 days on the abaxial as well as on the adaxial surface of intact leaves [31]. The development of CBB symptoms in the glasshouse after deposing infested faeces on cassava leaves, proved for the first time the transmission of Xam by *Z. variegatus*. The development of symptoms was especially favored by wounds. Eighty to hundred percent of wounded leaves showed angular leaf spots which developed to blight and later wilting of the leaves, whereas only 13.3 and 32.7% of the leaves showed symptoms when faeces were placed on the adaxial and abaxial surface of intact leaves, respectively [31]. However, Bani [54] did not obtain symptoms when Xam contaminated faeces were deposited in water drops on intact cassava leaves which may have been due to other environmental conditions. In the cassava field, *Z. variegatus* defecates on the adaxial surface of the leaves or on the soil surface, where faeces are moistened by rains or dew and a multiplication of Xam cells may be initiated. Rain splashing and wind could transport the cells to the lower and upper leaves. Also rain droplets could run down from the adaxial surface, reach the under-surface of the same leaf containing more stomata for bacterial entrance and may cause the development of symptoms.

2.2. Epidemiology and yield loss

Several mechanisms could be implicated in passive dispersal of Xam. These include mainly planting materials, weeds, soil debris, insects etc. The pathogen cells have been isolated from all of these sources which therefore may play a great role in the dissemination of the disease.

Epidemics can start from infected cassava cuttings which can act as an effective long distances dispersal mechanism when the infected cuttings coming from another region are used to install a new plantation. Xam can also be disseminated long distances by contaminated true seeds. Even though producers do not use true seeds to establish their cassava fields, seeds are widely used by cassava breeders to maintain and improve the germplasm and for the exchange of genotypes between countries and continents. The presence of the causal agent of CBB on and in cassava seeds was reported [28, 60–62]. It appears that cassava seeds are an

http://dx.doi.org/10.5772/intechopen.71527

inoculum source and can contribute to the dissemination of the disease. The pathogen survives epiphytically and multiplies on many weed species that are found in or close cassava fields. During the rainy season, inoculums can build up to high levels and Xam can be transported from weeds to cassava plants by wind and raindrops which are very important agents for short distance dissemination of bacteria. CBB pathogen has been proved to remain alive in cassava debris for long time when the debris is not decomposed. The contact of cassava leaves with the infected humidified debris on soil surface and rain splashing may favor the entrance of bacteria in the leaves through stomata on the abaxial surface and permit disease development. Grasshoppers (*Z. variegatus*) feeding on diseased cassava plants acquire the CBB bacterium that can be distributed within cassava field or in close cassava field.

Cassava bacterial blight (CBB) is one of the most severe diseases of cassava in several countries where the crop plays an important dietary and economic role. The disease is present in all cassava producing countries. Epidemics occur during the rainy season when high humidity and warm temperature favor the movement of bacteria and symptoms development. Recent surveys have revealed the prevalence of CBB in several West African countries with regional severe outbreaks [9, 10, 63]. The severity of symptoms varies widely with the cultivar, the ecology, the year and the virulence of the strain. When the development of CBB has been studied in the forest savannah transition and dry savannah zones using both resistant variety I30572 and susceptible variety BEN 86052 over 2 years, Fanou [31] observed that disease development in both varieties tested was more pronounced in the forest savannah transition zone than in the dry savannah. This may be explained by the different rainfall pattern in the two ecozones. Between the first inoculation until the beginning of the dry season, 4 months of wet period with a total of 410 mm rainfall were recorded in the forest savannah transition zone in the first year, but only 2 months with a total of 263.8 mm rainfall in the dry savannah. During the short rainfall period in the dry savannah, the establishment of the disease and its spread through the host plant was obviously restricted. After a long dry period, the survival of the epiphytic population of Xam might be lower. Thus, in the dry savannah the surviving residual population cannot induce a high number of leaf symptoms in the following cropping season compared to the high disease expression in the forest savannah transition zone after 12 months of vegetation despite of the important rainfall recorded in the dry savannah from March to end of July. The importance of rainfall for the development of CBB was also reported by Leu [64] who observed the occurrence of the disease in Taiwan from March to November when the weather was warm and wet. The variety BEN 86052 developed more severe symptoms than the variety I30572 in both ecozones and in both inoculated and non-inoculated variants [31]. Persley [65] also observed a higher disease development in a susceptible cultivar than in a resistant cultivar in the moist savannah zone (Ibadan) and in the dry savannah zone (Mokwa) in Nigeria. Comparing the varieties, BEN 86052 generally lost more root yield reflecting its susceptibility to the disease. The highest recorded loss after 12 months of up to 50% root yield observed in cultivar BEN 86052 occurred in the forest savannah transition zone in the second year when the study was repeated, with also the highest symptom severity, especially high percentages of individually evaluated spot and blight symptoms [31]. Using a susceptible cultivar in a trial, Leu [64] reported that loss caused by CBB in the Puli area of Taiwan differed from field to field and observed 10–15% root yield loss under 10–20% disease incidence and 25–30% root yield loss when disease incidence was 35% or more. Fifty percent or more yield loss due to CBB was reported in susceptible cultivars in Colombia (57%) [66], in Ibadan, Nigeria (58.2%) [67] confirming that CBB is a devastating disease and necessitates a particular attention.

2.3. Disease management

2.3.1. Sanitation measures

The rapid regional spread of CBB to areas where it did not exist before, indicating by the increase of CBB in the rainforest areas in Togo between the 1910s and one decade later [10] is a consequence of free movement of planting materials across ecozones and boundaries and indicates the weaknesses of existing quarantine systems in developing countries. Quarantine procedures are the first line of defence against CBB and should be reinforced by the governments to prevent the introduction of Xam strains to diseased-free regions. The causal agent of CBB is a stem- and seedborne bacterial pathogen and survives in planting materials for up to 30 months. The distribution of Xam in the stem may be continuous or discontinuous [31]. A symptomless plant from an infected field can harbor the pathogen, and also seeds from apparently healthy plants in a contaminated plantation can lodge Xam cells. Thus, to avoid the spread of the pathogen through the exchange of cassava stem cuttings and seeds to establish new plantations, or breeding purposes, planting materials should be collected from absolutely cassava bacterial blight-free fields. Consequently, the governments of each cassava producing countries should adopt the successful methods for producing bacteria-free planting material [24] and establish multiplication farms in disease-free areas from where educated producers could collect healthy planting material. This may prevent farmers to exchange infected planting materials among themselves and delay the dissemination of the disease from zone to zone. All cassava seed used for distribution should be subjected to thermal treatment using water at 60°C for 30 min or dry air at 65°C for 4 days [68].

2.3.2. Cultural measures

Cultural practices are successful to delay the spread of the CBB pathogen. Xam has no ability to survive freely for a long time in the soil [42]. Therefore, all cassava debris after harvesting should be removed from the field and burned or buried with deep ploughing, and the field should be planted with other crops or left under fallow.

Xam proved to survive epiphytically on many weed species [42], and bacterial cells may be transported by movement of men, insects, birds, and animals or wind-driven rains from contaminated weeds to cassava plants. So, all cassava fields should regularly be kept free of weeds. Bush fallow around cassava fields should be avoided to prevent epiphytic survival of Xam on weeds.

Several short-duration crops, such as maize, yam, sorghum, assorted vegetables and cowpea, are usually intercropped with cassava in the humid tropics of West Africa [69]. Intercropping was widely studied as a means to reduce pests and diseases [70–73], but not always with positive effect. Generally, intercropping has been reported as one of the measures to reduce cassava bacterial blight. Nyango [74], Terry [75], Ene [76] reported that cassava bacterial blight was significantly reduced by providing shade or intercropping cassava with maize or melon. The use of intercropping was proposed as means to reduce cassava bacterial blight in the dry savanna [77] and in the humid forest [78]. Significant reduction of cassava bacterial blight severity in cassava intercropped with cowpea and maize compared to cassava monoculture were observed in the forest savannah transition zone of Nigeria, with the highest disease reduction of 53% in a cassava-maize intercrop, without significant yield effect due to cropping system [31]. The latter author suggested that intercropping could have a barrier effect to inhibit the transport of the inoculum of Xam since bacterial diseases are generally disseminated in the field by rain splash and aerosols combined with wind. The effect of intercropping on cassava bacterial blight severity may vary with intercrops used and across ecozones. In our study in Benin, intercropping cassava-sorghum reduced cassava bacterial blight severity significantly up to 80% in three soil amendment treatments, at normal and late planting time in the forest-savannah transition zone and at normal planting time in the dry savanna zone, with few exceptions [79]. Also, the effect of intercropping cassava-maize and cassava-taro on cassava bacterial blight was investigated in Togo. Significant, but relatively low reductions of cassava bacterial blight severity were observed in cassava-maize intercropping in the forest savannah transition zone and in the wet savannah zone, and in cassava-maize and cassava-taro intercropping in the forest highland zone [80]. On the contrary, Sikirou [81] did not observe clear effects on cowpea bacterial blight when cowpea was intercropped with maize or cassava in the forest-savannah transition zone of West Africa. Although generally effects on root yield were not observed, the combination of late planting and intercropping in the dry savannah generally reduced cassava root yield. Cassava-sorghum intercropping generally had no effect on root yield compared to cassava monocropping with few exceptions in two sites (ecozones), making it a recommendable measure to reduce CBB, while intercropping with cowpea significantly reduced root yield by 52% compared to cassava monocropped, in the dry savannah site. On the contrary, a significant cassava yield loss due to intercropping cassava with maize was reported from the rainforest zone of Nigeria [82]. Okoli [69] reported significant cassava root yield losses up to 40% in susceptible and up to 35% in resistant cassava cultivars intercropped with cowpea, while Fanou [31] found no significant difference in cassava root yields between cassava-maize and cassava-cowpea intercropping and monocropping cassava. In maize-soybean intercropping, Mohta and De [83] reported increased total grain yield, whereas Crookston and Hill [84] observed no grain yield effect. Also, yields of intercropped soybean with maize were up to 32% lower than yields of soybean in monoculture, however, yield of intercropped maize was increased up to 53% compared with the yield of monoculture maize and compensated for the reduced yield of soybean [85]. Thus, the present results and studies of other authors show that intercropping may cause a yield reduction of the main crop, but, the additional yield gained by the intercrop has to be considered, which increases the land equivalent ratio [81].

Early and repeated removal of diseased cassava leaves slowed down the development of the disease during the investigations of Fanou and Wydra [86] and might prevent secondary infection. In an integrated CBB control system, when an infection appears despite the application of other successful methods, the diseased leaves should be removed early and subsequently buried. Thus, education of extension workers and farmers in the recognition of CBB symptoms should be part of the approaches in management of the disease. Regular inspection of the fields especially during the rainy season is needed to stop the expansion of the disease.

Accidental infection of cassava fields under integrated control measures should be prevented by installing the fields far away from old cassava fields or infected fields.

Among the agronomic measures to reduce disease epidemics, the shift of planting date to avoid the peak time of inoculum pressure during a susceptible stage of a crop is recommended. Also, for control of cassava bacterial blight, the shift to a late planting date was observed to reduce disease incidence and severity [87], and in our study, disease severity of bacterial blight was generally reduced by late planting in the last third of the rainy season with no effect on cassava root yield [79].

Rainy season generations of grasshoppers (*Zonocerus variegatus*) feeding on cassava plants in cassava bacterial blight-infected fields carry Xam cells on external and internal organs and in high quantity in the faeces [31, 59]. The role of grasshoppers in the spread of Xam during the rainy season proved that besides the external organs of the insect, the faeces also contribute to the distribution of the pathogen [31, 59]. Thus, it is concluded that the grasshopper is a vector for cassava bacterial blight. Therefore, control methods against high populations of the insect during the rainy season when CBB occurs would support the suppression of disease spread.

Resistance to cassava bacterial blight appears to be due to several genes mainly with additive effects, but also to some extent with non-additive effect. Difficulties in recommending suitable genotypes to farmers reside in high genotype-environment interactions for cassava bacterial. In our study, the results reveal the narrow basis for resistance to bacterial blight in local improved cassava varieties from Benin. Considering disease reaction and root yield across environments, only genotype TMS30572 was consistently moderately resistant to resistant and high-yielding in different environments [88]. Thus, genotype TMS30572 can be recommended to farmers. This genotype with a resistant reaction in the dry savannah in both years seemed to be specifically suitable to this ecozone. In Togo, Banito et al. [89] found that genotypes TMS30572 and TMS91/02316 with low disease severity and high root yield could be recommended to farmers. Continuous evaluation and further selection of resistant, high-yielding genotypes is necessary, also considering the observed development of genetically new strains which may overcome plant resistances [90, 91]. Therefore, an evaluation of plant reactions to identify genotypes with stable resistance to cassava bacterial blight should be performed under artificial inoculation with highly virulent strains from the area in order not to contribute to dissemination of strains in repeated years in several locations per ecozone. Additionally, inoculation with different pathotypes deriving from different regions [91] under controlled conditions in regions or countries where cassava is not grown is necessary to give a final evaluation of resistance of genotypes. Most of the IITA genotypes have been evaluated and continue to be evaluated for resistance to CBB and for their yield potential. In the screening studies of Fanou [31], eight genotypes (I89/00914, I30572, I89/00854, I89/02113, O83/00109, I50207, O88/01043 and I89/02078) of 23 screened ones proved to be resistant to CBB in 3 different ecozones, but efforts remain to be made in improvement of root yield of genotypes I89/02113, O83/00109, I50207, O88/01043 and I89/02078, which showed good resistance to symptoms, but low yield.

2.3.3. *Resistance mechanisms*

The role of leaf surface structures as first barriers to confer resistance to bacterial blight were elucidated by studying, leaf stomata and their occlusion with leaf waxes in cassava genotypes. Our results in Benin showed that differences in environmental conditions may have an influence on wax quantities and thereby, contribute to the high genotype x environment interactions in cassava [88]. Stomatal anatomy was reported to confer resistance to some varieties against certain of their bacterial pathogens [92]. Differences in thickness and permeability of cuticles, stomata, hydathodes and trichomes in varieties were observed by Schönherr and Baur [93]. Also, anti-microbial effects of epicuticular wax compounds such as terpenoides and flavonoides against bacteria or fungi were described [94]. But, Barthlott and Wollenweber [95] stated that the anti-microbial components of the epicuticular wax could be released or washed off after longer periods of rain making plants more susceptible to their pathogens. Additionally, we found that higher wax quantities specifically triterpenes were observed in the standard resistant genotype TMS 30572 compared to the susceptible Ben 86,052, and that waxes covered stomata on the abaxial leaf surfaces of both a susceptible and a resistant genotype, while the adaxial surfaces were not covered by wax, but wax was in crystalloid form. We also observed tendencies of lower stomata numbers on adaxial surfaces of the more resistant genotypes than of the susceptible genotype in combination with the lower wax quantities on this leaf side might therefore contribute to the resistance [96]. Also Cooper et al. [97], found adaxial stomata not being occluded by wax. Thus, Cooper et al. [97], reported that the abaxial leaf surface of cassava is nonwettable and seems unlikely as route of entry for Xam. Differences in nutrient availability through less foliar leaching of solutes diffusing across the wax-covered cuticle or direct effects of wax components influence microbial populations on leaf surfaces [98]. But, Fanou [31], observed in cassava a high level of epiphytic Xam populations on leaf surfaces of resistant and susceptible genotypes in different ecozones suggesting that cassava leaf waxes may have no significant effect on epiphytic bacterial populations. In bacterial leaf spot of tomato, stomatal frequency and morphology were shown to be associated with resistance to the disease [99].

In conclusion, cassava leaf surface wax and the number of adaxial leaf stomata might play a role in defence against bacterial blight, but seem not to be decisive for the resistance of genotypes. Lower stomata numbers and high wax quantities may be involved in reducing the number of bacteria invading leaves, but variations in wax quantities and the number of stomata in the tested genotypes were not or only tendentiously related to the described resistances. Comparing stomata distribution, the adaxial stomata are suggested to be portals of entry for the bacteria. Variability in wax quantities between genotypes and ecozones may be among the reasons for the observed high genotype x environment interaction of cassava.

Host plant resistance in cassava is described as polygenic and additively inherited, deriving from interspecific cross-breeding between *M. esculenta* and the wild relative *M. glaziovii* [100]. Genomic approaches demonstrated the induction of a high number of defence related genes in challenged cassava cell cultures with 26% of genes encoding for PR- or stress related proteins [101] or in inoculated cassava plants with 13% of analyzed transcript-derived fragments showing similarity to plant defense proteins [102]. Among biochemical mechanisms, the oxidative burst, phenylpropanoids, phenylalanin ammonia lyase and peroxidases were suggested to be involved in the resistance reaction of cassava [103, 104]. After infection with Xam, a resistant genotype reacted with lignin and callose deposits, and the production of tyloses and phenolic compounds associated with suberin within the infected vessels [29]. Quantitative trait loci (QTL) for resistance to bacterial blight strains from Latin America were identified, and molecular markers for breeding for resistance were developed [105, 106]. Among constitutive resistance mechanisms, a role of latex, produced abundantly after wounding, in defense is possible, indicated by its rapid coagulation and by its components such as lysozyme, chitinase, glucanase and protease [97, 107]. In addition, preliminary observations suggest a role of cell wall pectin in the resistance reaction, since pectin from young cassava leaves caused a synergistic rheological interaction with Xam lipopolysaccharides, while pectin from older, less susceptible leaves and pectins from other sources were not active [108]. A number of QTL for resistance to CBB, with major and minor effects as well as stable and unstable ones were detected. In 2000, Jorge and coworkers reported 12 QTL explaining 9–27% of the phenotypic variance. These QTL were detected in the F1 population using five Xam strains from Latin America, analyzing samples grown under greenhouse conditions. For the African strain ORST X-27 and one Colombian strain, resistance QTL appeared to be introgressions from a wild *Manihot* sp. and are located on one linkage group of the female-derived map, which has a large number of polymorphic markers and shows much lower recombination frequency than the rest of the genome. Eight novel QTL explaining between 7.2 and 18.2% of the resistance were identified under field conditions of natural disease pressure against four Colombian and one African strain and during two consecutive crop cycles in the BC1 population [106]. In our study, six QTL and five unlinked markers that explained between 16 and 33.3% of the phenotypic variance were characterized using quantitative data of symptom development after stem inoculation by the four African strains in the same BC1 population [91]. Nevertheless, some of these QTL and markers have to be confirmed by further studies because the population size was small, but they give some evidence that, with a larger sample size, we could be able to detect more QTL, especially in the CM8820 family. Our results suggest that several genes are involved in resistance to cassava bacterial blight. Among these QTL, two were located on linkage groups N and O, where we also found markers linked to resistance in the present study. More recently, two new QTL explaining 62% and 21% of the CBB resistance were identified to the Xam strains ClO151 and ClO121 [109], and two novel QTL which explain 10.9 and 12.6% of the field resistance to the disease, with four genes identified in the QTL intervals [110]. The genes code for a protein related to the vacuolar-sorting receptor, a serine protease carboxypeptidase, a C_2HC zinc finger-containing protein and for a core-2/i-branching beta-1,6-n-acetylglucosaminyltransferase protein. The low number of QTL detected in the case of the BC1 population in our study compared with the F1 population could be due to the number of markers selected for the BC1 mapping (121 markers) compared with the number selected for the F1 population (142 markers). Although the limited data did not allow the analysis of linkage between leaf resistance and markers, it may be speculated that different loci may be significant after leaf and stem inoculation. Thus, resistance based on strain-specific resistance can be improved by introducing the QTL underlying the resistance into a desirable genetic background or using them in gene pyramiding. Strain-specific resistance loci may contribute to explain the high genotype–environment interaction observed in selection of cassava genotypes for resistance to bacterial blight. The newly identified markers for cassava bacterial blight resistance can be used to increase the efficiency of identifying resistant genotypes for Africa. Incorporation of resistance loci in new lines by gene pyramiding and identification of additional resistance loci will contribute to selection of cassava genotypes with more effective and possibly durable resistance to Xam.

3. Conclusion

The review of studies from various authors and of two decades of our research reveals that an integrated control of cassava bacterial blight is possible. Application of sanitation measures and cultural methods as management strategies will reduce the disease impact and increase the crop yield potential. Further selection of bacterial blight resistant, high-yielding genotypes as well as continuous analysis of G × E and QTL × E interactions (Q × E), will allow estimating the impact of the environment over the QTL effect. Identification of other mechanism of resistance involved in plant defence and incorporation of resistance loci in new lines by gene pyramiding and identification of additional resistance loci will contribute to selection of cassava genotypes with more effective and durable resistance to *Xanthomonas axonopodis* pv. *manihotis*.

Acknowledgements

This publication was supported by Erfurt University of Applied Sciences, Erfurt, Germany.

Author details

André Antoine Fanou[1]*, Valerien Amégnikin Zinsou[1] and Kerstin Wydra[2]

*Address all correspondence to: andrefanou@gmail.com

1 Faculty of Agronomy, University of Parakou, Parakou, Republic of Benin

2 Erfurt University of Applied Sciences, Erfurt, Germany

References

[1] Bondar G. Una nova molestia bacteriana das hastes da mandioca. Characas e Quintais Saô Paulo. 1912;**5**(4):15-18

[2] Lozano JC, Sequeira L. Bacterial blight of cassava in Colombia: Etiology. Phytopathology. 1974a;**64**:74-82

[3] Hahn SK, Williams RJ. Investigation on Cassava in the Republic of Zaire. Report to the Commissaire d'Etat à l'Agriculture. Republic of Zaire, IITA, Ibadan, Mimeo. 1973. 12 p

[4] Maraite H, Meyer JA. *Xanthomonas manihotis* (Arthaud-Berthet) Starr, causal agent of bacterial wilt and leaf spot of cassava in Zaire. Pans. 1975;**21**:27-37

[5] Daniel JF. Le Dépérissement bactérien en Empire Centrafricain. Rapport ORSTOM. Congo: Centre de Brazaville; 1977

[6] Aklé J, Gnouhoué H. Cassava bacterial blight development in Benin. In: Cassava Bacterial Blight in Africa. Past, Present and Future. Report of an Interdisciplinary Workshop; 26-30 June 1978. IITA, Ibadan, Nigeria 1979. Editors: Terry ER, Persley GJ, Sheila CAC. p. 43

[7] Leu LS, Chen CT. Bacterial wilt of cassava (*Manihot utilisima* Pan1.) caused by *X. manihotis* (Arthaud-Berthet) Starr. Plant Protection Bulletin (Taiwan). 1972;**14**:17-26

[8] Bradbury JF. Guide to Plant Pathogenic Bacteria. Farnham House. Ferry Lane, Kew, Surrey, England: CAB International Mycological Institute; 1986. 332 p

[9] Wydra K, Msikita W. Overview of the present situation of cassava diseases in West Africa. In: Ekanayake IJ, editor. Proceedings of the. 6th Trienn. Symp. Intern. Soc. Trop. Root Crops Africa Branch (ISTRC-AB), 22-28 October 1995. Lilongwe, Malawi 1998. Editors. Akoroda MO. pp. 198-120

[10] Wydra K, Verdier V. Occurrence of cassava diseases in relation to environmental, agronomic and plant characteristics. Agriculture, Ecosystems and Environment. 2002; **93**:211-226

[11] Terry ER. Fear of cassava bacterial blight in Africa makes control imperative. World Crops and Livestock. 1977b;**29**(3):107-108

[12] Girard JC. Rapport de mission en République Populaire du Bénin sur les problèmes phytosanitaires du manioc 23 Avril au 7 Mai 1979 IRAT Montpellier 1979. 10 p

[13] Obigbesan GO, Matuluko EO. Effect of potassium and bacterial blight on the yield and chemical composition of cassava cultivars. In: Proceeding of 4th Symposium of the International Society for Tropical Root Crops. 1-7 August 1976. Cali, Colombia 1977. Editors: Cock J, MacIntyre R, Graham M. p. 185-188

[14] Dye DB. Genus *Xanthomonas* Dowson. A proposed nomenclature and classification for plant pathogenic bacteria. New Zealand Journal of Agricultural Research. 1978;**21**:153-177

[15] Vauterin L, Hoste B, Kersters K, Swings J. Reclassification of *Xanthomonas*. International Journal of Systematic Bacteriology. 1995;**45**:472-489

[16] Swings J, Vauterin L, Kersters K. The bacterium *Xanthomonas*. In: Swings JG, Civerolo EL, editors. Xanthomonas. London: Chapman & Hall; 1993. pp. 121-156

[17] Persley GJ. Epiphytic survival of *Xanthomonas manihotis* in relation to the disease cycle of cassava bacterial blight. Proc 4th Int. Conf. Plant. Path. Bact. INRA, Angers; 1978. p. 773-777

[18] Daniel JF, Boher B. Epiphytic phase of *Xanthomonas campestris* pv. *manihotis* on aerial parts of cassava. Agronomie. 1985a;**5**:111-116

[19] Perreaux D, Maraite H, Meyer JA. Detection of 3(methylthio)propionic acid in cassava leaves infected by *Xanthomonas campestris* pv. *manihotis*. Physiological and Molecular Plant Pathology. 1986;**28**:323-328

[20] Ewbank E. Etude du catabolisme d'acides aminés conduisant à la formation de phytotoxines par Xanthomonas campestris pv. manihotis (agent pathogène de la bactériose du manioc). [PhD Thesis]. Louvain-la-Neuve, Belgium: Université Catholique de Louvain; 1992. 191 p

[21] Persley GJ. Distribution and importance of cassava bacterial blight in Africa. In: Cassava Bacterial Blight. Report of an interdisciplinary workshop held at IITA, Ibadan, Nigeria, 1-4 November 1976. IDRC, Ottawa, Canada 1977a. Editors: Persley GJ, Terry ER, MacIntyre R. p. 9-14

[22] Wydra K, Banito A, Kpémou KE. Characterization of resistance of cassava genotypes to bacterial blight by evaluation of symptom types in different ecozones. Euphytica. 2007; **155**:337-348

[23] Daniel JF. Contribution à l'étude de la Biologie de *Xanthomonas campestris* Pathovar *manihotis* (Arthaud-Berthet et Bondar) Starr, Agent Responsable de la bactériose Vasculaire Du Manioc, *Manihot esculenta* Crantz. Thèse de Doctorat. Université de Paris-Sud Centre D'Orsay; 1991. 177 p

[24] Lozano JC, Wholey DW. The production of bacteria-free planting stock of cassava. World Crops. 1974;**26**:115-117

[25] Ikotun T. Survival of *Xanthomonas manihotis* in cassava tissues. Nigerian Journal of Plant Protection. 1977;**3**:31-36

[26] Ikotun T. Effect of *Xanthomonas manihotis* on cassava tissues. In: Cassava Bacterial Blight in Africa. Past, Present and Future. Report of an interdisciplinary workshop 26-30 June, 1978. IITA, Ibadan, Nigeria 1979. Editors: Terry ER, Persley GJ, Cook SA. p. 29-32

[27] Maraite H, Perreaux D. Comparative symptom development in cassava after infection by Xanthomonas manihotis or X. Cassavae under controlled conditions. In: Cassava Bacterial Blight in Africa. Past, Present and Future. Report of an interdisciplinary workshop 26-30 June, 1978. IITA, Ibadan, Nigeria 1979. Editors: Terry ER, Persley GJ, Cook SA. p. 17-24

[28] Daniel JF, Boher B. Etude des modes de survie de l'agent causal de la bactériose vasculaire du manioc, *Xanthomonas campestris* pv. *manihotis*. Agronomie. 1985b;**5**:339-346

[29] Kpémoua K, Boher B, Nicole M, Calatayud P, Geiger JP. Cytochemistry of defence responses in cassava infected by *Xanthomonas campestris* pv. *manihotis*. Canadian Journal of Microbiol. 1996;**42**:1131-1143

[30] Boher B, Brown I, Nicole M, Kpémoua K, Verdier V, Bonas U, Daniel JF, Geiger JP, Mansfield J. Histology and cytochemistry of interactions between plants and Xantho-monads. In: Nicole M, Gianinazzi-Pearson V, editors. Histology, Ultrastructure and Molecular Cytology of Plant-Micro-Organism Interactions. Kluwer Academic Plublishers; 1996. pp. 193-210

[31] Fanou AA. Epidemiological and Ecological Investigations on Cassava Bacteriel Blight and Development of Integrated Methods for its Control in Africa. [PhD Thesis]. Georg-August- University, Goettingen, Germany; 1999. Cuvillier Verlag. 199 p

[32] Kpémoua K. Etude Comparative Du Développement de Xanthomonas Campestris Pv. Manihotis Chez Des variétés de Manioc Sensibles et Résistantes. Approches Histologique, Ultrastructure et Cytochimique Des Mécanismes de la Pathogenèse. [Thèse de Doctorat]. Université de Nantes; 1995. p. 216

[33] Lambotte M, Perreaux D. Histopathology of cassava bacterial blight in susceptible and tolerant cultivars. In: Pro. Intern. Symp. On Diseases of Tropical Food Crops. U.C.L. Louvain-la-Neuve, Belgium; 1978. Editors: Maraite H, Meyer JA. p. 153-160

[34] Dinesen IG. Survival of pathogens. In: Klement Z, Rudolph K, Sands DC, editors. Methods in Phytobacteriology. Budapest: Akadémiai Kiadó; 1990. pp. 281-288

[35] Lozano JC, Sequeira L. Bacterial blight of cassava in Colombia: Epidemiology and control. Phytopathology. 1974b;**64**:83-88

[36] Terry ER. Cassava bacterial diseases. In: Proceedings of Cassava Protection Workshop; 7-12-11-1977 CIAT, Cali, Colombia; 1978. Editors: Brekelbaum T, Belloti A, Lozano JC. p. 75-84

[37] Otim-Nape GW. The effects of bacterial blight on germination, tuber yield and quality of cassava. In: Proceedings of the 10th International Congres, Brighton. British Crop Protection Council, London; 1983. p. 1205

[38] Ikotun T. Studies on the host range of *Xanthomonas manihotis*. Fitopatologia Brasileira. 1981;**6**:15-21

[39] Amusa T, Ikotun T, Hahn SK. Survival of *Xanthomonas campestris* pv. *manihotis* in soils under Mucuna fallow. In: Proc. 4th Symposium, ISTRC-AB, held in Kinshasa, Zaire, 5-08-12-1989. Editors: Akoroda MO, Arene OB. 1992, p. 193-196

[40] Dedal OI, Palomar MK, Napiere CM. Host range of *Xanthomonas manihotis* Starr. Annals of Tropical Research. 1980;**2**(3):149-155

[41] Marcano M, Trujillo G. Bacteriosis de la yuca; identificados algunos aspectos epidemiologicos del pathogeno que pueden utilizarse para disminuir los danos ocasionados par la enfernedad. Fonaiap Divulga [Venezuela]. 1982;**1**:12-13

[42] Fanou AA, Zinsou AV, Wydra K. Survival of *Xanthomonas axonopodis* pv. *manihotis* in weed species and in cassava debris: Implication in the epidemiology of cassava bacterial blight. International Journal of Advanced Research. 2017;**5**(4):2098-2112. DOI: 10.21474/IJAR01/4057

[43] Daniel JF, Boher B. Ecology of cassava bacterial blight: Epiphytic survival of *Xanthomonas manihotis* on aerial parts of the cassava plant. Proc. 4th Int. Conf. Plant Path. Bact., INRA, Angers; 1978. p. 763-771

[44] Elango FN, Lozano JC. Epiphytic survival of *Xanthomonas manihotis* on common weeds in Colombia. Proc. 5th Int. Conf. Plant Path. Bact., Cali, Colombia; 1981. p. 203-209

[45] Boher B, Verdier V. Cassava bacterial blight in Africa: The state of knowledge and implications for designing control strategies. African Crop Science Journal. 1994;**2**(4):505-509

[46] Fanou A, Wydra K, Zandjanakou M, Le Gall P, Rudolph K. Studies on the survival mode of *Xanthomonas campestris* pv. *manihotis* and the dissemination of cassava bacterial blight through weeds, plant debris and an insect vector. In: Root Crops in the 21st Century. Proceedings of the 7th Triennial Symposium of the International Society for Tropical Root Crops-Africa Branch (ISTRC-AB), Cotonou, Bénin 11-17 October 1998. Eds Akoroda MO, Ngeve JM. 2001. p. 569-575

[47] Hirano SS, Upper CD. Ecology and epidemiology of foliar bacterial plant pathogens. Annual Review of Phytopathology. 1983;**21**:243-269

[48] Rudolph K. Infection of the plant by *Xanthomonas*. In: Swings JG, Civerolo EL, editors. Xanthomonas. London: Chapman & Hall; 1993. pp. 193-264

[49] Persley GJ. Technical Report on IDRC Cassava Bacterial Blight Project. Ibadan Nigeria: IITA; 1977b. 18 p

[50] Ikotun T. The survival of *Xanthomonas manihotis* (Arthaud-Berther) Starr in the soil. Fitopatologia Brasileira. 1982;**7**:29-36

[51] Marcano M, Trujillo G. Perpetuacion de la bacteria *Xanthomonas campestris* pv. *manihotis* (Berthet y Bondar) dye. En el suelo a traves de resto de cosecha. Agronomia Tropical. 1984;**34**(4-6):7-19

[52] Thaveechai N, Leksomboon C, Kositratana W, Paradornutwat A. Rojanaridpiched C. Survival of *Xanthomonas campestris* pv. *manihotis* under natural field conditions. Kasetsart-Journal,-Natural-Sciences. 1993;**27**(1):25-32

[53] Terry ER, MacIntyre R. The international exchange and testing of cassava germplasm in Africa. Proc. Intern. Workshop held at International Institute of Tropical Agriculture, Ibadan, Nigeria. 1976. p. 35

[54] Bani G. Interactions *Zonocerus variegatus* (Orthoptère, Pyrgomorphidae) - Manioc Au Congo: Bioécologie d'un Ravageur Vecteur de la bactériose. [Thèse de Doctorat]. Université de Rennes 1; 1990. p. 195

[55] Modder WWD. Control of variegated grasshopper *Zonocerus variegatus* (L.) on cassava. African Crop Science Journal. 1994;**2**(4):391-406

[56] Givord L, Den Boher L. Insect transmission of okra mosaic virus in the Ivory Coast. The Annals of Applied Biology. 1980;**94**:235-241

[57] Whitney WK, Gilmer RM. Insects vectors of cowpea mosaic virus in Nigeria. The Annals of Applied Biology. 1974;**77**:17-21

[58] Daniel JF, Boher B, Nkouka N. Insect dissemination of *Xanthomonas manihotis* to cassava in the People's Republic of Congo. In: Tropical Root Crops: Research Strategies for the 1980s. Proc. of the 1st Trien. Root Crops Symp. of the Int. Soc. for Trop. Root Crops. AB, 8-12 Sept. 1980, Ibadan, Nigeria. Editors: Terry ER, Oduro KA, Caveness F. 1980. p. 66-68

[59] Zandjanakou-Tachin M, Fanou A, Le Gall P, Wydra K. Detection, survival and transmission of *Xanthomonas axonopodis* pv. *manihotis* and *X. axonopodis* pv. *vignicola*, causal agents of cassava and cowpea bacterial blight, respectively, in/by insect vectors. Journal of Phytopathology. 2007;**155**:159-169

[60] Persley GJ. Studies on the survival and transmission of *Xanthomonas manihotis* on cassava seed. Annals of Applied Biology. 1979b;**93**:159-166

[61] Elango F, Lozano JC. Transmission of *Xanthomonas manihotis* in seed of cassava (*Manihot esculenta*). Plant Disease. 1980;**64**:784-786

[62] Daniel JF, Boher B. Contamination of cassava flowers, fruits and seeds by *Xanthomonas campestris* pv. *manihotis*. In: Proc. 5th Int. Conf. of Plant Pathogenic Bacteria. Cali, Colombia 1981a. p. 614-617

[63] Banito A, Kpémoua K, Wydra K, Rudolph K. Bacterial blight of cassava in Togo: Its importance, the virulence of the pathogen and the resistance of varieties. In: DeBoer S, editor. Plant Pathogenic Bacteria. Dordrecht: Kluwer Academic Publishers; 2001. pp. 259-264

[64] Leu LS. Cassava bacterial blight in Taiwan. In: Diseases of Tropical Food Crops, Proc. of an Intern. Symp. U.C.L. Louvain-La-Neuve, Belgium, 1978. Editors: Maraite H, Meyer JA. 1978. p. 119-129

[65] Persley GJ. Studies on the epidemiology and ecology of cassava bacterial blight. In: Cassava Bacterial Blight in Africa. Past, Present and Future. Report of an Interdisciplinary Workshop held at IITA, Ibadan, Nigeria, 26-30-06-1978. Editors: Terry ER, Persley GJ, Sheila CAC. 1979a. p. 5-7

[66] CIAT. Annual Report 1973. Cali, Colombia: Centro Internacional de Agricultura Tropical; 1974. p. 254

[67] IITA. Annual Report 1976. Ibadan, Nigeria: International Institute of Tropical Agriculture. p. 1977

[68] Fanou AA, Wydra K. Physical and chemical treatment for the control of *Xanthomonas axonopodis* pv. *manihotis* in cassava seeds. Journal of Experimental Biology and Agricultural Sciences. 2015;**3**(1):54-59

[69] Okoli OO. Effect of planting dates and growth habits of cassava and cowpea on their yield and compatibility. Tropical Agriculture. 1996;**73**:169-174

[70] Trenbath BR. Intercropping for the management of pests and diseases. Field Crops Research. 1993;**34**:381-385

[71] Ahohuendo BC, Sarkar S. Partial control of the spread of African cassava mosaic virus in Benin by intercropping. Zeitschrift für Pflanzenkrankheiten und Pflanzenschutz. 1995; **102**:249-256

[72] Sikirou R, Wydra K. Effect of intercropping cowpea with maize or cassava on cowpea bacterial blight and yield. Journal Plant Diseases & Plant Protection. 2008;**115**: 145-151

[73] Fininsa C. Epidemiology of Bean Common Bacterial Blight and Maize Rust in Intercropping. [PhD thesis]. Uppsala, Sweden: University of Uppsala; 2001

[74] Nyango AK. Cassava bacterial blight in four regions of the United Republic of Tanzania. In: Cassava Bacterial Blight in Africa. Proc. Past, Present and Future Report of an Int. Workshop. IITA, Ibandan, Nigeria. 1979. p. 44-48

[75] Terry ER. Some epidemiological factors affecting the survival and dissemination of *Xanthomonas manihotis*. In: Workshop on Cassava Bacterial Blight in Nigeria. 1st. Umudike, Nigeria. Proc. Federal Agric. Res. and Training Station, Umudike; 1974. Ikpala EU, Cleeser HJ. (eds.). p. 39-48

[76] Ene LSO. Control of cassava bacterial blight (CBB). Tropical root and tuber crops. Newsletter. 1977;**10**:30-31

[77] Tabot S. Bacterial Blight of Cassava (Manihot Esculentus Crantz) in the Drier Savannas of Nigeria. [M.Sc. thesis]. Vol. 1995. Zaria, Nigeria: Ahmadu Bello University; 1995. p. 100

[78] Arene OB. Influence of shade and intercropping on the incidence of cassava bacterial blight. In: Proc. of Interdisciplinary Workshop on Bacterial Blight. IDRC Series 09C; 1976. p. 28-29

[79] Zinsou V, Wydra K, Ahohuendo B, Hau B. Effect of soil amendments, intercropping and planting time in combination on the severity of cassava bacterial blight and yield in two ecozones of West Africa. Plant Pathology. 2004;**53**:585-595

[80] Banito A. Integrated Control of Cassava Bacterial Blight in West Africa in Relation to Ecozones, Host Plant Resistance and Cultural Pactices. [PhD thesis]. Hannover, Germany: University of Hannover; 2003

[81] Sikirou R. Epidemiological investigations And Development of Integrated Control Methods of Bacterial Blight of Cowpea Caused by *Xanthomonas campestris* Pv. *vignicola* (Burkholder) Dye. [PhD thesis]. Göttingen, Germany: University of Göttingen; 1999

[82] Zuofa K, Tariah NM, Isirimah MO. Effects of groundnut, cowpea and melon on weed control and yields of intercropped cassava and maize. Field Crops Research. 1992;**28**:309-314

[83] Mohta NK, De R. Intercropping maize and sorghum with soybeans. The Journal of Agricultural Science. 1980;**95**:117-122

[84] Crookston RK, Hill DS. Grain yield and land equivalent ratios from intercropping corn and soybeans in Minnesota. Agronomy Journal. 1979;**71**:41-44

[85] Herbert SJ, Putnam DH, Poosfloyd MI, Vargas A, Creighton JF. Forage yield of intercropped corn and soybean in various planting patterns. Agronomy Journal. 1984;**76**: 507-510

[86] Fanou A, Wydra K. Removal of symptomatic cassava leaves as cultural practice to control cassava bacterial blight. international. Journal of Plant Pathology. 2014;**3**: 117-124

[87] Ambe JT. The effect of planting dates on three cassava diseases in Cameroon. International Journal of Pest Management. 1993;**39**:309-311

[88] Zinsou V, Wydra K, Ahohuendo B, Hau B. Genotype x environment interactions in symptom development and yield of cassava genotypes with artificial and natural cassava bacterial blight infections. European Journal of Plant Pathology. 2005;**111**: 217-233

[89] Banito A, Kpémoua KE, Wydra K. Expression of resistance and tolerance of cassava genotypes to bacterial blight determined by genotype x environment interactions. Journal of Plant Diseases and Plant Protection. 2008;**115**:152-161

[90] Assigbétsé K, Verdier V, Wydra K, Rudolph K, Geiger JP. Genetic variation of the cassava bacterial blight pathogen, Xanthomonas campestris pv. Manihotis, originating from different ecoregions in Africa. Plant Pathogenic Bacteria. Centre Adv. Study Botany, Univ. Madras, India. 1998. Mahadevan A. (ed.):223-229

[91] Wydra K, Zinsou V, Jorge V, Verdier V. Identification of pathotypes of *Xanthomonas axonopodis* pv. *manihotis* in Africa and detection of quantitative trait loci and markers for resistance to bacterial blight. Phytopathology. 2004;**94**:1084-1093

[92] Agrios GN. Plant Pathology. 4th ed. San Diego, USA: Academic Press; 1997

[93] Schönherr J, Baur P. Cuticle permeability studies: A model for estimating leaching of plant metabolites to leaf surfaces. In: Morris CE, Nicot PC, Nguyen C, editors. Aerial Plant Surface Microbiology. New York: Plenum Press; 1996. pp. 1-23

[94] Swain TJ. Secondary compounds as protective agents. Annal Review of Plant Physiology. 1977;**28**:479-501

[95] Barthlott W, Wollenweber E. Zur Feinstruktur, Chemie und taxonomischen Signifikanz epicuticularer Wachse und ¨ahnlicher Sekrete. Trop Subtrop Pflanzenwelt. 1981;**32**: 35-97

[96] Zinsou V, Wydra K, Ahohuendo B, Schreiber L. Leaf waxes of cassava (*Manihot esculenta* Crantz) in relation to ecozone and resistance to *Xanthomonas* blight. Euphytica. 2006;**149**:189-198

[97] Cooper RM, Kemp B, Day R, Gomez-Vasquez R, Beecing JR. Pathogenicity and resistance in Xanthomonas blight of cassava. In: Proceedings of the 10th Inter. Conf. on Plant Pathogenic Bacteria. DeBoerS.(Ed.) Charlotte town, Canada, Kluwer Academic Publishers, Dordrecht; 2001. The Netherlands. p. 319-323

[98] Knoll D, Schreiber L. Influence of epiphytic microorganisms on leaf wettability: Wetting of the upper leaf surface of *Juglans regia* and of model surfaces in relation to colonization by micro-organisms. The New Phytologist. 1998;**140**:271-282

[99] Ramos LJ, Narayanan KR, Macmillan RT. Association of stomatal frequency and morphology in *Lycopersicon* species with resistance to *Xanthomonas campestris* pv. *vesicatoria*. Plant Pathology. 1992;**41**:157-164

[100] Hahn SK, Terry ER, Leuschner K, Akobundu IO, Okali C, Lar R. Cassava improvement in Africa. Field Crops Research. 1979;**2**:193-226

[101] Kemp BP, Beeching JR, Cooper RM. Pathogenicity and Resistance in *Xanthomonas* Blight of Cassava. (Abstract and Poster) 2001. Available from www.danforthcenter.org/media/video/cbnv/session8/S8-08.htmaccessed in March, 2005

[102] Santaella M, Suarez E, Lopez C, Gonzalez C, Mosquera G, Restrepo S, Thome J, Badillo A, Verdier V. Identification of genes in cassava that are differentially expressed during infection with *Xanthomonas axonopodis* pv. *manihotis*. Molecular Plant Pathology. 2004;**5**:549-558

[103] Beeching JR, Han Y, Gomez-Vasquez R, Day RC, Cooper RM. Wound and defense responses in cassava as related to post-harvest physiological deterioration. In: Romeo JT, Downum KR, Verpoorte R, editors. Phytochemical Signals and Plant-Microbe Interactions. NewYork: Plenum Press; 1998. pp. 231-249

[104] Gomez-Vasquez R, Day R, Buschmann H, Randles S, Beeching JR, Cooper RM. Phenylpropanoids, phenylalanine ammonia lyase and peroxidases in elicitor-challenged cassava (*Manihot esculenta*) suspension cells and leaves. Annals of Botany. 2004;**94**:87-97

[105] Jorge V, Fregene M, Duque MC, Bonierbale MW, Thome J, Verdier V. Genetic mapping of resistance to bacterial blight disease in cassava (*Manihot esculenta* Crantz). Theoretical and Applied Genetics. 2000;**101**:865-872

[106] Jorge V, Fregene M, Velez CM, Duque MC, Thome J, Verdier V. QTL analysis of field resistance to *Xanthomonas axonopodis* pv. *manihotis* in cassava. Theoretical and Applied Genetics. 2001;**102**:564-571

[107] Flood J, Cooper RM, Deshappriya N, Day RC. Resistance of cassava (*Manihot esculenta*) to *Xanthomonas* blight in vitro and in planta. Aspects. Applied Biology. 1995;**42**: 277-284

[108] Wydra K, Agbicodo E, Ahohuendo B, Banito A, Cooper RMC, Dixon AGO, Jorge V, Kemp B, Kpémoua K, Rudolph K, Verdier V, Witt F, Zandjanakou M, Zinsou V. Integrated control of cassava bacterial blight by (1) combined cultural control measures and (2) host plant resistance adapted to agroecological conditions, and (3) improved pathogen detection, In: Akoroda MO (Ed.). Ninth Symposium of International Society of Tropical Root Crops—Africa Branch. 2003. p. 506-515

[109] Lopez C, Quesada L, Bohorquez A, Duque M, Vargas J, Tohme J, et al. Mapping EST-derived SSRs and ESTs involved in resistance to bacterial blight in Manihot Esculenta. Genome. 2007;**50**:1078-1088. DOI: 10.1139/ G07-087

[110] Soto Sedano J, Mora Moreno RE, Calle F, López Carrascal CE. QTL identification for cassava bacterial blight resistance under natural infection conditions. Acta Biológica Colombiana. 2017;**22**(1):19-26. DOI: http://dx.doi.org/10.15446/abc.v22n1.57951

Chapter 3

Cassava Superelongation Disease in the Caribbean

Angela T. Alleyne

Additional information is available at the end of the chapter

http://dx.doi.org/10.5772/intechopen.71509

Abstract

An important economic constraint to the growing cassava industry in the Caribbean islands is the disease caused by the fungal pathogen *Sphaceloma manihoticola*, synonym *Elsinoë brasilensis* (Bitancourt & Jenk). One hundred percent incidence has been recently observed on some farms in the Caribbean islands. The fluctuation in individual farming practices such as lack of fertilizing and irrigation schemes may play a role in the level of health and disease resistance of the plants, which in turn may affect the severity of the disease and levels of incidence. Severe elongation may be seen of the internodes in mature plants but primary symptoms include small yellow leaf spots, leaf curling, stem and petiole scab-like lesions and defoliation. The use of disease-free planting material, fungicide pre-treatment of nodal stem cuttings and germplasm maintenance of *in-vitro* stocks of high performing varieties is suggested. However, new molecular tools for disease diagnosis and analysis of the pathogen population dynamics are required to adequately manage the disease in the region.

Keywords: cassava, super-elongation, gibberellin A4

1. Introduction

Manihot esculenta Crantz (cassava)—a woody shrub of the family *Euphorbiaceae* native to South America, is extensively cultivated as an annual crop in tropical and subtropical regions such as West Africa, Southern and Central America, and South East Asia [1, 2], for its edible, starchy, tuberous root. While production showed increased growth in Africa and Asia in the 1990s, production levels of cassava in Latin America and the Caribbean were relatively stable [2]. Since 2005 however, Latin America and the Caribbean has seen a resurgence in cassava production as the implication of threats to food security have emerged and new markets for secondary cassava products continue to grow and develop [2]. Consequently, cassava has been recognized as a diverse crop for development of primary and secondary agricultural products in the Caribbean region. From 2004 to 2014, cassava production has increased from 20 to 400% from Trinidad and Tobago in the south to Bahamas in the north (**Table 1**) [3]. However, with increased production of cassava, Superelongation disease (SED) has re-emerged as a significant threat to its cultivation in the region [4, 5].

Superelongation disease of cassava is caused by the fungal pathogen *Sphaceloma manihoticola* (Bitancourt & Jenk) synonym *Elsinoë brasilensis* and has been responsible for crop losses of up to 80% in several Latin American countries, such as Columbia, Brazil, Venezuela [4, 6, 7]. It is therefore considered to be of major of economic importance due to its potential to severely lower yields [8, 9].

The genus *Sphaceloma* de Bary (Melanconiales) is composed of over fifty fungal species [6, 10]. The majority of the species thrive in tropical or subtropical regions. Conidia are small, unicellular and hyaline, formed either in an acervulus-like structure or on continuous fertile layers of densely packed phialidic conidiophores [6, 11–13]. Different species have the ability to form large spindle-shaped, septate spores which may be pigmented with a thick cell wall. This characteristic growth has been referred to as the "fawcetti" conidia and allows it to be carried long-distances by the wind, aiding in dissemination of the pathogen [7, 14, 15].

Normally cassava is planted from stem or nodal cuttings known as stakes, which are at least 10 months old [16–19]. The cuttings should be healthy; otherwise the plants produced will bear diseases that were infecting the stake. The cultivation of cassava through the planting of disease-free stakes is therefore designed to reduce the effect of potential diseases such as SED [20].

In the English speaking Caribbean islands, there are several local varieties of cassava grown; with growers in each island having their particular preference; these include local names and descriptions such as: Sugarloaf, Butterstick, Redstick, Maracas Blue Stick, Maracas Black Stick, Green stem, and Guyana Sweet among others [19]. However, improved cassava varieties obtained from the International Centre for Tropical Agriculture (CIAT) are generally higher yielding and have been bred and maintained *in vitro* for tolerance to particular pests and diseases such as bacterial blight, anthracnose, SED, and thrips [6, 19]. Despite the introduction of improved cassava varieties, it was observed that in Brazil many of these resistant

Country	**Productivity per year in tonnes**[a]		
	2004	**2008**	**2014**
Bahamas	155	413	938
Barbados	317	466	553
Trinidad	575	2746	3194
Jamaica	16,758	14,991	16,549

[a]Data taken from FAOSTAT 2014. FAO Database, Food and Agriculture Organization of the United Nations. Rome, Italy. http://www.fao.org/faostat/en/

Table 1. Growth in cassava production from 2004 to 2014 among three Caribbean islands.

varieties are not adopted by the farmers, so these benefits may not be transferred to farmers who may have their own criteria for choosing cassava planting material [21]. A similar response to varieties exists in some Caribbean islands. Interestingly, farmers may also lack knowledge of the scientific names and nomenclature of a specific variety being planted, being only familiar with a common description which remains in the local discourse.

This preferential planting of cassava varieties by the farmer therefore narrows the genetic base, resulting in genetic erosion [22] and increases exposure to endemic diseases such as SED, which is prevalent in the region. Tracking of cassava varieties for cultivar identification through the use of genetic markers [22, 23] is therefore currently a necessary exercise for the cassava industry in the Caribbean islands.

2. Disease description

Epidemics of SED result in reduced root size and poor quality tubers, besides dramatic yield losses [7, 19]. An outbreak of the disease was reported for the first time in the Tolima Valley of Colombia in 1972 and 1976 [24]. In 1994, SED was reported in Brazil in cassava crops near Manaus, Maues in the Amazon region [7]. Superelongation disease was also reported in Brazil again in 1994 in Sao Paulo where the disease was observed in cassava crops in the Valle de Paranapanema [25]. In 2007, the disease was observed on cassava in fields of north central and southern Trinidad and Tobago [4] and is also widespread in Barbados [5], the Dominican Republic and Panama [4]. In an island wide survey of 2015–2016 in Barbados, an overall incidence of SED of 72–88% and severity levels of approximately 49% was seen on the island.

The disease affects the leaves, petioles and stems of the cassava plant. Early symptoms of the disease appear on expanding leaves, juvenile stems and floral tissues. Leaf spots are chlorotic and present as small, circular to irregular discolorations, approximately 0.5–5 mm in diameter, lightly colored and sometimes necrotic with a yellow halo [4, 7, 26]. Abundant spots may eventually deform the leaf causing sharp curvatures of one or more of the leaf lobes. This leaf curl causes the lower leaf surface to face upwards, resulting in severe defoliation [19]. Infrequently, chlorotic spots with necrotic centers are seen on the leaf lamina which when dried, produce a "shot hole" appearance. In the later stages of the disease, raised corky cankers appear as lesions on the petioles, leaf veins, and stems [5, 7, 13]. The stem cankers are usually hypertrophic which may coalesce to produce large elliptical to fusiform lesions [26] (**Figure 1**).

Secondary and advanced symptoms from which the disease takes its name, comprise of exaggerated internode elongation in severely infected young stalks, in susceptible cassava cultivars [4, 7, 12, 25]. Rapid elongation caused by SED results in weak plants because the stem is unable to support mature healthy growth, and is frequently followed by die-back and extensive apical defoliation. Internode elongation is prevalent in seasons during which cassava is actively growing. Therefore, during the dry season internodes do not usually elongate, even if numerous stem lesions are present [12].

Figure 1. Symptoms of superelongation disease of cassava caused by *S. manihoticola*. (A) Yellow discolorations and spots necrotic lesions surrounded by a yellow halo and (B) fusiform scab-like lesions on cassava petiole.

Thus SED is considered a disease with economic impact because it not only affects the crop but the planting material [7, 13]. The stem elongation observed is a result of the over production of Gibberellins, specifically gibberellin A4 (GA_4), produced by the fungus *S. manihoticola* [7, 15, 26, 27].

3. *Sphaceloma manihoticola*: the pathogen

Sphaceloma manihoticola when cultured exhibits varied morphological characteristics depending on the media it is grown on [6, 7]. *S. manihoticola* was first observed and documented by Bitancourt and Jenkins [14], when they observed the pathogen on cassava and considered it a new species based solely on its symptomology and host species. Fungal colonies are usually observed as a yellow mucoid mass but there may be other colors such as black and orange, all dependent on the strain and growth conditions [6, 7]. *Sphaceloma manihoticola* has such variable colony morphology that Zeigler et al. [26] considered this feature to be limited in its usefulness to distinguish species.

The morphology of young colonies ranges from a yeast-like growth of budding, unicellular propagules and short hyphal strands to a distinctly mycelial form [10, 28]. It was noted that as colonies matured, they became raised, convoluted and bound in a gelatinous matrix. The fungus is known to synthesize large amounts of exopolysaccharides which account for the difficulty of removing the mycelium from the culture media [29]. The colonies studied by Zeigler and Lozano [13] were pulvinate or raised and deeply fissured, gummy to occasionally mucoid on agar media. Colony color ranged from orange to yellow or orange to bright red, rust and brown on Potato Dextrose agar (PDA) [6]. Yellow or orange colonies frequently formed small red sectors [6] and colony color changed based on the growth medium used. Cassava leaf agar supplemented with glucose generally produced a mixture of bright red, orange and black colonies, while the same colonies on Czapek Dox agar (CZA) were uniformly orange in color and produced no aerial mycelium [13]. Another study on PDA showed colony color from nearly purely white through yellow to deep reddish purple and black [26].

Reeder et al. [4] consistently isolated *S. manihoticola* from infected planting material by plating on potato carrot agar. The colonies that formed were slow growing, pulvinate, fissured and bright red to tomentose in color. Conidiophores were phialidic, conidia were hyaline, non-septate and ellipsoid forming a continuous layer [4]. Morphology of the pathogen as previously observed by Zeigler [10] and others [7, 25] is therefore highly diverse.

Four different morphologies were observed when isolates were grown on CZA in Barbados (**Figure 2**). The range in colors is due to the presence of elsinochromes which are red/orange pigments that are produced by *Elsinoë* spp. and *Sphaceloma* spp. [30]. These pigments have been shown to vary based on the available nutrients in the growth medium such as glucose [28]. Elsinochromes contain perelenequinone which is a non-host virulent factor and causes lesions during fungal infections such as in citrus scab [30]. These pigments cause lipid peroxidation and electrolyte leakage into the infected leaves [31]. The presence of this phytotoxin containing pigment in *S. manihoticola* and its interaction with reactive oxygenic species may account for the lesions and necrosis observed in SED. In addition, given that these pigments were expressed on CZA, in which sucrose is the carbon source, suggests that pigment production might have been stimulated by that carbon source. Further studies are therefore necessary on isolates in the Caribbean islands to explain their yellow pigment.

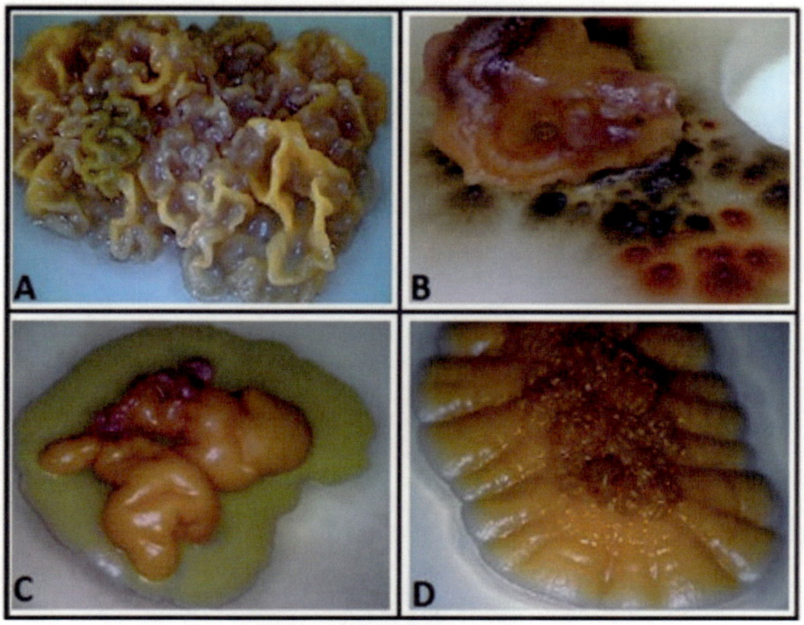

Figure 2. Gross morphology of *S. manihoticola* on various agar media: (A) Czapek Dox agar, (B) Cassava Leaf agar supplemented with Glucose, (C) twenty-eight day colony on Potato Dextrose agar, (D) Potato Dextrose agar.

4. Gibberellin A4

Gibberellins are a group of at least 136 different diterpenoid compounds of plant or fungal origin [32, 33]. These molecules are synthesized from acetyl CoA via the mevalonic acid pathway.

The production of gibberellin GA_4 by *S. manihoticola* promotes the growth and elongation of cells, stimulates rapid stem and root growth, induces mitotic division in leaves and increases seed germination rate [27, 29, 34]. Thus, the symptoms of SED are a result of the production of GA_4 by *S. manihoticola* [7, 26, 29]. In standard incubations it was determined that the wild-type strain of *S. manihoticola* (DSM1638 from the Leibniz Institute DSMZ-German Collection of Microorganisms and Cell Cultures) produced GA_4 in concentrations of up to 7 mg per liter of culture filtrate. *Sphaceloma manihoticola* does not produce the commercial gibberellins of GA_3, GA_1 or GA_7 which are also synthesized by *F. fujikuroi*, thus with GA_4 being the main gibberellin produced by *S. manihoticola* it can be isolated in a pure form from culture filtrates of the pathogen [26, 27, 35].

Although plants and fungi produce structurally identical gibberellins, the biosynthetic steps in the pathway for the formation of gibberellins differs significantly [36]. A major contrast in the biosynthetic pathway is the stage at which the hydroxyl groups are introduced. Fungal gibberellic acid biosynthesis requires only cytochrome P450 monooxygenases, while the formation of plant gibberellic acid requires both membrane bound cytochrome P450 monooxygenases and soluble 2-oxyglutarate dependent dioxygenases [32]. In *S. manihoticola* the production of gibberellic acid is controlled by the GA biosynthetic gene cluster.

The gene cluster has been characterized and consists of the genes *SmP450-2* (AM886290.1), *SmP450-1* (AM 886288.1) and (AM 886289.1) *SmP450-4*. These three genes are surrounded by two open reading frames; the bi-functional ent-copalyl diphosphate synthase/ent-kaurene synthase and geranylgeranyl diphosphate synthase [5, 29] (**Figure 3**).

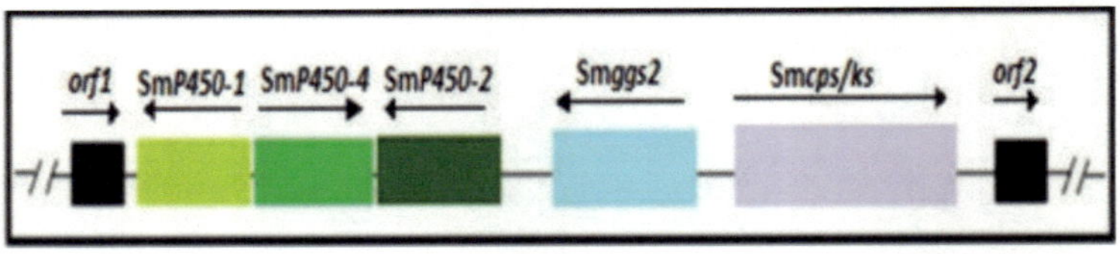

Figure 3. The gibberellin biosynthesis gene cluster of *S. manihoticola* (adapted from Bömke et al. [29]).

5. Detection of Superelongation disease in cassava

Classic methods used in the identification of pathogen infections depend on the observations of seasonally variable elongation and subsequent laboratory confirmation by fungal growth of cultures. However, this can take weeks due to the slow growing nature of the pathogen [5, 13]. With SED, because symptoms are variable and may depend on environmental conditions, visual observations of these symptoms alone are not a reliable indicator of SED infection [12]. For example, although in Trinidad a high incidence was recorded in 2008, elongation of the internodes was not observed [4]. This was also apparent in Barbados in 2014, but hyper-elongation was evident in several cassava growing fields severely infected with *S. manihoticola* in 2015 [28]. Krausz [12], stated it is not peculiar for plants affected with SED not to exhibit elongation of internodes especially in the dry season. Therefore, variability in the consistency and severity of secondary symptoms of the disease decreases the reliability of visually confirming the disease by field observations alone.

Currently there is a disease severity scale based on symptom appearance in infected cassava. The scale was introduced by the Cassava Pathology Program at CIAT [24]. Infected cassava plants are assigned a numerical value for disease severity. As signs and symptoms of SED increased in plants, they are assigned a higher numerical value correlating with the symptoms presented, e.g., a value of 1 is assigned to plants that have no sign of SED, 2 is used for the development of spots or cankers on leaves or petioles, 3 used for signs of cankers on leaves, petioles and stems with severe leaf distortion and 4 is assigned to plants displaying elongation, cankers on leaves, petioles and stems, severe leaf distortion and scorching [20]. Additionally, the use of molecular techniques such as polymerase chain reaction (PCR) has the ability to surpass many of the shortcomings of measuring the disease severity using a disease rating scale. These methods present advantages of being specific, accurate and are faster than traditional techniques.

For analysis of the gibberellic acid gene cluster organization in *S. manihoticola* Bömke et al. [29] synthesized SMP primers. The SMP primers developed were specific for particular SMP transposons responsible for the regulation of the gibberellic acid gene cluster organization. Bömke et al. [29] utilized the SMP primers in order to characterize the gibberellin biosynthetic cluster in *S. manihoticola*. However, the SMP primers were not used to characterize the fungus, *S. manihoticola*. Since then, primers have been developed for amplification of the *SmP450-2* gene (Gen Bank Accession AM 886290) which serves for detection of SED in asymptomatic field and artificially inoculated cassava leaves, with SED [5]. They have also been used to quantify varying disease severity levels in cassava leaves and stem lesions [37].

In 2000, Alvarez and Molina characterized *S. manihoticola* by targeting the internal transcribed spacer (ITS) region of ribosomal DNA using the PCR primers: ITS4 and ITS5 [25]. Brazilian isolates of *S. manihoticola* could also be distinguished from those infecting milkweed and RAPD molecular markers designed for SED were able to distinguish variation in the pathogen population from South central Brazil and Colombia [38]. They suggested that pathogen variation may be determined by geographic location and even smaller locales such as a municipality, indicating centers of pathogen diversity [38].

Despite these established techniques to estimate disease severity and identify the disease, pathogen quantification remains one of the main challenges in the disease management of crops and moreso in SED in cassava.

High quality cassava planting material has a key role in the maintenance of genetic purity and plants free from pathogens and disease. Cassava farmers are constantly faced with the issue of generational build-up of diseases and pathogens through the use and reuse of infected planting material.

The differences in morphological diversity among isolates of the pathogen when cultured on varying media, and the use of molecular markers to identify the pathogen might also suggest that molecular differentiation is present in the pathogen population in the Caribbean islands, as suggested for Colombia and Brazil [6, 38].

6. Disease resistance

Both the SPM-1 and SPM-9 primer pairs designed from the gene *SmP450-2* [5] demonstrated the capability for early detection of the *SmP450-2* in the local varieties Butterstick, White stick, Red stick and the CIAT cultivar CM6604, 24 hours post-inoculation [28]. Contrastingly, the SPM-1 and SPM-9 primers did not detect the *SmP450-2* gene until 14 days' post-inoculation in the MCOL-22 and BRA-383 CIAT cassava varieties [28]. This variability in detection of the *SmP450-2* gene by the SPM primers could be applicable to investigate susceptibility and resistance among local varieties. Thus, the cassava varieties in which the SPM primers detected the expected amplicon 24 hours post-inoculation were possibly more susceptible to SED infection.

The CIAT library of cassava cultivars describes the BRA-383 cultivar as a landrace derived from Brazil and commonly referred to as, "Vassourao." The BRA-383 variety grows to a height of 200 cm and has linear-pandurate leaf lobes with white roots. Titus et al. [19] in their report on commercial cassava production described BRA-383 as being susceptible to SED infection. Additionally, the MCOL-22 variety, also known as "Uvita" originated from Colombia and was described to have white roots and grows to a height of 150 cm with the leaf lobes being straight or linear. Interestingly, the MCOL-22 variety has also been recognized as susceptible to SED infection [6], but is widely cultivated in some islands.

Although both the BRA-383 and the MCOL-22 varieties have been previously described as susceptible to SED in the region, investigations using the *SmP450-2* gene as a determinant of SED susceptibility demonstrated both varieties expressed the GA4 gene much later compared to the other local varieties examined. This may suggest differential susceptibility in local varieties when compared to well-described CIAT varieties. Zeigler et al. [6] examined several morphological features of cassava that could correlate with SED resistance and suggested that stem and leaf cuticle could account for resistance, because the water-borne inoculum cannot adhere to susceptible leaf tissues with thick cuticles. Physiological characteristics such as plant architecture, apical maturity and growth rate were also suggested to affect the resistance of cassava varieties to SED infection [6]. Thus, it is possible that the cassava varieties demonstrating susceptibility to SED infection in the Caribbean islands have physiological characteristics which aid or support in *S. manihoticola* infection.

It is also possible that the local varieties in the Caribbean islands are no longer the same varieties as described by the CIAT database. This raises the question for further investigation into the origins and characteristics of currently disseminated cassava stakes. Locally it is known that mixing and exchange of cassava stakes among farmers are common practices in most farming communities and could result in mis-identification of cassava varieties. Traditionally, tolerant varieties against SED such as Mex 55 and Mex 23 which were imported from CIAT are no longer tolerant to SED in Barbados, and are no longer distinguishable from locally grown varieties [19, 20], this could further account for the consistency observed in early detection of the *SmP450-2* gene in susceptible local varieties. Studies are ongoing for the resistance performance of current CIAT varieties in the island.

The presence of cassava varieties showing such a high level of susceptibility to SED within the region could be the result of weak pathogenic specialization due to crossing two highly resistant cassava varieties with high physiological resistance. The result is a loss of modifying factors ultimately causing the progeny to have traits which are likely to be overcome by the highly variable pathogen [6].

7. Disease management

In cassava production, and farming in general, healthy planting material is the first step to having a disease-free crop. Farmers sometimes plant more than one variety of cassava per plot and the different varieties also have differences in disease resistance. Over 20 cassava varieties have been shown to express a level of resistance to Superelongation [39].

The most effective means of preventing infection and the spread of SED in cassava is by the planting of disease-free stakes in areas where weeds are scarce [19]. As a precautionary measure, stakes can be treated with a broad spectrum fungicide [19, 20]. In areas where SED is endemic, plants affected by the disease should be removed from the field and burnt. In the Caribbean islands, research by Chandler [20] produced a fact sheet indicating that to eradicate SED, Captafol® ($C_{10}H_9Cl_4NO_2S$) (Chevron™) at 400 ppm could be used as a dip for pre-treating planting stakes. Recommended weed treatment included: pendimethalin (Herbadox®, BSAF, Chile) at 4.5 liters per hectare (ha) or pendimethalin (Herbadox® 45 CS) at 4 liters per ha together with diuron (Karmex®DF (DuPont™)) at 1.5 kg per ha, or Karmex®DF at 1.5 kg per ha together with alachlor (Lasso™) at 3 liters per ha. The recommended practice was to inter-row directed sprays with Gramoxone® (Syngenta™).

Titus et al. [19] conducted research on hot water treatment of cassava as a method of disease management. The treatments used in the study included: thermotherapy, stem cuttings placed in water bath at 49°C for 49 minutes; stem cuttings immersed for 5 minutes in the commercial fungicide Kocide® 3000 ($Cu(OH)_2$) (DuPont™) and stem cuttings immersed for 5 minutes in the commercial fungicide Score® ($C_{19}H_{17}Cl_2N_3O_3$) (Syngenta™). The research determined that the treatment using Kocide® recorded the lowest average AUDPC score of 6.3.

In addition to SED being spread via infected stake cuttings, there has been increasing emphasis on the role of weeds as hosts of *S. manihoticola* infection. Chandler [20] described weed control as being critical during the first few months before the lead canopy closed over. The most common weeds occurring in cassava fields are grasses such as: *Imperata cylindrical* (Spear grass), *Cynodon dactylon* (Bermuda grass); *Panicum maximum* (Guinea grass), and *Pennisetum polystachion* (feathery pennisetum) [40]. *Sphaceloma manihoticola* and *Elsinoë* species are also common pathogens on weedy and ornamental plants related to cassava in Central and South America. These species of plants affected by *S. manihoticola* include *Jatropha curcas* (physic nut), *Jatropha aconitifolia* Muell., *Manihot glaziovii* (ceara rubber), *Euphorbia pulcherrima* Willd., and weeds such as *Euphorbia brasiliensis*, *Euphorbia heterophylla* L., *Euphorbia hypericifolia* L., and *Euphorbia prunifolia* [6, 41–43].

There are multiple challenges to integrating effective and economical weed control. Rapidly growing weeds in cassava farms will cover the ground almost completely and utilize the nutrients and water from the soil, limiting these materials for cassava plant growth [42]. Weed growth results in shading of the cassava plant and therefore decreases sunlight exposure to young cassava plants. Therefore, it is almost impossible to remove the weeds without affecting the growing cassava.

Moreover, observations of SED infections in Costa Rica occurring immediately after planting in grassy fields may further suggest the possibility of grass species being alternate hosts of *Sphaceloma* species (Alvarez, E. personal communication).

Bermuda grass is one of the most common weeds found in cassava production [44] and this weed among others harbors pests that later infect cassava. Bermuda grass has been recently found to be a reservoir host of *S. manihoticola* in Barbados by using the *Spm1* molecular marker [45]. *Cynodon dactylon* may therefore be a potential reservoir host of *S. manihoticola* in the Caribbean.

One mechanism by which fungi spread, even globally, is via wind so at a local level since tomentose spores are observed in *S. manihoticola* they may easily be carried by the wind from the Bermuda grass to cassava found nearby.

A combination of these factors satisfies the parameters of a disease triangle; cassava fields throughout the island—the host, a weed that harbors the pathogen and others that weaken the plants' robustness and suitable environmental conditions favorable to pathogen growth. This finding of alternate hosts is therefore important in improving disease management of SED in local communities in the Caribbean islands.

8. Conclusion

As cassava production has increased in the region the impact of SED has also risen. The long-term consequences for productivity and success of the cassava industry are yet to be measured. Disease mitigation strategies are use of clean stakes as planting material, use of SED or tolerant resistant varieties, weed control, crop rotation, and fungicide treatments. Disease awareness in cassava cropping area is also required so continuous famer training in disease recognition is important in mitigating the devastation of this disease in cassava planting areas, especially in the Caribbean islands.

Author details

Angela T. Alleyne

Address all correspondence to: angela.alleyne@cavehill.uwi.edu

Department of Biological and Chemical Sciences, Faculty of Science and Technology, The University of the West Indies, Cave Hill Campus, Bridgetown, Barbados

References

[1] El-Sharkawy MA, Cock JH, Lynam JK, del Pilar Hernandez A, Cadavid LLF. Relationships between biomass, root yield and single leaf photosynthesis in field grown cassava. Field Crops Research. 1990;**25**(12):183-201

[2] International Fund for Agricultural Development (IFAD), Food and Agriculture Organization of the United Nations. The World Cassava Economy: Facts, Trends and Outlook. Rome, Italy: FAO; 2000. pp. 1-41

[3] FAOSTAT/Food and Agriculture Organization of the United Nations. Available from: http://www.fao.org/faostat/en/

[4] Reeder R, Kelley PL, St. Hill AA, Ramnarine K. Superelongation disease, caused by *Elsinoë brasilensis*, confirmed on cassava in Trinidad and Tobago. Plant Pathology. 2009;**58**(4):800

[5] Alleyne AT, Gilkes JM, Briggs G. Early detection of Superelongation disease in Manihot esculenta Crantz (cassava) using molecular markers for gibberellic acid biosynthesis. European Journal of Plant Pathology. 2015;**141**:27-34

[6] Zeigler RS, Alvarez E, Lozano JC. Characteristics of cassava resistance to Superelongation disease (*Elsinoë brasilensis*). Tropical Pest Management. 1983;**29**(2):148-158. DOI: 10.1080/09670878309370791

[7] Alvarez E, Molina ML. Characterising the *Sphaceloma* fungus causal agent of Superelongation disease in cassava. Plant Disease. 2000;**84**(17):423-428

[8] Takatsu A, Fukuda S, Hahn SK, Caveness FE. Current status of cassava disease in Brazil. In: IITA, editor. Workshop on Global Status of and Prospects for Integrated Pest Management of Root and Tuber Crops in the Tropics; 25-30 October, 1987; Ibadan, Nigeria. Nigeria: IITA; 1990. pp. 127-131

[9] Hillocks RJ, Wydra K. Cassava: Biology, Production and Utilization. Oxon, U.K.: CABI Publishing; 2002

[10] Zeigler RS. The Superelongation disease of cassava: Pathogen taxonomy and physiology, and some characteristics of host resistance [dissertation]. New York: Cornell University; 1982. 133p

[11] Jenkins AE, Bitancourt AA. Revised descriptions of the genera *Elsinoë* and *Sphaceloma*. Mycologia. 1941;**33**(3):338-340

[12] Krausz J. The Superelongation disease of cassava [dissertation]. New York: Cornell University; 1976

[13] Zeigler RS, Lozano JC. The relationship of some *Elsinoë* and *Sphaceloma* species pathogenic on cassava and other *Euphorbiaceaae* in Central and South America. The Phytopathology. 1983;**73**(2):293-300. DOI: 10.1094/Phyto-73-293

[14] Bitancourt A, Jenkins AE. Estudos sobre as Miringiales. Vinte Novas especies de Elsinoaceas Neotropicais. Arquivos do Instituto Biologico. 1950;**20**:1-28

[15] Fan XL, Barreto RW, Groenwald JZ, Bezerra JDP, Peirira OL, Cheewangkoon R, et al. Phylogeny and taxonomy of the scab and spot anthracnose fungus *Elsinoë* (*Myriangiales, Dothiddeomycetes*). Studies in Mycology. 2017;**87**:1-41

[16] Lozano JC, Toro J, Castro A, Belotti A. Production of Cassava Planting Material. Centro International de Agricultura Tropical (CIAT). Cali, Colombia: CIAT; 1977. 27p

[17] International Center for Tropical Agriculture (CIAT). Cassava: The Latest Facts about an Ancient Crop. Cali, Colombia: CIAT; 1996

[18] EL-Sharkawy MA. Cassava biology and physiology. Plant Molecular Biology. 2004;**56**: 481-501. DOI.org/10.1007/s11103-005-2270-7

[19] Titus P, Lawrence J, Seesahai A. Commercial cassava production. Technical Bulletin, Caribbean Agricultural Research and Development Institute (CARDI). 2011;**5**:1-16. St. Augustine, Trinidad

[20] Chandler F. Caribbean Agricultural Research and Development Institute (CARDI). Commercial Cassava Production. Factsheet. 1992; Order no.: CP-F/16-89

[21] Ospina B, Poats S, Henry G. Integrated Cassava research and development projects in Colombia, Ecuador and Brazil: An overview of CIAT's experiences. In: Dufon D, O'Brien GM, Best R, editors. Cassava Flour and Starch: Progress in Research and Development. Colombia: CIAT; 1996. p. 271

[22] Fregene MA, Suarez M, Mkumbira J, Kulembeka H, Ndedya E, Kula A, et al. Simple sequence repeat marker diversity in cassava landraces: Genetic diversity and differentiation in an asexually propagated crop. Theoretical and Applied Genetics. 2003;**107**(6): 1083-1093

[23] Rabbi I, Kulakow PA, Manu-Aduening JA, Danky AA, Asibuo JY, Parkes EY, et al. Tracking crop varieties using genotyping by sequencing markers: A case study using cassava (*Manihot escuclenta* Crantz). BMC Genetics. 2015;**16**:115. DOI: 10.1186/s12863-015-0273-1

[24] International Center for Tropical Agriculture (CIAT). Production Systems of Cassava. Annual Report. Cali, Colombia: CIAT; 1973

[25] Alvarez E, Meija JF. Molecular and pathogenicity characterization of *Sphaceloma manihoticola* isolates from South-Central Brazil. Plant Disease. 2003;**87**(11):1322-1328

[26] Zeigler RS, Powell LE, Thurston HD. Gibberellin A4 production by *Sphaceloma manihoticola* causal agent of Superelongation disease. Phytopathology. 1980;**70**(7):589-593. DOI: 10.1094/Phyto-70-589

[27] Rademacher W, Grabe JE. Gibberellin A4 produced by *Sphaceloma manihoticola*, the cause of the Superelongation disease of cassava (*Manihot esculenta*). Biochemical and Biophysical Research Communications. 1979;**91**(1):35-40

[28] Gilkes JM. Quantification of Superelongation disease in *Manihot esculenta* using *Spm* molecular markers [thesis]. Bridgetown, Barbados: The University of the West Indies, Cave Hill Campus; 2015. 180 pp

[29] Bömke C, Rojas MC, Gong F, Hedden P, Tudzynski B. Isolation and characterization of the gibberellin biosynthetic gene cluster in *Sphaceloma manihoticola*. Applied and Environmental Microbiology. 2008;**74**(17):5325-5339

[30] Liao HL, Chung KR. Genetic dissection defines the roles of elsinochrome phytotoxin for fungal pathogenesis and conidiation of the citrus pathogen *Elsinoë fawcetti*. Molecular Plant Microbe Interactions. 2008;**21**(4):469-479

[31] Wang LY, Bau HJ, Chung KR. Accumulation of elsinochrome phytotoxin does not correlate with fungal virulence among *Elsinoë fawcetti* isolates in Florida. Journal of Phytopathology. 2009;**157**(10):602-608

[32] Hedden P, Phillips A. Gibberellin metabolism: New insights revealed by genes. Trends in Plant Science. 2000;**5**:523-530

[33] Hedden P, Thomas SG. Gibberellin biosynthesis and its regulation. Biochemical Journal. 2012;**444**:11-25

[34] Rademacher W. Occurrence of gibberellins in different species of the fungal genera *Sphaceloma* and *Elsinoë*. Phytochemistry. 1992;**31**(12):4155-4157

[35] Bhattacharya A, Kourmpetli S, Ward DA, Thomas SG, Gong F, Powers SJ, et al. Characterization of the fungal gibberellin desaturase as a 2-oxoglutarate-dependent dioxygenase and its utilization for enhancing growth. Plant Physiology. 2012;**160**(2):837-845

[36] Malonek S, Rojas MC, Hedden P, Gaskin P, Hopkins P, Tudzynski B. The NADPH-cytochrome P450 reductase gene from *Gibberella fujukuroi* is essential for gibberellin biosynthesis. The Journal of Biological Chemistry. 2004;**279**(24):25075-25084

[37] Alleyne AT, Gilkes JM. Quantification of disease severity in field and laboratory grown cassava by application of molecular markers for measuring Superelongation disease. In: Phytopathology. St. Paul, Minnesota, USA: American Phytopathology Society; 2014

[38] Alvarez E, Llano G, Loke JB, Meija JF. Applying biotechnology tools to control plant diseases. Fitopatología Colombiana. 2005;**28**:93-97

[39] Kawano K, Umemura K, Kano Y. Field assessment and inheritance of cassava resistance to Superelongation disease. Crop Science. 1983;**23**:201-205

[40] Melifonwu A. Weeds and their control in cassava. African Crop Science Journal. 1994;**2**(4): 519-530

[41] Swart L, Crous PW, Kang JC, Mchau RA, Pascoe I, Palm ME. Differentiation of species of *Elsinoë* associated with scab disease of *Proteaceae* based on morphology symptomatology, and ITS sequence phylogeny. Mycologia. 2001;**93**:365-379

[42] Agumagu A, Adesope OM, Mathews-Njoku EC. Perception of infestation problems on cassava farms and preference for weed management practices in humid agro-ecological zone of Nigeria. Agriculture Conspectus Scientificus. 2008;**73**(2):115-120

[43] Valadatilde D, Santos JB, Carvalho FP, Barraos-Silva E, Sebastiatilde J, Concenccedil G. Competitive capacity of cassava with weeds: Implications on accumulation of dry matter. African Journal of Agricultural Research. 2013;**8**(6):525-531

[44] Melifonwu A, James B, Aihou K, Weise S, Amah E, Gbhaduidi B. Weed Control in Cassava Farms. Cotonou, Benin: International Institute of Tropical Agriculture (IITA); 1997. 13p

[45] Alleyne AT, Gilkes JM. Occurrence of the cassava pathogen *Sphaceloma manihoticola* in a reservoir host, Bermuda grass (*Cynodon dactylon*) in Barbados. Phytopathology. 2014; **104**(S3):6

Section 3

Improving Cassava Production

Chapter 4

Storage Root of Cassava: Morphological Types, Anatomy, Formation, Growth, Development and Harvest Time

Luiz JCB Carvalho, Josefino F. Filho,
James V. Anderson, Priscila G. Figueiredo and
Songbi Chen

Additional information is available at the end of the chapter

http://dx.doi.org/10.5772/intechopen.71347

Abstract

Cassava (*Manihot esculenta*, Crantz) is considered a starchy root crop that provides staple food for millions of people in tropical and subtropical regions of the world. Research efforts are directed toward genetic breeding and cultivation of cassava to improve cassava storage root starch production, nutritional values, and industrial utilization. Cassava storage root (CSR) is a vegetative storage organ with indeterminate type of growth that has a central cylinder (edible part) originated by the swelling of primary root and crown roots. Comprehensive studies on thickened primary root (secondary growth) are rare, incomplete, and to a certain extent, missing. In this chapter, we review and forward studies that move our knowledge on cassava storage root (CSR). CSR generally forms up to 12–14 storage root (SR) per plant, which can originate from three sources of propagating plant materials as well as being induced *in vivo* and *in vitro*. Types of storage root (morphologically defined), CSR physiology, tissue anatomy/histology (secondary growth), chemical composition of the edible part, biochemical features, gene expression and proteomics as secondary growth proceeds are of major importance in order to breed cassava plant for agriculture utilization. Storage root morphology varies in shape from cylindrical to globular. Time to initiation of storage root formation varies from 45 to 90 days after planting (DAP), depending on the leaf auxiliary bud position in the vegetative propagating material at the plant source. Storage root growth, starch accumulation, and nutrient contents are largely dependent on genotypes. Storage root anatomy can be identified by eight characteristics common to a root with secondary growth and starch reserve variants. Histological characterizations can be used to identify cell types of primary and secondary meristems, procambium, vascular cambium, phellogen, phelloderm, primary and secondary xylem and phloem, storage parenchyma and sclerenchyma. Three types of meristematic cell differentiations occur as secondary growth proceeds; one due to cork cambium with plane perpendicularly oriented cell division, second due to plane longitudinally oriented cell division in the root apex, and third longitudinally oriented in the epidermal cells. Chemical composition of the storage root varies in the central cylinder (edible part) depending on the sample position in the root and the plant genotype. Therefore, biochemical characteristics are known to change with tissue age as secondary growth proceeds. Moreover, the composition of stored starch varies with tissue age across the central cylinder and may be used as a physiological indicator for bulk storage root maturation and storage root harvest time.

Keywords: storage root, secondary growth, physiology, development, maturation

1. Introduction

Cassava (*Manihot esculenta,* Crantz) is a starchy root crop that provides a staple food source for millions of people in tropical and subtropical regions of the world. Worldwide, research efforts are directed toward genetic breeding and cultivation of cassava to improve cassava storage yield, root starch production, nutritional quality, and industrial utilization. Cassava storage root (CSR) is an indeterminate, vegetative storage organ that results from the swelling of primary root crown root, with the central cylinder as the edible part. Studies on secondary growth of cassava storage root (CSR) are rare, incomplete, and to a certain extent, missing. In this chapter, we review our comprehensive studies related to (CSR) morphology, storage root (SR) formation, SR physiology (growth analysis, development and maturation), anatomy/histology (secondary growth), and biochemical (carbohydrate, carotenoids, proteins, and gene expression) characteristics as secondary growth proceeds in order to understand yield of CSR.

2. Storage root of cassava features

2.1. Storage root utilization, shape and diversity

The practical utilization of CSR can be described in relation to 11 features that vary in importance, depending on the end use. These important characteristics are ranked (**Table 1**) in relation to their utilization for fresh consumption and industrial use (two most common uses of cassava by mankind).

2.2. Storage root of cassava diversity

Diversity in CSR morph types (**Figure 1**) is considered important cassava breeding traits when considering mechanical harvest.

Diversity in central cylinder of CSR (**Figure 2**) for carotenoids (**Figure 2A**), and carbohydrate and starch iodine staining pattern (**Figure 2B**) indicate a large genetic [1] and are the most popular traits used for genetic breeding proposes [2–4].

2.3. Cassava storage roots formation and induction

A cassava plant can form up to 14 storage roots per plant, depending on the genotype. Storage root can initiate from three distinct sources (**Figure 3**) of plant propagating material. These include direct embryonic root formation at the seed germination event (**Figure 3A**) to form a single-tap SR (**Figure 3B**), the leaf axillaries bud in stem cuttings forming a single SR

Storage root features/utilization	Fresh consumption	Industrial cassava
Storage root format	+++++	+++++
Early harvest storage root	+++++	+++++
Storage root HCN content	+++++	+
Storage root color of central cylinder	+++++	+
Storage root high fiber content	+	++
Storage root high starch content	++	+++++
Storage root starch quality	+++++	+++
Storage root vitamins	+++++	+
Storage root easy peel off	+++++	+++++
Storage root cooking	+++++	–
Storage root rotting	+++++	+++++

Table 1. Features of cassava storage root and its importance ranked in association with practical utilization by mankind.

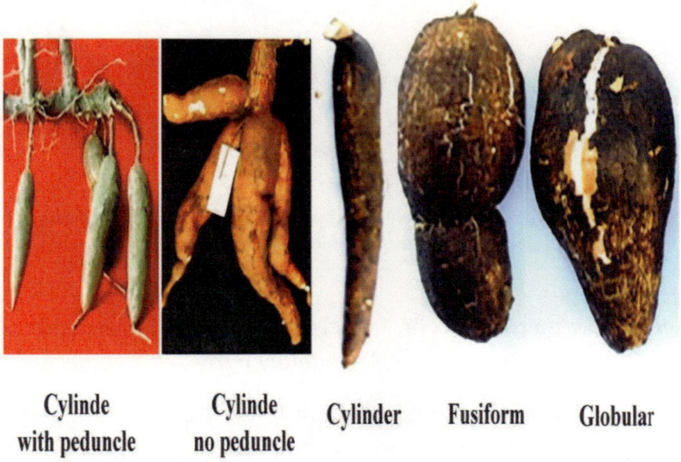

Figure 1. Cassava storage root morphological types.

(**Figure 3C**), and a number of nodal callus from the bases of stem cuttings forming more than one SR (**Figure 3D**), and buried nodes at the base of stem cuttings forming SR or induced *"in vitro"* plants [5]. The effect of leaf bud position on the stem cutting from a 1-year old mother plant is observed in **Table 2.**

2.4. Storage root anatomy and histology features

The anatomy of cassava storage root was first described by Rateaver [7] and more recently at [6]. From the basic secondary growth of CSR shown in **Figure 4**, it is possible to recognize at least 12 cell types in the storage root associated to secondary tissues including primary

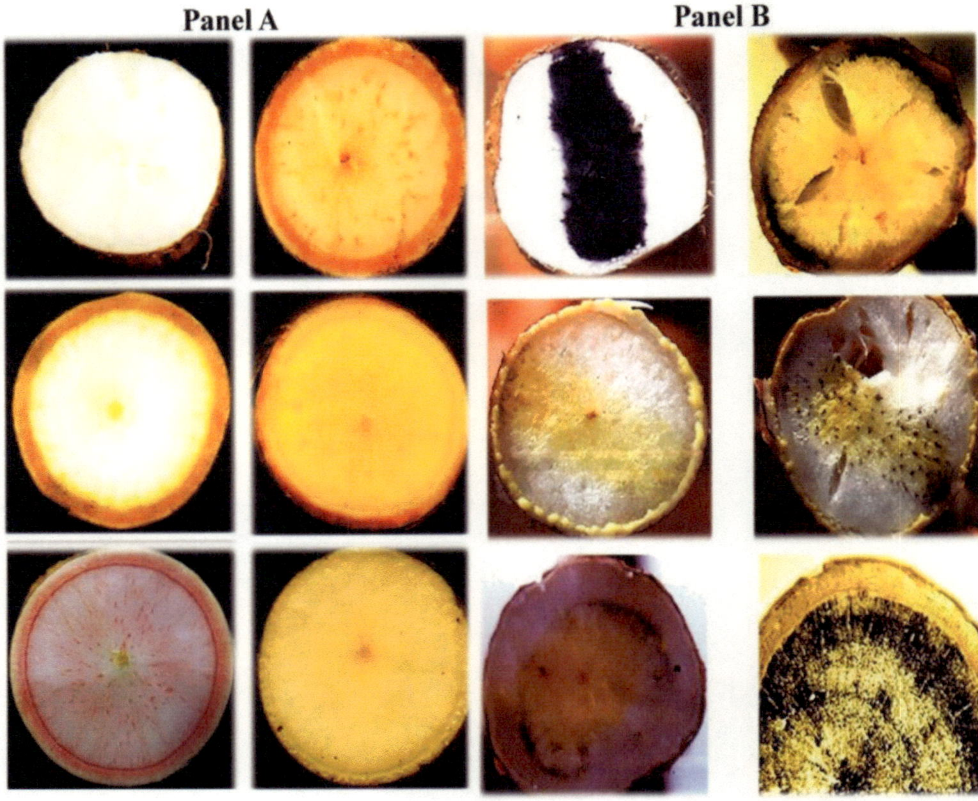

Figure 2. Diversity of cassava storage root in the central cylinder (edible part) related to carotenoid types and content (Panel A) and carbohydrate types as stained with iodine solution (Panel B).

meristem cells, secondary meristem cells, vessels, primary xylem, secondary xylem, primary phloem, secondary phloem, parenchyma cells, sclerenchyma, and epidermal cells.

Cassava storage roots grow in length from the apical meristem forming new cells continually, as generally observed in other plants carrying root secondary growth. In addition to the primary tissues (**Figure 4F)**, cassava storage root has secondary tissues that add thickness to a primary root (**Figure 4B–E**). Secondary tissues develop from two types of meristems. Based on these observations, from fibrous root, we defined six stages of CSR growth (**Figure 4G**). The cork cambium, originates beneath the epidermis, generally by pericycle dedifferentiation, producing cork cells and pushes them toward the outside of the root. As the cork expands outward, the endodermis, cortex, and epidermis die and peel off. The cork replaces them and becomes the outer covering of the root. The other secondary meristem, the cambium, lies between the primary xylem and the primary phloem. It produces secondary xylem cells toward the center of the root, and secondary phloem cells toward the outside. Qualitatively (**Figure 5**) and quantitatively (**Figure 6**), this pattern of tissue and cell type distribution in CSR over DAP as secondary growth proceeds indicates that CSR peel (secondary phloem, phellogen, and phelloderm), vascular cambium, and secondary xylem showed in **Figure 6A**, and

Figure 3. Source of storage root from cassava planting material. Germinating seeds (A) forming single-tap storage root (B). Leaf bud in the stem cutting from plant material forming single-tap storage root from leaf axillary bud germination (C) and callus on the stem cutting forming multiple storage roots (D).

Stem cutting bud position (bottom to top)	Counting bud age (DAP)	Fibrous root (S1)	Swelled root (S2)	Swelled root (S3)	Swelled root (S4)
StP1	30	32	1.2 (cm)	3.9 (cm)	11.3 (cm)
StP2	45	66	1.9 (cm)	3.0 (cm)	0
StP3	52	28	1.0 (cm)	3.2 (cm)	0
StP4	59	42	0.65 (cm)	0	0
StP5	66	14	0	0	0
StP6	73	23	0	0	0
StP7	87	50	0	0	0
StP8	94	21	0	0	0
StP9	115	29	0	0	0
StP10	122	0	0	0	0

Table 2. Cassava storage root formation in relation to leaf bud position in the stem cuttings from a 1-year old mother plant. Number of storage roots formed at leaf axillary bud from stem cuttings of the plant material. Initial fibrous root and defined stage of storage root S1. S2, S3, and S4 (as shown in **Figure 4G**) were based on root diameter (cm) starting 30 days after planting (DAP).

central cylinder (vessels and parenchyma cells in secondary xylem) shows opposite fashion. While secondary xylem peels, as well as vessels decrease with DAP, the secondary xylem and secondary parenchyma cells increases. Based on this analysis, we developed a tissue layer sampling system (**Figure 7**) and used the procedure for studies on biochemical features such as

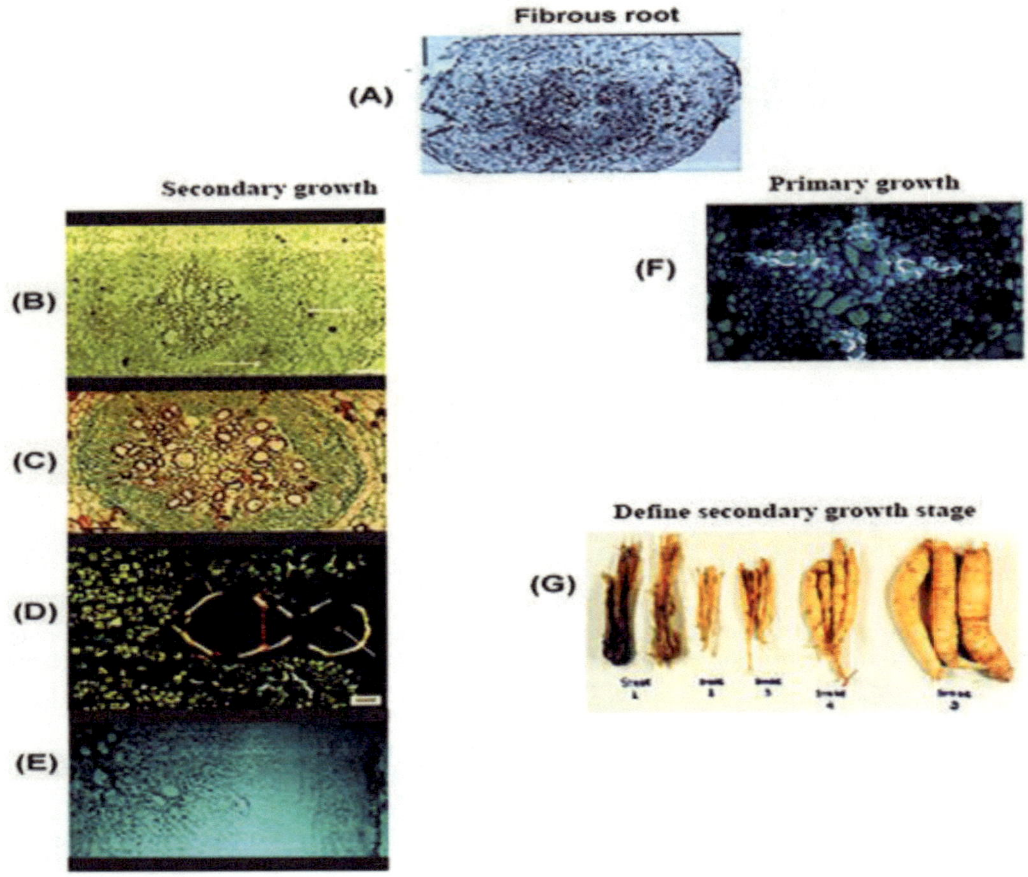

Figure 4. Recognizing storage root anatomy change initiation and advanced secondary growth stages in storage root of cassava. (A) refers to the initial fiber root; (B) refers to the initial pro cambium differentiation in fibrous root with pericycle dedifferentiation; (C) refers to the early events of secondary growth initiation; (D) refers to the complete secondary tissue formation with mature vessels; (E) refers to full secondary tissue formation; (F) refers to primary growth in fibrous root; and (G) defined six stages of storage root formation based on SR diameter.

carbohydrate (single sugar and starch) content [3], amylose percent variation [4], protein content variation [7], carotenoid content and type variation [1], and gene expression analysis [8–10].

2.5. Storage root growth, development, and physiological maturation

Storage root growth analysis was performed based on sampling SR at different time points after stem cuttings were planted in field plots at EMBRAPA Cerrados (Lat 15°35,769°) (Long 47°42,664°) and (Alt 977m) for a crop season of up to 170 days after planting (DAP) using genotypes for industrial use (cv.436) and fresh consume (cv. 982). Developmental stages of storage root (SR) were defined based on SR diameter (cm), SR length (cm), carbohydrate, carotenoid composition and content, protein content, fiber content, and fiber/starch ration to

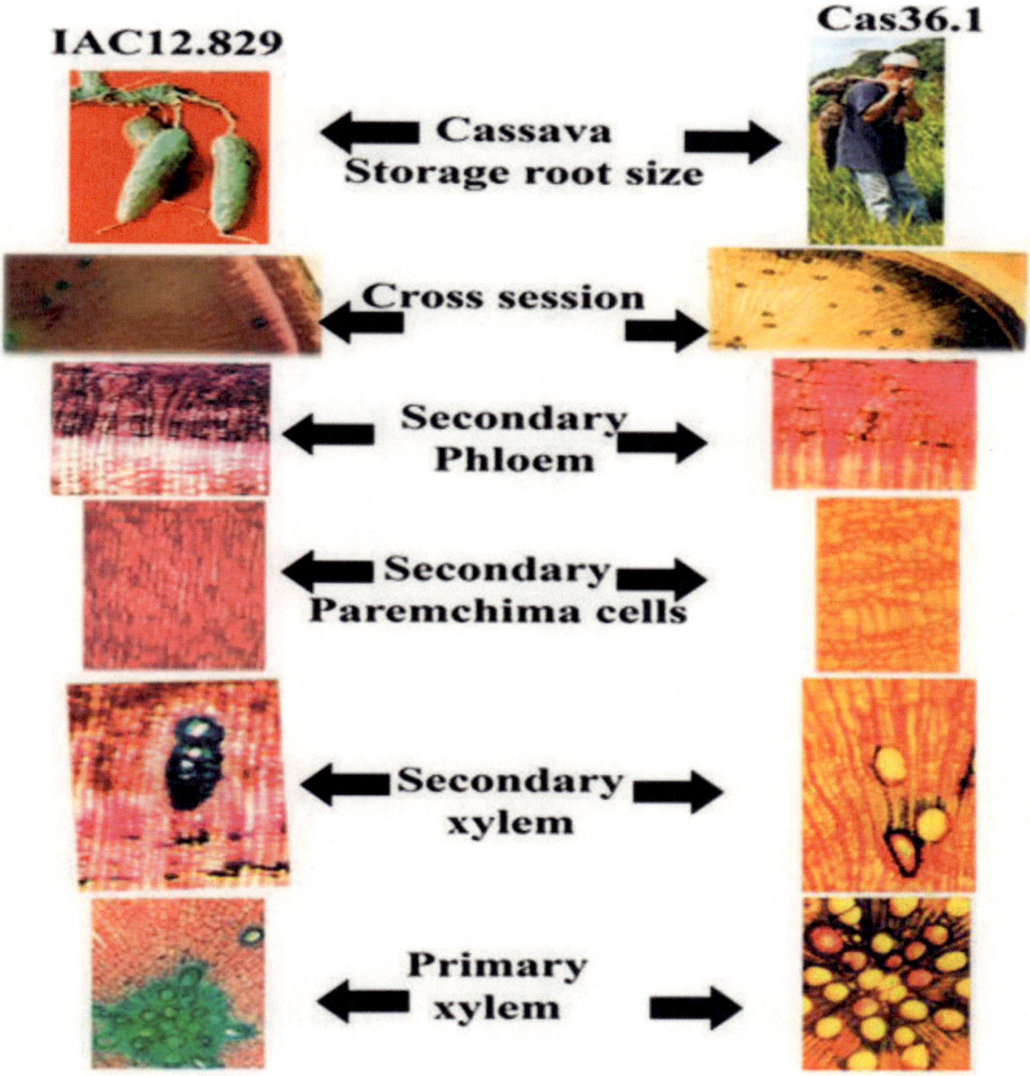

Figure 5. Visualization of cassava storage root morphology type (root size and shape) from two contrasting cassava genotypes. Cultivar IAC12.829 refers to commercial cultivar with the traditional type of storage root. Landrace Cas36.1 refers to a sugary cassava with giant storage root. Storage root tissues distinctions are observed. Cross session shows pattern of different stain with toluidine blue stain (traditional cassava) and iodine stain (sugary cassava). Microscopic observation for the major tissue types in both cassava types. Tylosis formation is observed only in sugary genotype.

accomplish harvest time (physiological maturation). Results shown in **Figure 8** indicate that CSR formation initiates 30 DAP, reaching a maximum number of SR (12–14) by 90 DAP, SR diameter increased linearly up to 170 DAP, while SR length reach a plateau around 40–70 DAP **(Figure 8 Panel A**) depending on the genotype. Either SR dry matter (%) or SR dry weight (gram/plant), and starch accumulation (gram/plant) extended up to 170 DAP and is largely dependent on the genotype (**Figure 8 Panel B**).

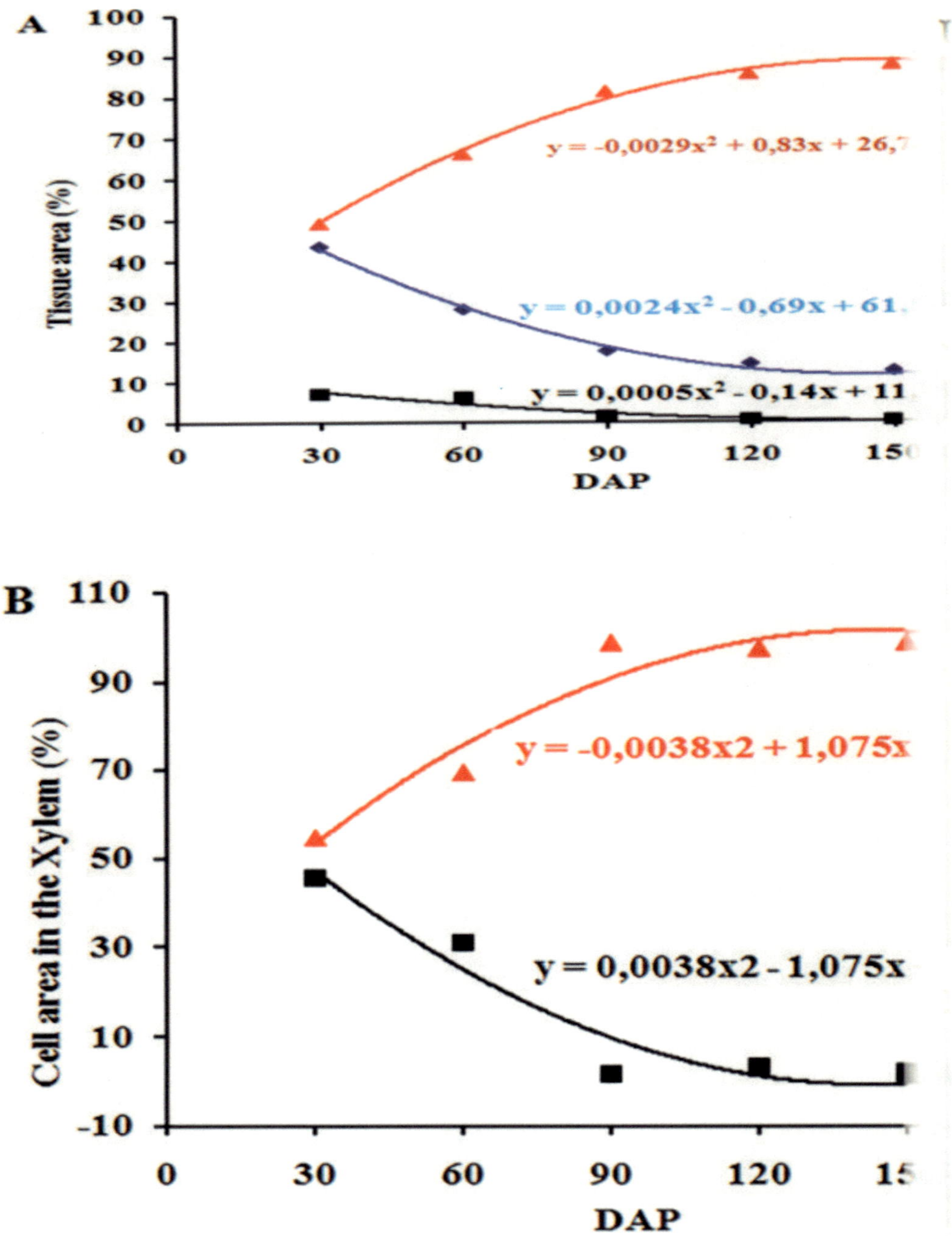

Figure 6. Changes in proportion distribution of tissue and cells type in cassava storage roots as secondary growth proceeds. (A) Refers to tissue of peel (secondary phloem, phellogen, and phelloderm), vascular cambium, and secondary xylem. (B) Refers to vessels and parenchyma cells in secondary xylem.

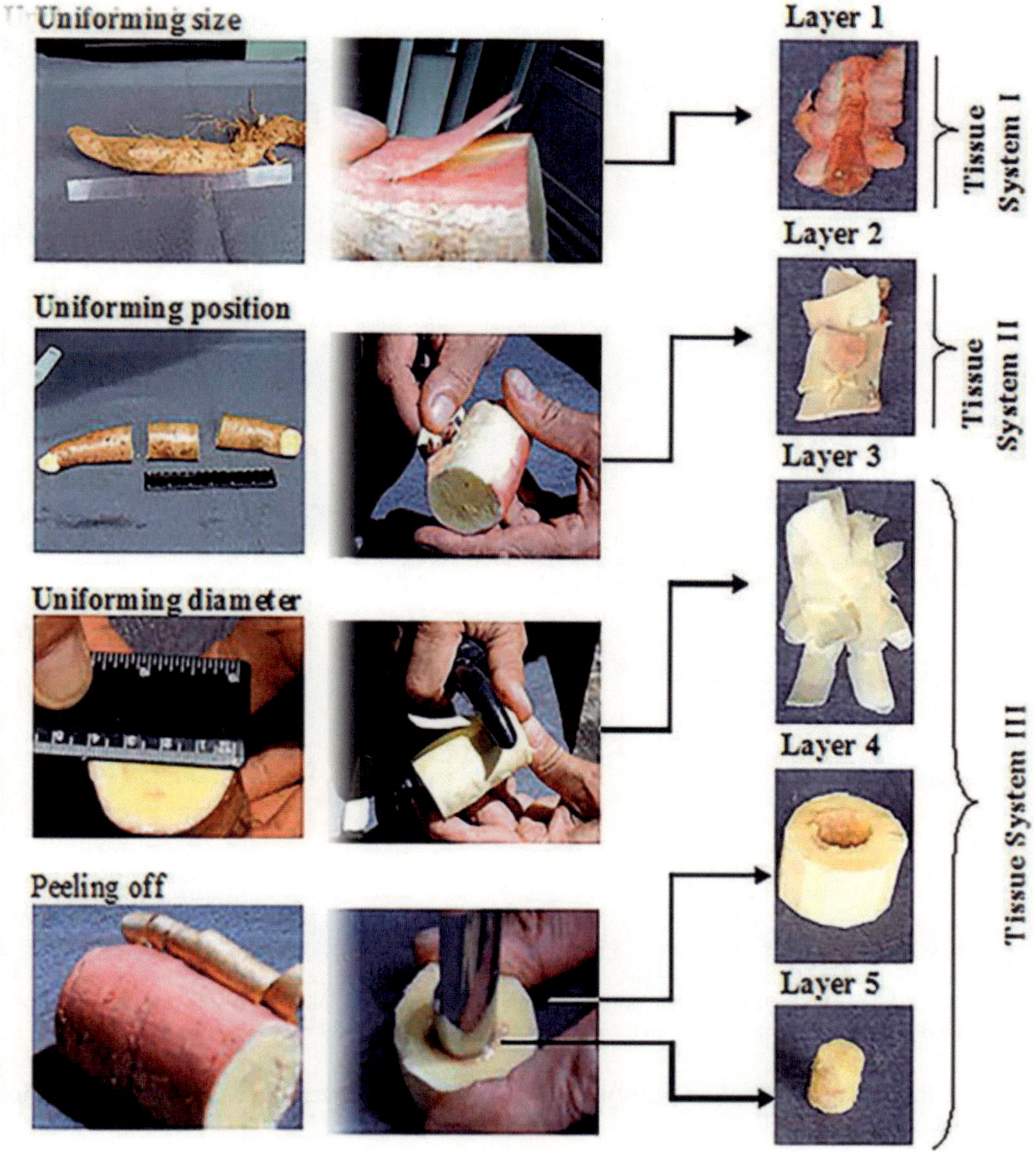

Figure 7. Step by step for storage root tissue sampling system used to further biochemical feature studies of cassava storage root as secondary growth proceeds. Tissue sample I (layer 1), tissue sample II (layer 2), and tissue sample III (layer 3, layer 4, layer 5). Tissue cell compositions are as described in **Figure 5**.

The SR maturation (physiological maturation), as taken by the rate of CSR growth, starch accumulation, and crude fiber accumulation, and crude fiber/starch ratio vary in relation to conventional utilization of the crop (**Table 3**). The major differences occurring are early harvest

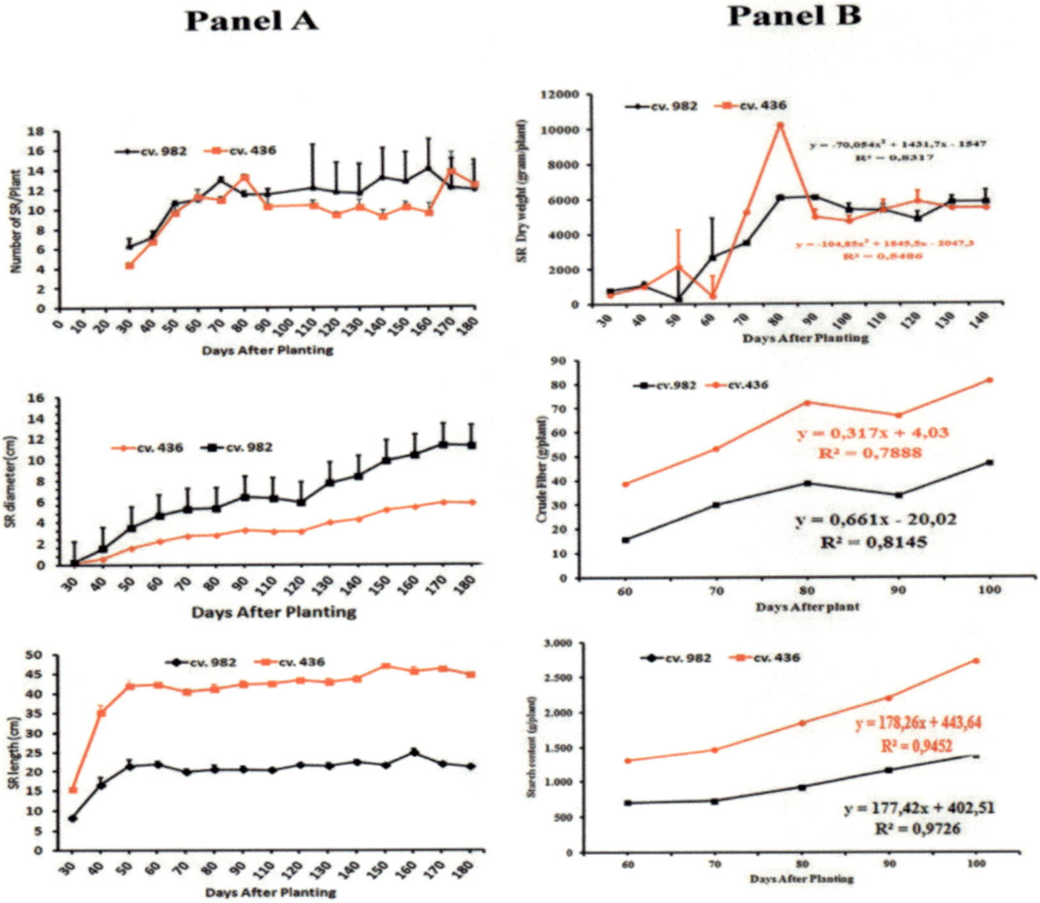

Figure 8. Storage root formation, growth, and development analysis. Panel A—storage root formation, referring to number of storage root per plant, storage root central diameter, and storage root length. Panel B—referring to total dry matter, starch, and crude fiber accumulation over time. Plants were grown at EMBRAPA Cerrados (Latitude 15°35,769°) (Longitude 47°42,664°), and (Altitude 977 m) for a crop season of up to 170 days after planting (DAP).

time for the fresh consumption genotype (cv. 982) and late harvest time for the industrial use genotype (cv.436).

The overall chemical composition of CSR has recently been reviewed [1]. The major conclusions indicate that fresh peeled cassava storage roots are rich in carbohydrates (30–35%), low in protein (1–2%), and fat (<1%). In addition, CSR has nutritionally significant amounts of calcium (50 mg/100 g), phosphorous (40 mg/100 g), and vitamin C (25 mg/100 g), and poisoned values of cyanogenic glycosides upon the hydrolyses of linamarin [11, 12, 15, 16]. In this chapter, we forward our knowledge on nutritional values of CSR based on three major biochemical features that lead to more precise natural variation in the composition and accumulation of carbohydrates (free sugar and starch), carotenoids (type and content), and proteins (content and exploratory functionalities) in the CSR central cylinder.

http://dx.doi.org/10.5772/intechopen.71347

Storage root growth parameters/genotype	cv. 982	cv. 436
Days to form storage root (DAP)	60	90
Root growth rate (g/plant/day)	1431.70	1845.5
starch accumulation rate (starch gram/root/day)	177.42	178.26
fiber accumulation rate (fiber gram/root/day)	0.661	0.317
Fiber/starch ration	0.0037256	0.0017783
Harvest time	Early season	Late season

Table 3. Storage root growth and development parameters of an early (cv. 982) and late season (cv.436) harvest time in cassava crop. Plants were grown at EMBRAPA Cerrados (Latitude 15°35,769°) (Longitude 47°42,664°) and (Altitude 977 m) for a crop season up to 170 days after planting (DAP).

2.6. Storage root biochemical features and natural genetic variation

This chapter focus on the identification of spontaneous mutations in two biochemical pathways (sucrose/starch conversion and carotenoid biosynthesis), as well as mechanisms of carotenoid and proteins accumulation, and gene expression analysis.

Carbohydrate composition, content, and genetic variation: *Sugary* cassava is a unusual SR phenotype as observed in **Figure 2** (**Panel B**) for the cross session of SR stained with iodine solution,

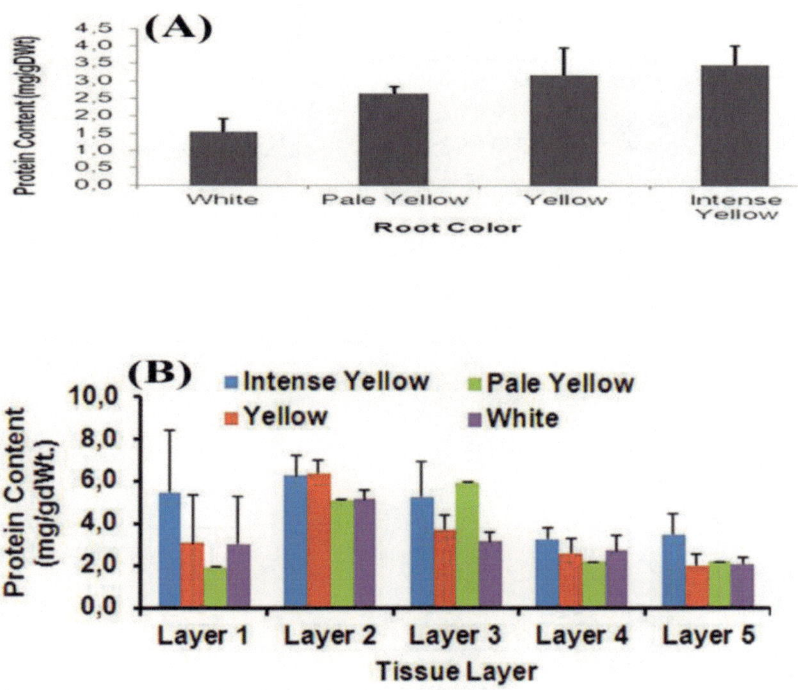

Figure 9. Variation in total protein content of storage roots (mg/gDWt) in relation to four categories of central cylinder color genotypes (A) and tissue age (B).

cells morphology, free sugar composition, and sucrose/glucose content in relation to normal genotypes and SR tissue age [3].

Carotenoid biosynthesis, accumulation, and genetic diversity: Landraces diversities (**Figure 2 (Panel B)**) have been studied to understand carotenoid biosynthesis [6, 12] mechanisms of carotenoid accumulation [1, 7, 13], identification of mutants [13], and breeding commercial varieties [2].

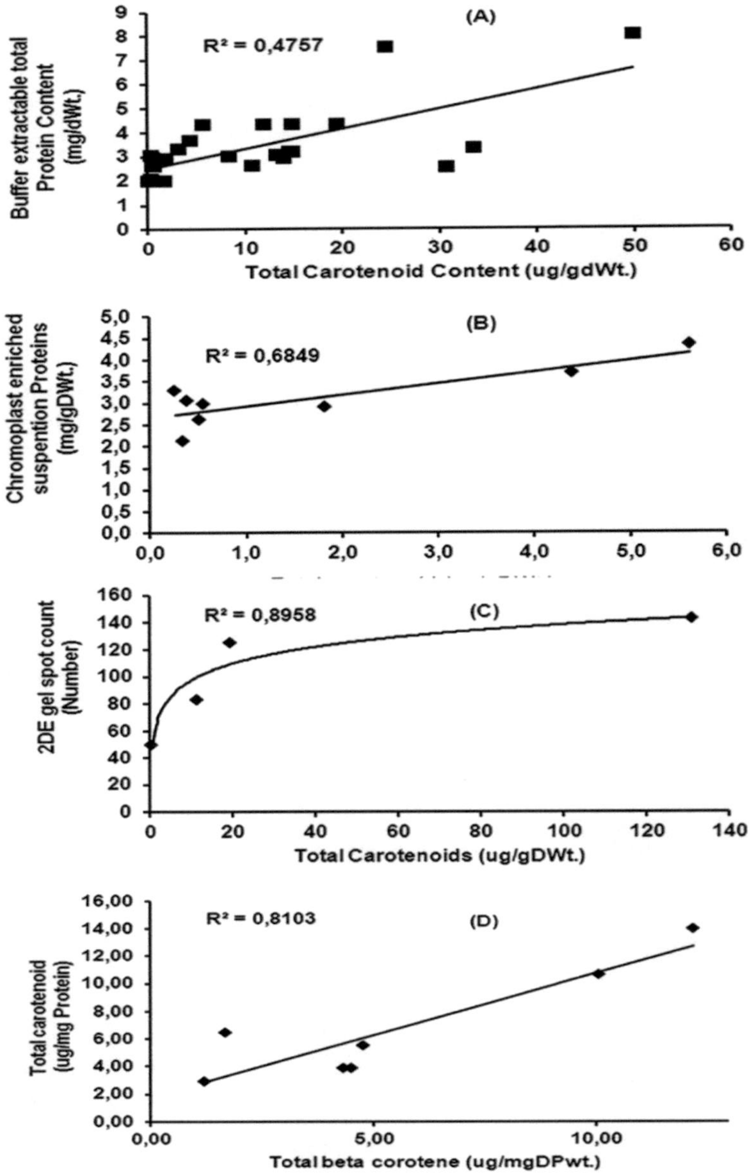

Figure 10. Correlations of total carotenoids (μg/mgDWt) and (A) buffer extractable proteins content, (B) chromoplast suspension proteins (mg/gDWt), (C) counting number of proteins in 2DE gel separated, and (D) total β-carotenoid content in cassava storage roots.

The major achievements, includes the discovery of a putative mutant for the gene LYCb that leads to the accumulation of solely lycopene in the landrace CAS51 and the discovery of a mutant for the gene HYDb that leads to accumulate mainly β-carotene in the landrace CAS64. Discovery of a single point mutation on the gene coding for protein SHSP that lead to the sequestration specifically of β-carotene in landrace CAS64. Six new commercial varieties were developed, registered, and protected in 5 years instead of 15 years as it is ordinarily done. It has been reported that sampling variation among plants and roots from the same plant is responsible for 20–25% [13] that causes uncertainty of values used for selection of clones in a breeding program. The sampling tissue system based on tissue age, as discussed above, could improve the accuracy of quantification of total carotenoid content for this propose.

Protein content and exploratory functionalities: Cassava storage root protein content variations predicted functionalities, patterns of distribution in source and sink organs, and post-harvest physiological deterioration studies using PROTEOMIC's technologies.

Cassava storage root proteins content in relation to color categories of genotypes (**Figure 9**): Similar to carbohydrate, protein content varies in two ways. One, higher protein content is observed in pigmented cassava rather than in white cassava (**Figure 9A**). Second, protein content varies according to tissue type and age across the central cylinder by decreasing from layer 3 to layer 4 to layer 5 (**Figure 9B**). In addition, protein content is strongly correlated with total carotenoid content (**Figure 10**). Heat shock proteins (HSPs) are the most abundant proteins types [13] in cassava storage root and are closely associated to accumulation of total carotenoid, with small shock proteins (SHSPs) being the major type of HSP [13].

3. Synthesis and conclusions

The studies discussed in this chapter highlight the importance of natural variation in landraces previously unknown for the cassava community in several ways. 1. Accurate estimation of the genetic of traits in landraces derived from alteration in two major metabolic pathways (starch and carotenoid) of great relevance for the two recognized practical utilization of CSR by using physiological concepts and sampling strategy. 2. Describing a CSR sampling procedure specific for CSR to estimate traits of agronomic importance for the two major practical utilization of CSR to improve product quality. 3. Incorporation of those genetic variants in a conventional breeding program, which reduced the time for obtain new commercial varieties. 4. Discovery of three putative mutants in the CSR. Further researches to dissect transcriptome and proteome of CSR are under way using the sampling system proposed in this chapter to elucidate molecular mechanisms regulating CSR formation, growth, development, and physiological maturation.

Acknowledgements

We would like to acknowledge the financial support provided by the Rockefeller Foundation (RF96010#25 and RF9707#26 in the search for landrace diversity); Conselho Nacional de

Desenvolvimento Cientifica e Tecnológico – CNPq (grant #480410.2001-1 for functional genomic work); Programa Nacional de Pesquisa em Biotecnologia – CENARGEN (Project No 1 060302058 for carbohydrate analysis); IAEA (contract #13188 for funds supplied to gene expression analysis); National Special acknowledgements to Nature Science Foundation of China (NSFC grant #31271776), and NSFC-CGIAR International (Regional) Cooperation and Exchange Programs (grant #31361140366) for providing funds for this publication.

Source of Funds

National Nature Science Foundation of China (NSFC grant number 31271776).

Author details

Luiz JCB Carvalho[1]*, Josefino F. Filho[2], James V. Anderson[3], Priscila G. Figueiredo[4] and Songbi Chen[4]

*Address all correspondence to: luiz.castelo@embrapa.br

1 EMBRAPA Genetic Resources and Biotechnology, Brasilia-DF, Brazil

2 EMBRAPA Cerrados, Brasilia-DF, Brazil

3 USDA-ARS, Sunflower and Plant Biology Research Unit, Fargo, ND, USA

4 Tropical Crops Genetic Resources Institute, Chinese Academy of Tropical Agricultural (CATAS), Hainan, China

References

[1] Carvalho LJCB, Agustini MA, Anderson JV, Vieira EA, de Souza CR, Chen S, Silva JP. Natural variation in expression of genes associated with carotenoid biosynthesis and accumulation in cassava (*Manihot esculenta* Crantz) storage root. BMC Plant Biology. 2016;**16**(1):133

[2] Carvalho LJCB, Vieira EA, Fialho J d F d. A genomic assisted breeding program for cassava to improve nutritional quality and industrial traits of storage root. Crop Breeding and Applied Biotechnology (Online). 2011;**11**(4):289-296 ISSN 1984-7033. http://dx.doi.org/10.1590/S1984-70332011000400001

[3] Carvalho LJCB, Souza CRB, Cascardo JMC, Bloch Junior C, Campos L. Identification and characterization of a novel cassava (*Manihot esculenta* Crantz) clone with high free sugar content and novel starch. Plant Molecular Biology. 2004;**56**:634-659

[4] Daiuto ER, Cereda MP, Carvalho LJCB. Características e propriedades de amidos extraído de camadas da raízes de mandioca cv. Mico (*Manihot esculenta* Crantz). Brazilian Journal of Food Technology. 2002;**5**:2176-2223

[5] Ribeiro MC, Figueiredo LFA, Loureiro ME, Cabral GB, G.B. Cabral, Carvalho LJCB. Study on the induction of storage root formation in cassava grown under in vitro culture and nutrient solution. Proceedings of The Second International Scientific Meeting Of CBN. Bogor-Indonesia. August 22–26, 1994; 1995. p. 795-803. (Working Document # 150)

[6] Lowe SB, Mahon JD, Hunt LA. Early development of cassava (*Manihot esculenta*). Canadian Journal of Botany. 1982;**60**(12):3040-3048

[7] Carvalho LJCB, de Souza CRB, Cascardo JC Mattos, Agostini MAV, Vieira EA Anderson JV, Lippolis J. 2009. Natural genetic variation in cassava (*Manihot esculenta* Crantz) landraces: A tool for gene discovery. In: QY Shu (editor). Induced Plant Mutations in the Genomics Era. Food and Agriculture Organization of the United Nations, Rome, 313-316

[8] de Souza CRB, Carvalho LJCB, Almeida ERP, Gander ES 2003. A cDNA sequence coding for glutamic acid protein is differentially expressed in cassava storage root. Protein and Peptide Letters, p. 167-170

[9] de Souza CRB, Carvalho LJCB, Cascardo JMC. Comparative gene expression study to identify genes possible related to storage root formation in cassava. Protein and Peptide Letters. 2004;**11**(6):577-582

[10] de Souza CRB, Carvalho LJCB, de Almeida ERP, Gander ES. A cassava gene coding for a glutamic acid rich protein expressed during storage root tissue development and related to the latex allergen family; 2004

[11] Safwan II, Mohammed UA. Review on the nutritional value, cultivation and utilization potential of some minor and under-utilized indigenous root and tuber crops in Nigeria. International Journal of Advanced Research. 2016;**4**(3):1298-1303

[12] Montagnac JA, Davis CR, Tanumihardjo SA. Nutritional value of cassava for use mutant in cassava (*Manihot esculenta* Crantz). Journal of Agricultural and Food Chemistry. 2009;**56**(16):7215-7222

[13] Carvalho LJCB, Lippolis J, Chen S, Batista DSC, Vieira EA, Anderson JV. Characterization of carotenoid-protein complexes and gene expression analysis associated with carotenoid sequestration in pigmented cassava (*Manihot Esculenta* Crantz) storage root. Open Biochemistry Journal. 2012;**6**:116-130

[14] Duque ALC, Escobar AF. Sampling strategies for proper quantification of carotenoid content in cassava breeding. Journal of Plant Breeding and Crop Science. 2011;**3**(1): 14-23

[15] Burns AE, Gleadow RM, Zacarias AM, Cuambe CE, Miller RE, Cavagnaro TR. Variations in the chemical composition of cassava (*Manihot esculenta* Crantz) leaves and roots as

affected by genotypic and environmental variation. Journal of Agricultural and Food Chemistry. 2012;**60**:19

[16] Islamiyat Folashade Bolarinwa, Moruf Olanrewaju Oke, Sulaiman Adebisi Olaniyan and Adeladun Stephen Ajala. A review of cyanogenic glycosides in edible plant. In: Sonia Soloneski and Marcelo L. Larramendy. A Review of Cyanogenic Glycosides in Edible Plants ISBN 978-953-51-2717-8, Print ISBN 978-953-51-2716-1216 pages. InTech. Chapters published 26-10-2016 under CC BY 3.0 license. DOI: 10.5772/62600

Chapter 5

Computer Simulation of Cassava Growth

Velayudhan Santhakumari Santosh Mithra,
A.R. Seena Radhakrishnan and Divya K. Lekshmanan

Additional information is available at the end of the chapter

http://dx.doi.org/10.5772/intechopen.71510

Abstract

To achieve higher productivity in a shorter time, growth simulation models are used in countries like Thailand to provide agro advisory to cassava farmers. Crop model helps to study the response of the crop to any environmental change. Computation of yield potential is important to adopt proper management practices to maximize the yield. Many simulation models of cassava have been developed in different parts of the world. Most of the deficiencies in earlier models were well taken care of in the cassava simulation model GUMCAS [40]. Latest process model, SIMulation of CASsava (SIMCAS), describes cassava growth and yield with good accuracy. This model emphasizes working out the sink capacity and source potential of the plant because the balance between them is a critical requirement for determining the final economic yield of the plant SIMCAS was developed with the aim of applying it for agro advisory purposes. The location- and variety-specific potential tuber yield under given weather conditions are calculated by the model. The most suitable planting time to achieve maximum yield from cassava in a particular locality can also be found out from this model.

Keywords: crop model, crop simulation, SIMCAS, cassava, simulation model

1. Introduction

The problem faced by developing world agriculture is often described as the need to produce more from less and solutions to this problem are aimed at producing more rather than reversing the trend that causes less. While the production of more is important, global efforts must focus on halting and reversing the trends that lead to diminishing agricultural lands and related natural resources. To achieve this, a greater understanding of cross sectional impacts must be achieved at all levels of agriculture and therefore the spread of information must be wider. Crop production must be increased to meet the rapidly growing food demands through sophisticated agricultural process, while it is important to protect other natural resources and the environment. New agricultural research is needed to provide additional information to farmers, policy makers and other decision makers on how to accomplish sustainable agriculture over the wide variations in climate change around the world. Therefore many researchers have over the years shown interest in finding ways to estimate the yield of crops before harvest [1]. In this context plant growth and development models should be elaborated to supply a basis for planning and managing crop production [2].

Dynamic crop growth simulation is a relatively recent technique that facilitates quantitative understanding of the effect of these factors and agronomic management factors on crop growth and productivity. Crop models are computer program that mimic the growth and development of crops [3] or is a simple representation of crop [4] used to study crop growth and to calculate growth responses to the environment [5]. Model simulates or imitates the behavior of the real crop by predicting the growth of its components or organs or parts. A crop growth simulation model not only predicts the final state of the production or harvestable yield, but also contains quantitative information about major processes involved in the growth and development of the crop.

Crop model is a simple mathematical representation of the crop, helps to study the response of the crop to any environmental change. Simulation is the study of the behavior of a system represented by a dynamic model. A particular growth model can be suitable for a particular crop but may be inappropriate for another crop. The user and developer must carefully formulate the model before using, adapting and refining it. SIMulation of CASsava (SIMCAS) is the cassava growth simulation model which describes growth and yield of cassava with good accuracy.

2. SIMCAS

Cassava (*Manihot esculenta* Crantz) is the world's fourth most important source of energy [6] and is a climate resilient crop which is likely to be the best bet under the changing global climatic condition. It is the primary staple food for more than 800 million people in the world and represents a household food bank [7]. By achieving higher productivity in a shorter time, poverty alleviation efforts can be hastened. To achieve this target, growth simulation models are used in countries like Thailand to provide agro advisory systems to cassava farmers [8]. Computation of yield potential is important to adopt proper management practices to maximize the yield. Several studies have been reported on the relationship of environmental factors on plant parts and tuber yield of cassava.

Many simulation models of cassava have been developed in different parts of the world. The model developed by Cock et al. [9] helps to predict the performance of the crop under different combinations of parameters such as branching time, leaf age, etc. This model is of great importance to breeders to determine parameters having maximum influence on tuber yield. Crop Growth Rate (CGR) is assumed to be a constant function of Leaf Area Index (LAI) and it is a serious limitation of this model that the performance of the crop under different solar radiation and temperature conditions cannot be studied. Other assumptions of fixed life span and maximum leaf size add further restrictions to this model in its use under different environmental conditions. Fukai and Hammer [10] developed a model based mainly on empirical relationships derived from datasets collected under Australian conditions. To use this model under different environments, the relationships in the model itself must be recalculated. Most of the deficiencies in earlier models were well taken care of in the cassava simulation model GUMCAS [11]. Maximum potential CGR, a varietal character, is included in this model for calculating dry matter production. The effect of stress due to solar radiation, temperature and water deficit on CGR has been computed with the help of multipliers, as in the case of the Fukai and Hammer model. Diffused and direct components of solar radiation and its differential entrapment by leaves are not included in the model. Although leaf size is one of the most important plant parameters determining dry matter production and final tuber yield, it is computed as a function of time. Thus, the environmental influence on final tuber yield is considerably restricted in this model. The fraction of dry matter allocated to the stem and leaves is calculated as a linear function of developmental time. Hence, this model cannot predict a drastic reduction of tuber yield, which sometimes happens under conditions of excessive vegetative growth. To predict branching, optimum photoperiod and photoperiod sensitivity were derived from the datasets of Keating et al. [12], which were collected for the variety M Aus 10 under Australian conditions. To use this model for a different variety or under a different environment, these values should be recalculated empirically. Dynamic partitioning of dry matter into shoot and storage roots is very important to reflect real field conditions, but has not been done in previous cassava models [13]. In the latest process model, SIMulation of CASsava (SIMCAS), most of the shortfalls observed in the earlier models were rectified and adopted some concepts and methodologies used in them. This model describes cassava growth and yield with good accuracy. This model emphasizes working out the sink capacity and source potential of the plant because the balance between them is a critical requirement for determining the final economic yield of the plant [6].

SIMCAS was developed with the aim of applying it for agro advisory purposes. The location- and variety-specific potential tuber yield under given weather conditions are calculated by the model. The most suitable planting time to achieve maximum yield from cassava in a particular locality can also be found out from this model. For this purpose, historic weather data for the locality should be analyzed to find out the general weather situation of the locality. The model can be run using known weather data and different dates of planting to find out the most suitable planting time. Once the potential yield is found out, proper management practices should be followed to maximize the yield. This model helps to reduce moisture, nitrogen and potassium wastages, and to maximize yield by applying the required amounts at the proper time. The model also calculates stresses due to shortages of moisture, nitrogen and potassium on crop growth and yield.

3. Model description

The driving variables of the models are weather parameters, crop parameters and nutrient parameters.

Crop phenology: In this model phenological development of cassava was calculated on each day in terms of growing degree days (GDD). The optimum temperature (TOPT), base temperature (TBASE) and maximum temperature for the growth of cassava were computed to be 28.5–30, 13.0 and 36–40°C, respectively [14].

Cassava is a long-day plant [4]. Photoperiod influences different processes in cassava like time of root initiation [15], leaf area and branching [12]. Photoperiodic effect in this model is calculated using the method followed in GUMCAS. *P_GDD* is calculated, by multiplying this photoperiodic effect with GDD, to simulate leaf growth, branching, fibrous root growth, etc., which are reported to be influenced by photoperiod [16, 17].

Sprouting: In cassava, sprouting is influenced by temperature. Under conditions of constant and alternating temperatures, sprouting response is found to be the same [18]. A minimum soil temperature of 12–17°C, optimum soil temperature of 28.5–30°C and maximum soil temperature of 36–40°C are required for the sprouting of cassava stems [14]. In this model the varietal parameter Temergence, is defined as GDD at which sprouting begins.

Stem growth: The number of lateral branches and sympodial branches (forks) on each lateral branch are important characters in determining the source potential of cassava. Leaves formed on these parts are major components of the source potential. Cassava forms new leaves and stem simultaneously with filling of storage roots. Stem growth has priority over tuber growth under the conditions of limited carbohydrate supply [9, 19]. Cock et al. and Fukai et al. observed a direct relationship between assimilate supply and stem growth. Among the weather factors, higher temperature favors stem and branch growth [20]. Increase in stem length is calculated in terms of GDD in this model [18].

Leaf area: Leaf Area Index (LAI) is an important component determining the final yield of cassava. It is determined by the parameters like number of active apices, rate of leaf formation per apex, leaf size, and leaf life. Leaf formation per apex is decreased at lower temperatures and increased over the temperature range 20–28°C. Time to reach maximum leaf size also increases at lower temperature. Similarly, leaf life decreases as temperature increases, but it remains more or less constant at 28°C [21]. Keating et al. [12] reported that a longer photoperiod promotes the production of leaf area and thereby a high LAI. Shading is an important factor reducing LAI by its role in enhancing leaf shedding [19]. Leaf shedding plays a key role in determining the LAI and is related more to leaf aging and mutual shading than to temperature [20]. Tan and Cock [22] reported that plants with single stem have large leaf area and that the leaf life increased when number of apices reduced. Short days increase leaf life and long days promote rate of leaf formation and thereby high LAI [17]. In this model leaf formation (*dLFi/dt*) on i DAP is computed as a function of TMEAN. Potential life of cassava leaf ($Life_{POT}$) in days is a cultivar character [23]. This model calculates the number of days (Life) left in the life of individual leaves on each day since its formation, as a function of *TMEANi*. Leaf shedding is simulated in this model as a function of mutual shading. An algorithm to calculate mutual shading is developed and used in this model. The algorithm is described below.

a. Calculate the number of days left in the life of leaf 'k' on i DAP ($Life_{k,i}$)

b. In cassava, leaves are arranged on the stem spirally. Five leaves complete one circle and the sixth leaf and first leaf will be in the same line. The five leaves, which complete one circle, will occupy part of the area of a circle with radius (PLLN), which is equal to the sum of petiole length and the length of the mid rib of each leaf.

Light received by the first circle of leaves are not shaded and receive full sunlight, i.e., $light\_leaf_1 = 100$.

$$shade\_leaf_c = \frac{\sum_{k=n-(5c+1)}^{n-(5c-3)} LA_k}{\pi.PLLN^2}.100, \tag{1}$$

Where,

n is the total number of leaves retained, c is the circle in which the leaf exists, LA_k is the area of the leaf k, $shade\_leaf_c$ is the shade on the leaves of the circle c.

Light received by the leaves are calculated as:

$$light\_leaf_c = light\_leaf_1.\left(1 - \prod_{k=1}^{c-1} \frac{shade\_leaf_k}{100}\right) \tag{2}$$

Life k,i is modified by incorporating the effect of shading as:

$$Life_{k,i} = \frac{28.0.Life_{POT}}{TMEAN_i}.\frac{(light\_leaf_c - shade\_leaf_c)}{light\_leaf_c} \tag{3}$$

Branching: Branching has important role in terms of canopy development and dry matter partitioning [9, 22]. New branches are produced under a good supply of carbohydrates [19]. Long photoperiod conditions are found to promote branching [24]. Temperature is also important in determining the branching habit of cassava. Irikura et al. reported that branching is a varietal character and is delayed by reduced temperature or high temperature [21]. For cassava, the critical threshold value of photoperiod that promotes branching is 12–13 h [12]. The number of nodes produced before the first fork in cassava has genotypic character [24]. Under low fertility conditions, branching is very poor. Similarly, lateral branching occurs under conditions of good illumination and soil fertility. In this model, two conditions are set for emerging branches:

a. There should be a fixed number of nodes before forking stems

b. If total dry matter produced by the plant should be more than the potential requirement

Fibrous roots: Fibrous roots are very important in determining the capacity of the plant to absorb water and nutrients. Short day length and reduced light adversely affect fibrous root growth. Fibrous root growth is sensitive to reduced carbohydrate supply [16]. In this model, initiation of fibrous roots is simulated when P_GDDi reaches a fixed value, which is variety specific. Number of fibrous roots (***n_Froots***) is also set as varietal parameter in this model.

Tubers: Tubers are storage organs and are an economically important part of cassava in most parts of the world. Time of storage root initiation is determined by the interaction of photoperiod and temperature, and is delayed by low temperature conditions, which generally prevail in the subtropics [15]. Boerboom [25] proposed that the plant should attain a dry weight before the storage root initiation happens. This concept is adopted in this model to simulate tuber initiation.

Solar radiation absorbed by the plant is calculated using the method suggested by Johnson et al. [26].

Calculation of photosynthesis: Rate of leaf photosynthesis (PLi, g CO_2 m^{-2} of leaf area day^{-1}) on i DAP, gross canopy photosynthesis (Pci, g CO_2 $m^{-2}$of leaf area day^{-1}) and total dry matter (Pn′i, g m^{-2} of leaf area $plant^{-1}$), is calculated using standard methods Penning de Vries et al. [4] and Johnson et al. [26].

Partitioning of dry matter: Storage roots, stem and growing leaves constitute the sink portion of cassava and active leaves constitute the source portion [23]. In cassava, leaf area and tubers develop simultaneously. Hence, there always exists competition for assimilates between different plant parts [6]. Aboveground plant organs always have a preference over root growth [9] and, hence, in this model dry matter is first partitioned between stem, leaves and fibrous roots and whatever is remaining will be stored in the tuber. In this model the dry matter produced on i DAP (Pn'_i) is calculated using leaf area on previous day. Before tuber initiation, whatever dry matter remaining after allocating to leaves and stem will be partitioned to fibrous roots and after tuber initiation this will be transported to tubers. In this model it is assumed that after tuber initiation, fibrous roots do not receive any dry matter.

4. Stress estimation

Under field conditions, stress due to shortage of many factors essential for the growth of the crop limits the crop from achieving its potential yield. Moisture, nitrogen and potassium are the three most important factors essential for the growth of cassava. This model calculates the stresses due to shortages of moisture, nitrogen and potassium on growth and yield of cassava as well as the uptake of N and K.

4.1. Effect of drought stress on crop growth

Reduction in dry matter production due to drought stress (WS) is depending on the stage at which the crop experiences stress. Alves [23] reported that for cassava first 5 months after planting (MAP) are the most critical periods of drought stress. The effect of drought stress on crop growth is computed using the method suggested by Allen et al. [27]. Since the data on wind speed is not commonly available, Priestly-Taylor method [27] is used in this model for calculating reference evapotranspiration (ET_0), which is used in the calculation of drought stress.

4.2. Effect of nitrogen deficit on crop growth

From the change in biomass, crop demand for nutrients during a given time period can be calculated [28]. In this model stress experienced by the crop due to the shortage of nitrogen

(N_{stress}) is the ratio of Npot (nitrogen required to produce the dry matter at potential rate) and the actual nitrogen uptake (N_{uptake}) [18].

4.3. Effect of potassium deficit on crop growth

Potassium is an important nutrient for cassava because of its role in photosynthesis and the translocation of photosynthates [29]. Stress experienced by the crop due to the shortage of K (K_{stress}) is calculated in this model as the ratio of K_{pot} (potassium required to produce the dry matter at potential rate) and the actual potassium uptake (K_{uptake}) [18].

5. Growth simulation with SIMCAS model

The model requires three data files for simulating the growth and to predict the phenology, dry matter production and distribution of cassava crop.

5.1. Plant parameters required to run the model

Plant parameters required to run the model are:

a. ***n_leaf***: leaves at the time of emergence

$$n\_leaf = \frac{LF_n{}^*28}{\sum_{i=1}^{n} 28 - |Tmean_i - 28|} \tag{4}$$

Where,

LF_n is number of leaves on n DAP,

$Tmean_i$ is mean temperature on ith DAP

b. ***n_stem***: Describes growth rate of stem (0C.cm^{-1})

$$n\_stem = \frac{GDD_i}{HT_i} \tag{5}$$

Where,

HT_i is the height of the stem on ith DAP

GDD_i = GDD on ith day after planting

c. ***d_roots***: P_GDD at the time of initiation of fibrous roots

d. ***dwt_rt_MAX***: It is the maximum dry weight, which the fibrous root of cassava can attain (g)

e. ***dwt_rt_EMERG***: The dry weight of fibrous roots of cassava at the time of their emergence (g)

f. ***n_F roots***: the number of fibrous roots of cassava

g. *SLA*: Describes the specific leaf area(g*(cm^2)$^{-1}$)

h. *stwt_MAX*: Describes the maximum dry matter that can be apportioned to the unit length of the stem (g/cm^{-1})

i. $LAREA_{MAX}$: Describes the maximum area that an individual leaf can attain (cm^2)

j. $LAREA_{START}$: Describes the area of individual leaves at the time of its full emergence (cm^2).

k. $life_{POT}$: Describes potential life of cassava leaf (days).

l. *node_branch*: Describes number of nodes, which should be produced on the stem since last branching to production of new branch

m. *st_rt_plant weight*: Describes initial plant weight, i.e., the weight of the stem, leaves, and fibrous root together, at which tuber initiates.

n. *n_S_tubers*: Describes number of tubers produced by a variety

o. *n_shoots*: Describes the number of shoots produced by the variety. The user can set its value so that the software will do the simulation for that number of shoots. If its value is set as −99 the software will find out the actual number of shoots produced by the plant under the experimental conditions.

5.2. Weather parameters required to run the model

a. Maximum temperature (°C)

b. Minimum temperature (°C)

c. Solar radiation (MJ/day) or Sunshine hours (h)

d. Maximum relative humidity (%)

e. Minimum relative humidity (%)

f. Precipitation (mm)

Daily weather data of the cropped area is required.

5.3. Fertilizer parameters required to run the model

a. Nitrogen (kg ha^{-1})

b. Phosphorous (kg ha^{-1})

c. Potassium (kg ha^{-1})

These data files should be prepared in CSV format to run the model.

The estimated values of the crop parameters for the two varieties, Sree Vijaya, and H 226, are given in **Table 1**.

Plant parameters	Sree Vijaya	H 226
n_leaf	1.121636	1.10964
n_stem	12.53228	12.817006
d_roots	85.7309	109.13437
dwt_rt_MAX	7.762	14.065
dwt_rt_EMERGENCE	0.053832	0.0694762
n_F roots	23.64993	21.597397
SLA	0.002288	0.0024023
stwt_MAX	0.553091	0.42715
Larea MAX	467.3508	305.10788
Larea START	3.206739	2.1434482
Life POT	95	115.66667
node_branch	76.6	102.66667
st_rt_pltwt	285.561	134.967
n_S_tubers	9.373267	9.76
shoots	2.2	2

Table 1. Mean plant parameter values of Sree Vijaya and H 226.

6. Evaluation of the model

The model accuracy could only be tested against limited datasets. To evaluate the accuracy of predictions, simulated tuber yield for the varieties Sree Visakam, Sree Shya and M4 were collected from different locations and compared with the observed values. It is observed that the predicted values are reasonably close to the observed values. Mean absolute percentage deviations between observed and predicted values for the said varieties were 13.2, 17.0 and 3.4%, respectively. This shows that model's predictions are very close to the real field situation (**Figure 1**).

7. Sensitivity analysis

The sensitivity of the final output (t ha^{-1}) of the three varieties to the perturbations in each plant parameter was computed. Sensitivity (β) (**Table 2**) was estimated as the ratio of the fractional change in yield to the fractional change in a particular parameter [11].

An average 5% change on either side of the parameter value was used to compute the β (**Table 2**). Leaf area is the most sensitive parameter in determining tuber yield. The parameters, ***Life$_{POT}$*** and ***node_branch***, which play a major role in determining the canopy size of the plant and thereby are important in determining the source potential of the plant, show high β values. High β values of these parameters for all three varieties confirm the importance of vegetative growth in determining the final yield.

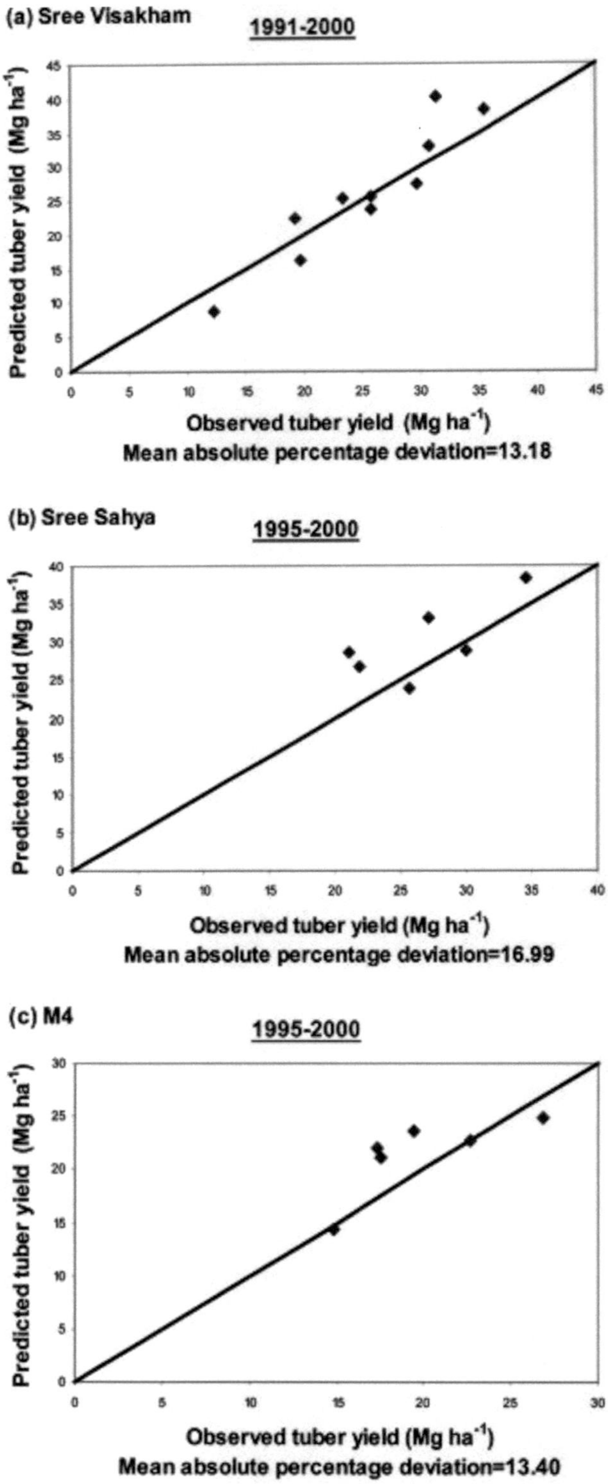

Figure 1. Observed vs. predicted tuber yield (t ha^{-1}) for the varieties: (a) Sree Visakham, (b) Sree Sahya, and (c) M4.

Parameter	β		
	Sree Visakham	Sree Sahya	M4
n_leaf	0.09	0.13	0.37
n_stem	0.51	0.47	0.48
d_roots	0.00	0.08	0.11
dwt_rt_MAX	0.00	0.00	0.00
dwt_rt_EMERG	0.00	0.00	0.00
n_Froots	0.00	0.00	0.00
SLA	0.20	0.19	0.18
stwt_MAX	1.23	0.47	0.48
LAREAMAX	0.34	0.33	0.97
LAREASTART	0.02	0.06	0.08
LifePOT	1.41	1.29	2.64
node_branch	0.56	0.67	0.72
st_rt_pltwt	0.35	0.29	0.34
n_S_tubers	0.57	0.00	0.01

Source: Ref. [18].

Table 2. Results of sensitivity analysis of plant parameters for the cassava varieties Sree Visakham during 6.06.1994 to 22.03.1995, Sree Sahya during 16.06.1992 to 22.03.1993 and M4 during 16.06.1996 to 22.03.1997.

8. Discussion

The GUMCAS model predicts branching based on photoperiod alone [11]. In this model, a new method is followed for predicting branching. Influence of assimilate supply on branching is widely reported. Fukai et al. [19] reported that new branches are produced when there is a good supply of carbohydrates. The assumption that branching occurs when the ratio of supply to demand of assimilates is >3 is modified and adopted in this model. Branching was found to be less under high competition for assimilates [30]. This model predicts branching based on the varietal parameter, ***node_branch***, also. Because branching is a process, which marks the beginning of reproductive phase, production of nodes is calculated as a function of P_GDD, whereby the interaction between photoperiodic effect and temperature is the basis for determining phenological development. When the number of nodes produced on the stem since the previous branching crosses ***node_branch***, the model simulates the production of another branch. The number of new branches produced from this stem is dependent on the assimilate supply. If the assimilate supply is more than that required for the stem, three branches are produced, otherwise two branches are only produced. Unlike the GUMCAS model, there is no fixed limit for the maximum number of branches. According to this model, branching is delayed by low temperature and short day conditions, because under these conditions leaf formation and dry matter production are reduced. Thus the findings of Irikura et al. [21] and Veltkamp [17] are fulfilled by this model.

The stem is an important sink and stands in strong competition to tubers for dry matter [15]. In this model, stem growth is calculated as a function of GDD. Here, the stem growth is promoted more under summer than under winter conditions. This hypothesis is supported by Manrique [20].

All factors determining LAI, which is one of the most important components of dry matter production, are carefully included in this model. To predict leaf production, the findings of Irikura et al. [21] are adopted in this model by assuming that the rate of leaf production is lowered as temperature moves from 28°C. The longer photoperiod, which results in high P_GDD, causes a large increase in leaf area, thereby substantiating the findings of Keating et al. [12]. The time to reach maximum leaf size will be longer at low temperatures [12].

The primary cause of leaf fall is shading [19]. A new algorithm to calculate shading is proposed in this model. The angle at which the leaf is held on the stem is not considered here.

Tuberization is simulated with the help of the parameter ***st_rt_pltwt*** (total plant weight at which storage root production starts), as proposed by Boerboom [30, 31]. Accumulation of plant weight is dependent on many weather parameters, which support the development of leaf area. The plant weight accumulation at low temperature becomes slow. This will result in delayed tuberization [12, 15].

Source–sink relationships are given due importance in this model. In the first phase of the model, source and sink capacity are calculated based on weather parameters and dry matter production is calculated using the existing source, i.e., the source potential of the previous day. So, dry matter production becomes more realistic. In the second phase, dry matter produced is allocated to the sink already calculated. If the dry matter produced is not sufficient to meet the potential sink capacity, its size will be proportionally reduced. In such situations, there will be no stored dry matter in the main storage organ, i.e., tubers. Whereas, if the dry matter produced is more than the sink capacity, all the sink portions will grow to their potential capacity and if some dry matter remains it will be stored in the storage organs.

Potential dry matter production on each day is calculated by this model based on the prevailing weather conditions. This is further modified by the stress experienced by the crop. This model calculates water, nitrogen and potassium stresses only. Each stress is calculated separately and the reduction from the potential yield is calculated. More elaborate algorithms for finding out stresses may improve the accuracy of the model predictions. Algorithms to calculate stresses due to shortage of P, other micronutrients, pests and diseases, etc., are not included in this model. Extensive research on impact of various stresses on the growth of cassava will definitely help to improve the accuracy of model predictions. This model helps in giving advice to farmers to reduce stresses and to realize the potential yield. The precise time and quantity of application of water, N and K help the farmers to reduce deficits and obtain maximum yield from the crop. Before giving a comprehensive advisory, extensive validation of the model is required and the economics also should be worked out.

9. Conclusion

SIMCAS, growth simulation model of cassava is composed of various submodels/routines for predicting phenology, productivity and assimilate partitioning. These models were built mainly using data reported in literature, as well as data collected by conducting field experiments. Water, potassium and nitrogen are important stress factors included in the model. The predicted values while compared with field observations shows a good agreement between predicted and observed tuber yield even though the stem and leaf weight predictions are deviating from the observed values. This model is based on sufficiently studied physiological processes and extensive validations under field conditions are necessary before its field application. Improvements in the estimation of drought, nitrogen and potassium stresses are required to give more accurate predictions. By considering the detailed dynamics of water, nitrogen and potassium in soil and plant, the estimation of stress due to their shortage will be improved. However, SIMCAS can be considered as a good tool for advising the farmers about the potential yield of cassava at a given locality and to develop strategies for maximizing the yield by managing irrigation and N and K fertilizations.

Author details

Velayudhan Santhakumari Santosh Mithra[1]*, A.R. Seena Radhakrishnan[2] and Divya K. Lekshmanan[3]

*Address all correspondence to: vssmithra@gmail.com

1 ICAR-Central Tuber Crops Research Institute, Thiruvananthapuram, India

2 University of Kerala, Thiruvananthapuram, Kerala, India

3 ICAR-CPRS, Ooty, India

References

[1] Rauff KO, Bello R. A review of crop growth simulation models as tools for agricultural materiology. Agricultural Sciences. 2015;**6**:1098-1105

[2] Durado Neto D, Teruel DA, Reichardt K, Nielsoen DR, Frizzone JA, Bacchi OOS. Principles of crop modeling and simulation: 1. Uses of mathematical models in agricultural science. Science Agriculture Piracicaba. 1998;**55**:46-50

[3] USSDA 2007

[4] Penning de Vries FWT, Jansen DM, ten Berge HFN, Bakema A. Simulation of Ecophysiological Processes of Growth in Several Annual Crops. Wageningen (The Netherlands): IRRI Los Bonos and Pudoc; 1989

[5] Migleitta F, Bindi M. Crop growth simulation model for research, farm management and agrometeorology. EARSeL Advances in Remote Sensing. 1993;**2**(2):148-157

[6] Cock JH. Cassava. In: Goldsworthy PR, Fisher NM, editors. The Physiology of Tropical Field Crops. New York: Wiley-Interscience; 1984

[7] Lebot V. Tropical Root and Tuber Crops. Cassava, Sweet Potato, Yams and Aroids; 2008. www.cabi.org

[8] Sarawat V, Ratanasriwong S, Puangprakon K, Sringam P, Jintrawet A. MunThaiDSS: A decision support system for cassava production. In: Zazueta F, Ninomiya S, Chitradon R, editors. Proceedings of the Joint Conference of the 4th International Conference of the Asian Federation of Information Technology in Agriculture and the 2nd World Congress on Computers in Agriculture and Natural Resources; August 9–12. Bangkok, Thailand; 2004. pp. 136-139

[9] Cock JH, Franklin D, Sandoval G, Juri P. The ideal cassava plant for maximum yield. Crop Science. 1979;**19**:271-279

[10] Fukai S, Hammer GL. A simulation model of the growth of cassava crop and its use to estimate cassava productivity in Northern Australia. Agricultural Systems. 1987;**23**:237-257

[11] Matthews RB, Hunt LA. GUMCAS: A model describing the growth of cassava (*Manihot esculenta* L. Crantz). Field Crops Research. 1994;**36**:69-84

[12] Keating BA, Evanson JP, Fukai S. Environmental effects on the growth and development of cassava (*Manihot esculenta* Crantz). I. Crop development. Field Crops Research. 1982;**5**: 271-281

[13] Gray VM. A comparison of two approaches for modeling cassava (*Manihot esculenta* Crantz) crop growth. Annals of Botany. 2000;**85**:77-90

[14] Keating BA, Evanson JP. Effect of soil temperature on sprouting and sprout elongation of stem cuttings of cassava (*Manihot esculenta* Crantz). Field Crops Research. 1979;**2**:241-251

[15] Ekanayake IJ. Physiology of cassava. In: Ekanayake IJ, Osiru DSO, Porto MCM, editors. IITA Research Guide. Ibadan (Nigeria): IITA; 1998. 55 p

[16] Aresta RB, Fukai S. Effects of solar radiation on growth of cassava (*Manihot esculenta* L. Crantz). II. Fibrous root length. Field Crops Research. 1984;**9**:361-371

[17] Veltkamp HJ. Physiological Causes of Yield Variation in Cassava (*Manihot esculenta* L. Crantz). Agricultural University Wageningen papers. Wageningen (The Netherlands): Agricultural University; 1986

[18] Santhosh Mithra VS, Sreekumar J, Ravindran CS. Computer simulation of cassava growth: A tool for realizing the potential yield. Archives of Agronomy and Soil Science. 2013;**59**(4): 603-623. DOI: 10.1080/03650340.2011.653681

[19] Fukai S, Alcoy AB, Llamelo AB, Patterson RD. Effects of solar radiation on growth of cassava (*Manihot esculenta* Crantz). I. Canopy development and dry matter growth. Field Crops Research. 1984;**9**:347-360

[20] Manrique LA. Plant morphology of cassava during summer and winter. Agronomy Journal. 1990a;**82**:881-886

[21] Irikura Y, Cock JH, Kawano K. The physiological basis of genotype temperature interaction in cassava. Field Crops Research. 1979;**2**:227-239

[22] Tan SL, Cock JH. Branching habit as a yield determinant in cassava. Field Crops Research. 1979;**2**:281-289

[23] Alves AAC. Cassava botany and physiology. In: Hillocks RJ, Thresh JM, Bellotti AC, editors. Cassava: Biology, Production and Utilization. Wallingford (UK): CAB International; 2002

[24] Hunt LA, Wholey DW, Cock JH. Growth physiology of cassava (*Manihot esculenta* L. Crantz). Field Crop Abstracts. 1977;**30**:77-91

[25] Van Kraalingen DWG, Stol W, Uithol PWJ, Verbeek M. User Manual of CABO/TPE Internal Communication. Wageningen (Netherlands); 1991

[26] Johnson IR, Riha SJ, Wilks DS. Modelling daily net canopy photosynthesis and its adaptation to irradiance and atmospheric CO_2 concentration. Agricultural Systems. 1995;**50**:1-35

[27] Allen RG, Pereira LS, Raes D, Smith M. Crop Evapotranspiration – Guidelines for Computing Crop Water Requirements. FAO irrigation and drainage paper. Rome (Italy): FAO; 1998. 56 p

[28] Van Noordwijk M, Van de Geijn SC. Root, shoot and soil parameters required for process-oriented models of crop growth limited by water or nutrients. Plant and Soil. 1996;**183**:1-25

[29] Rajendran N, Nair PG, Mohankumar B. Potassium fertilization of cassava in acid laterite soils. Journal of Root Crops. 1976;**2**:35-37

[30] Tsay JS, Fukai S, Wilson GL. The response of cassava to spatial arrangement and to soybean intercrop. Field Crops Research. 1987;**16**:19-31

[31] Boerboom BWJ. A model of dry matter distribution in cassava (*Manihot esculenta* Crantz). Netherlands Journal of Agricultural Science. 1978;**26**:267-277

Chapter 6

Domestication Syndrome in Cassava (*Manihot esculenta* Crantz): Assessing Morphological Traits and Differentially Expressed Genes Associated with Genetic Diversity of Storage Root

Luiz Joaquim Castelo Branco Carvalho,
James V. Anderson, Songbi Chen, Chikelu Mba and
Münevver Doğramaci

Additional information is available at the end of the chapter

http://dx.doi.org/10.5772/intechopen.71348

Abstract

Cassava (*Manihot esculenta* Crantz) provides a staple food source for millions of people in tropical and subtropical world regions. Brazil is the major center of diversification for species of the *Manihot*, and a center for domestication of the cultivated species originated from wild ancestral *M. esculenta subsp. flabellifolia.* Genetic breeding of cassava depends on landraces. Molecular phylogenetic technologies used to study genetic traits selected by mankind in crops, are likely to predict proposed "domestication syndrome." Phylogenetic trees use DNA sequences alignment to infer on gene historical events. A study on regulatory and structural complexity that dictates gene/protein function, will add non-sequence information to predict a more complete understanding of functional evolution. Transcriptional profile contains critical information on when and where a gene is manifested. These regulatory properties could explain functional genes diversity achieved within gene families across closely related species such as cassava and its ancestor. Microarray technologies measure transcriptional response of gene to a given environmental or genetic factor. Integration of genomic and transcriptomic data provides more detailed picture of molecular evolution. This chapter describes comprehensive study using the wild relative of cassava ancestor, recognition of natural morphological trait changes during domestication, and gene expression of cassava storage root.

Keywords: *Manihot, M. esculenta subsp flabellifolia, M. esculenta subsp esculenta,* cassava, domestication syndrome, genetic diversity, gene expression

1. Introduction

Although obtaining space for the intentional cultivation of edible plants often starts with the clearing of forests and modification of landscapes, it was the ability to domesticate plants that made agriculture possible in the first place. Domestication consists of a set of consecutive stages that begins with the original set of plant traits and evolves through the increase in selection frequency for desirable traits (the domestication traits). In the genus *Manihot*, the geographical occurrence and species relationships provided the first source of candidate species to be domesticated, which culminated in the emergence of cultivated plants adapted to both human needs and a cultivated environment that may be directly associated to the type of crop **(Table 1)**. A subset of traits that collectively form the morphological and physiological differences between the cultivated plant and their wild progenitors (the domestication syndrome) is specific for a plant adapted to human needs. For the case of cassava (*Manihot esculenta subsp. esculenta*), we selected a set of observable traits which suffered intense human-driven selection in relation to its ancestor (*M. esculenta subsp. flabellifolia*). These traits include early stages of environmental change resulting from the transition from forest (shade) to open-field cultivation. To determine what regulatory genes differ during the change in growth habit traits, such as vine-type (ancestor) to shrub-type (cultivated) growth, thickening of fibrous root (ancestor) into storage root (domesticated) and flowering set reduction as domestication progressed. The present chapter contemplate, the actual knowledge on the issue of cassava domestication report our current and forward studies on the evolutionary suite [1, 2] of genetic diversity in cassava landraces using genomic, transcriptomic, and proteomic technologies for cassava storage root, as the major domestication trait in cassava. Based on gene expression analysis, we identified a set of exploratory regulatory gene networks associated with diversity among minimum mankind artificial interference on the domestication of the major domestication trait (storage root formation) in cassava crop in relation to their ancestor originated in Brazilian Amazon (a major center for domestication of cassava). Finally, the pattern of exploratory regulatory gene networks linked to genotypes diversity was used to predict the early steps in the domestication process of

Domestication index	Other crops	Cassava
Removal of the forest	Yes	Yes
Seed dispersion at campground	High/low	Refractive
Human migration	High/fair	Fair
Gardening crops	High/fair	Fair
Discarded undesirable genotype	None/low	Low
Storage root	None/high	High
Genetic diversity	Moderate/low	Moderate
Plasticity	Low/high	High
Genetic bottleneck	Moderate/low	High

A set of hypothetical domestication index used to rank for their utilization by mankind and its effect on genetic processes.

Table 1. Crop domestication processes: a suite of procedures performed by humans to domesticate plants for food production.

the cassava ancestor as mankind first removed the forest (low-light shade environment) causing a clear open space (abundant light environment) with relevant landscape alterations that promoted alteration in the cassava ancestor plant as observed in its growth habit, storage root formation, and flowering set pattern.

2. Cassava wild relatives and the ancestral species

Mexico and Brazil are considered two relevant centers of diversification for *Manihot* species [3–5]. In Brazil, the geographical distribution for the genus *Manihot* is presented in **Figure 1**. Current systematic working models for botanical descriptions and classifications of species in the genus *Manihot* [6–9] contrasts in two major areas compared with an earlier classical monograph [9]. The early classical monograph suggests a total number of 98 species for the genus *Manihot*, which adopts a section classification system for grouping morphological closeness of species [9, 10] and proposes a compiled species concept to explain the origin of a single cultivated species (*M. esculenta subsp. esculenta*). The current model avoids the section classification system, reduces the number of species by using synonymies, uses an evolutionary approach, and permits a single ancestor species (*M. esculenta subsp. flabellifolia*) to explain the origin of the cultivated species (*M. esculenta subsp. esculenta*) of the genus *Manihot* [5–8]. In this chapter, we first contrast these two approaches by using a phylogenetic analysis of **ribosomal RNA internal transcribed spacer (ITS)** for 17 *Manihot* species from Brazil and Mexico to identify those most closely related to *M. esculenta subsp. esculenta* (cassava), which is recognized as the only cultivated species of the genus.

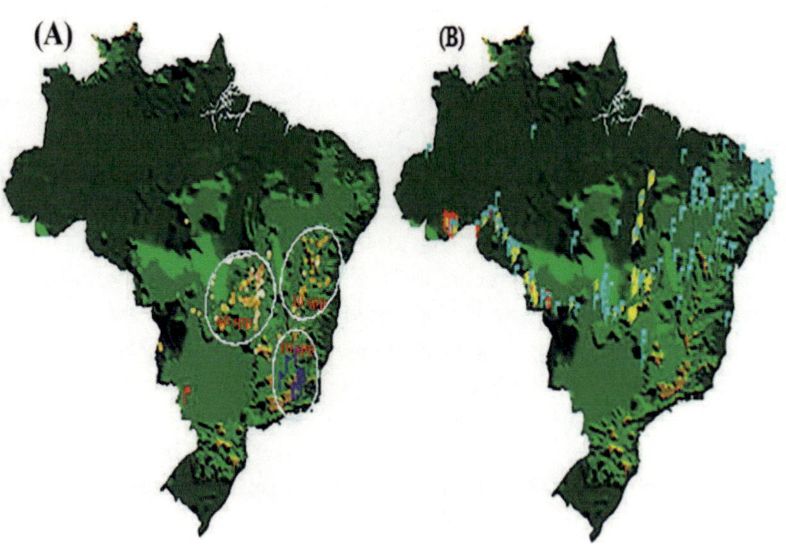

Figure 1. Occurrence and geographical distribution of *Manihot* genus in Brazil. (A) Refers to distribution of 60 species (larger circle) of the genus in the central part of Brazil, 10 species (upper small circle) of the genus in the east part, and 10 species (lower small circle) of the genus in southeast part of Brazil. (B) Refers to distribution of *M. esculenta subsp. flabellifolia* from Northwest Amazon to Northeast in the central part (yellow flag) of Brazil and *M. esculenta subsp. esculenta* from Northwest Amazon to Northeast of Brazil (blue flag).

Results indicate that species from Brazil, including *M. esculenta subsp. flabellifolia*, are phylogenetically related to the cultivated species *M. esculenta subsp. esculenta* rather than species from Mexico [3, 4]. These analyses indicated that cassava probably did not originate from Mexico. Therefore, it is not the result of a compiled species but instead possibly has a Brazilian single species ancestor originally named *M. esculenta subsp. flabellifolia* [5–7]. A gene pool analysis [8] for the cultivated species (*M. esculenta subsp. esculenta*) and its wild relative identify two gene pools involving 13 *Manihot* species in gene pool 2 (GP2) and 4 *Manihot* species in gene pool 1 (GP1), which has *M. esculenta subsp. flabellifolia* based on overall similarity of several morphological characteristics indicates *M. esculenta subsp flabellifolia* as the closest alias of cassava [8].

3. Domestication as evolutionary processes

Phylogenetic techniques used for determining the molecular evolution of a crop have relied predominantly on sequence information to model the evolutionary history that determines plant speciation and domestication. Phylogenetic trees are based on alignment of DNA or protein sequences, from which evolutionary distances between genes can be inferred. However, transcriptional behavior of a gene is poorly represented by DNA sequence data alone. A gene's transcriptional profile may contain critical functions, including when and where a gene is expressed, and the conditions under which gene expression is manifested. This chapter addresses questions on how function transcriptional profiles vary due to changes in the environment (light intensity) and due to genetic diversity of landraces and commercial breed varieties.

3.1. Molecular evolution of a crop species

Factors involved in regulation of expressed genes or gene sets could be crucial in explaining the key functional differences between related genes whose function, during selection (natural and artificial), cannot be distinguished from DNA sequence alone [11–19]. Attempts to predict expression patterns of genes using sequence information [20] have typically been limited by the complexity and diversity of factors influencing genes. Thus, sequence-based prediction of a gene's regulation remains a premature goal. However, transcriptomic approaches, for example, using microarray chip or RNA-Seq technology, allow for a direct, quantitative measurement of global transcriptional responses to a given environmental or genetic factor and are useful experimental sources for obtaining large-scale gene expression data [21, 22]. Genomic data sets spanning a wide selection of the cassava ancestor (*M. esculenta subsp. flabellifolia*), landraces from the Amazon, and breeding cultivars (cv.) are publicly available [23–33], providing a ready source of data for studying several aspects of gene transcription behavior. The integration of genome and transcriptome data [34–44] provides an increasingly detailed picture of molecular evolution by incorporating regulatory behavior into models of the evolution of gene expression and function. Here, we report steps toward understanding changes in gene expression to model gene evolutionary function using our cassava domestication syndrome hypothesis. Specifically, the cassava domestication syndrome hypothesis considers changes in traits such as plant growth habit, storage root formation and flowering from the cassava ancestor (*M. esculenta subsp. flabellifolia)* to becoming the cultivated species (*M. esculenta subsp. esculenta*). **Figure 2** illustrates these variables as observed from field trips to the Amazon and recorded images.

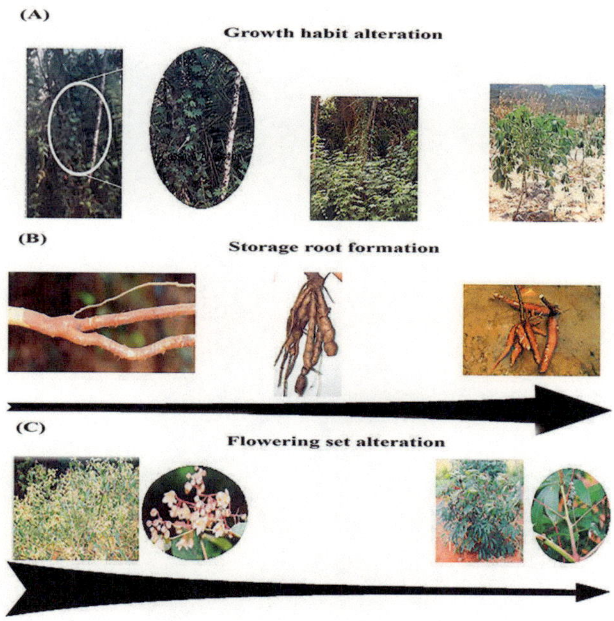

Figure 2. Diagram representing the early step of domestication of cassava crop. Field observed phenotype pattern differences between *M. esculenta subsp flabellifolia* (ancestor) in the forest (shade environment) and open field (full light environment, after removal of the forest) growth of the cultivated species *M. esculenta subsp esculenta*. (A) Growth habit change from vine type of growth (ancestor) to shrub type of growth (cultivated cassava). (B) Storage root formation from fibrous normal root (ancestor) in the forest (shade environment) and open field (C) flowering set pattern alteration between ancestor and cultivated cassava.

3.2. Differentially expressed genes

A cDNA microarray chip designed for Euphorbiaceae [24] was probed with total RNA extracted from storage root (31 samples total) of cassava with diverse storage root traits. This chapter documents a total of 569 genes which were identified as differentially expressed (*p-value* of 0.005) between storage root of *M. esculenta subsp. flabellifolia* (ancestor) and various *M. esculenta subsp. esculenta* landraces and cultivars. Hybridization intensity values were statistically analyzed to further identify groups among the differentially expressed genes (DEG). The complexity of the experimental design and analyses for screening groups of DEG was achieved using two statistical strategies [34, 35]. First, principal component analysis (PCA) was performed to observe the number of DEG among each group. The PCA results identified four groups among the DEG. Considering the cassava domestication syndrome hypothesis described above, the questions to address with the available data are (i) do these results occur due to differences in expression of genes per se or, (ii) in part, due to the selected genetic backgrounds in the experimental design? Therefore, the second approach used recursive partitioning to obtain tentative conclusions about the grouping patterns [34, 35], as shown in **Figure 3**.

3.3. Ontology and functional classification of differentially expressed genes (DEG)

Analysis of the DEG identified 22 distinct groups among gene ontologies and functional classifications. The groups (**Figure 4**) highlighted in yellow (i.e., "Protein with Binding Function

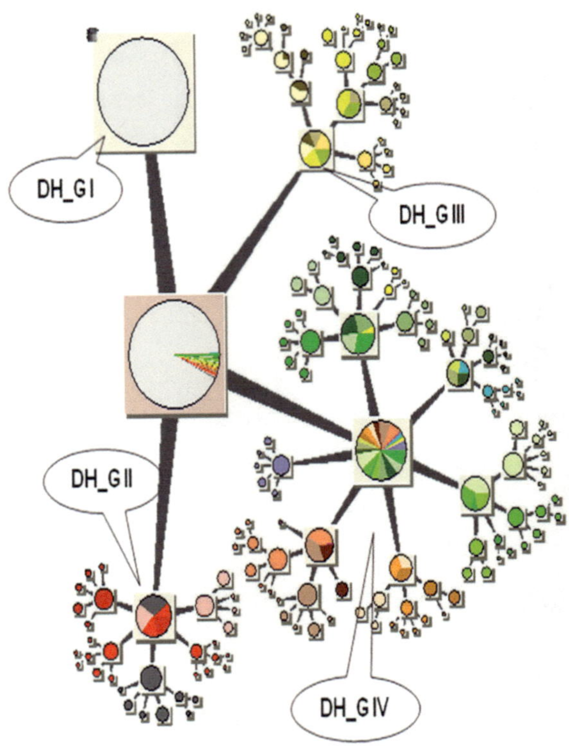

Figure 3. Partitioned pattern for a set of 569 differentially expressed genes (DEG) in storage root between cassava ancestor (*M. esculenta subsp flabellifolia*) and cultivated species (*M. esculenta subsp esculenta*). Group DH_G1 partitioning pattern showed no statistically significant further groups formation, while DH_G2, DH_G3 and DH_G4 formed variable numbers of subgroups and sub-sub groups of DEG.

or Cofactor Requirement," "Regulation of Metabolism and Protein Function," and "Cellular Transport, Transport Facilities and Transport Routes") were targeted to elucidate candidate genes involved in regulation of these key pathway networks.

3.4. Exploratory pathway networks and candidate regulatory genes

The program Pathway Studio [43] was used to conduct subanalysis (SNEA) to identify potential regulatory networks from transcriptome data obtained in this study and available databases, as previously described [36–42]. The results on statistics (shown in **Tables 2** and **3**) and visualization of gene networks (shown in **Figures 5** and **6**) took into consideration three types of molecular interaction mechanisms (expression target, protein binding, and protein modification).

These results indicate node operating gene/hub, edge genes which are regulated (activated or silenced), and their expression level—increased abundance (blue color) or decreased abundance (pink color) among genes visualized in the pathways; regulatory genes such as transcription factors and other gene products modulating functionality (protein binding and modification) were observed. The node/hub gene regulates the network and genes, while on the edge are

http://dx.doi.org/10.5772/intechopen.71348

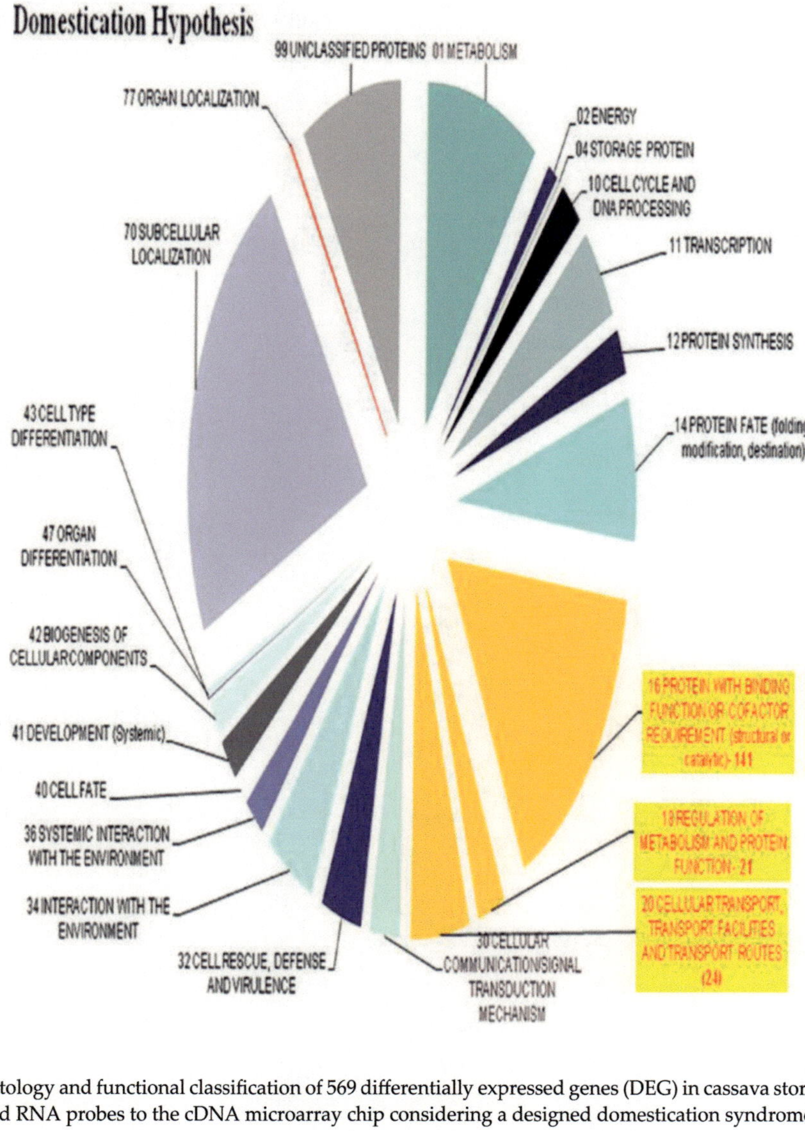

Figure 4. Ontology and functional classification of 569 differentially expressed genes (DEG) in cassava storage root using 31 hybridized RNA probes to the cDNA microarray chip considering a designed domestication syndrome hypothesis.

regulatory genes of a particular network. **Table 4** summarizes the list of nodes/hubs in the networks unique to each class of landraces based on comparisons to the cassava ancestor and the cv. IAC 12-829.

3.5. Identification of regulatory gene sets

Exploratory pathways network common among cassava ancestor (*M. esculenta subsp flabellifolia*), cassava cultivars (*M. esculenta subsp esculenta*)), and cassava landraces (*M. esculenta subsp esculenta) were based on the analysis of DEG using* Pathway Studio software to conduct the Gene Set Enrichment Analyses (GSEA) algorithm and establish the level of significance (*p-value*) of

Name	Gene set root	Entities	Neighbors	p-Value
Comparison of FLA vs. a representative group from the World Collection (cv. MTAI16/1, cv. CM2177–2/1, cv. Reina1 and cv. Veronica1)				
Expression targets	Calmodulin	8	7	0.0032
	Ubiquitin	7	6	0.0076
	CRY1	9	8	0.0229
	CTR1	7	6	0.0270
	Proteasome	9	8	0.0453
Binding partners (protein interaction)	PHYB	9	8	0.0004
	PHYA	7	6	0.0324
	PRL1	6	5	0.0366
	14–3-3	13	12	0.0464
Comparison of FLA vs. cv. IAC12.829 (white-type storage root)				
Expression targets	Ubiquitin	7	6	0.0074
	Calmodulin	8	7	0.0367
	AXR3	6	5	0.0404
Binding partners (protein interaction)	PHYB	9	8	0.0063
	SCF	7	6	0.0125
	WAVE1	6	5	0.0280
	PRL1	6	5	0.0325
	PHYA	7	6	0.0386
	TIR1	6	5	0.0391
Comparison of FLA vs. landrace Cas51 (pink-type storage root)				
Expression targets	ABI1	6	5	0.0027
	Ubiquitin	7	6	0.0192
	CRY1	9	8	0.0324
	FLC	10	9	0.0452
	Proteasome	9	8	0.0477
Binding partners (protein interaction)	PHYB	9	8	0.0006
	PHYA	7	6	0.0147
	SCF	7	6	0.0307
	TIR1	6	5	0.0318
	COP1	6	5	0.0414
Comparison FLA vs. Landrace Cas36.1 (sugary-type storage root)				
Expression targets	Calmodulin	8	7	0.0042
	Ubiquitin	7	6	0.0073
	CTR1	7	6	0.0247

Name	Gene set root	Entities	Neighbors	p-Value
Binding partners (protein interaction)	PHYB	9	8	0.0008
	PRL1	6	5	0.0349
	SLY1	6	5	0.0429
Comparison of FLA vs. Landraces CAS36.7, Cas64, Cas56, and Cas62 (yellow-type storage root)				
Expression targets	Ubiquitin	7	6	0.0043
	CRY1	9	8	0.0230
	Calmodulin	8	7	0.0331
	EIN2	13	12	0.0396
Binding partners (protein interaction)	PHYB	9	8	0.0006
	SCF	7	6	0.0369

Comparison between cassava ancestral *M. esculenta subsp. flabellifolia* (FLA) and genetic diversity of cultivated species *M. esculenta subsp. esculenta* (cassava).

Table 2. Sub network enrichment of classes of genes and their level of statistically significance (*p-value*).

Common to all genotype	Landrace (pink)	Landrace (*sugary*)	Commercial cv. (white)	Landrace (yellow)	World core collection (white)
Ubiquitin(ET)	FLC(ET)	SLY1(PI)	PRL1(BP)	EIN2((ET)	14–3-3(PI)
PHYB(PI)	TIR1(PI)	PRL1(PI)	WAVE1((PI)	SCF(PI)	
	ABI1(ET)				

Cassava classes of genotype include commercial cv. IAC 12–829 (white type), landrace Cas51 (pink color), landraces CAS36.7, Cas64, Cas56, and Cas62 (yellow), landrace Cas36.1 (sugary), and world core collection representatives cv. MTAI16/1, cv. CM2177–2/1, cv. Reina1, and cv. Veronica1 (white type). Abbreviation refers to gene name as in **Table 4**. Abbreviation (in parenthesis) accounts for molecular interaction mechanism as expression target (ET), protein interaction (PI), and protein binding (PB) regulatory gene function.

Table 3. Exploratory regulatory node/hub genes unique to different classes of cassava genotype compared to cassava ancestral (*M. esculenta subsp. flabellifolia*).

gene groups on the identification of functional gene according to the Sub-Network Enrichment Analysis (SNEA) for biological processes, cellular components, and molecular functions using based on annotation of cassava genes to *Arabidopsis* database. The gene sets were grouped per their function and selected for retrieving and visualizing regulatory networks or pathways they form. Together these results add new knowledge about the potential functionality of gene products previously unknown in cassava storage root, their potential roles in the domestication trait, as well as in the flowering set trait. As an example of these analyses, we propose a hypothetical hormonal gene regulatory model (**Figure 7**) to represent the effect of environmental light changes likely caused due to removal of ancestral cassava from the forest as illustrated in **Figure 2**. However, it is important to clarify that Gene Set Enrichment Analyses (GSEA) and Sub-Network Enrichment Analysis (SNEA) is based on annotation of cassava genes to *Arabidopsis*. Thus, for our proposed models, we are assuming that the cassava gene products are performing similar functions as their Arabidopsis homologues (all *Arabidopsis* annotations used in this chapter, and their known functions can be obtained from [44]).

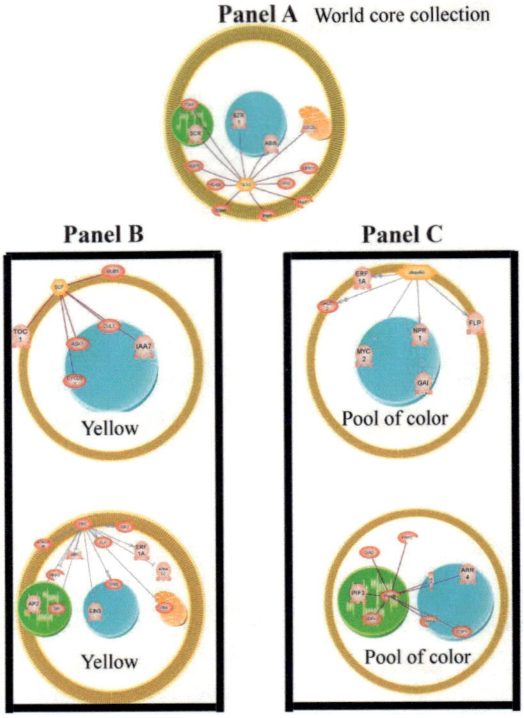

Figure 5. Diagram showing exploratory pathway networks for regulatory genes related to cassava ancestral (*M. esculenta subsp flabellifolia*) domestication storage root trait and its genetic diversity in the cultivated species (*M. esculenta subsp esculenta*). Panel A. Differentially expressed regulatory gene node/hub and regulated genes of ancestral and world core collection representatives. Panel B. Differentially expressed regulatory gene node/hub and regulated genes of ancestral and the yellow color storage root landraces. Panel C. Differentially expressed regulatory gene node/hub and regulated genes of ancestral and the pool of colors (white, yellow, and pink) storage root landraces genotypes. Blue and pink colors symbols are up and down regulated genes. Arrow (⟶) refers to positive regulation of the gene and block line (…⊣…) refers to silencing the gene.

The removal of ancestral cassava from a shaded forest environment would be expected to alter regulatory networks and pathways involved in light perception and signaling, as highlighted in **Figure 5** (Panel C) and **Figure 6** (Panel A) (the Brazilian collection). Further, altering light quantity and quality or selection for storage root traits during the domestication process also appears to have differentially impacted gibberellic acid (GA) signaling regulation of DELLAs, as indicated by **Figure 5** (Panel C) (GAI) and **Figure 6** (Panel B) (SLY1, RGA1, RGL2, GAI). These alterations are likely to have impacted known regulatory networks involving interactions between DELLAs and PIF3/PIF4 [39, 40]. Because light perception (lack of shade) would be expected to reduce the positive impact that GA signaling has on inhibiting DELLA, the function of DELLAs in reduced GA-induced elongation likely resulted in dwarfed and bushy phenotypes [41, 42]. In aboveground photosynthesizing tissues, shifts in expression of genes linked to the GA/DELLA regulatory pathway would also be expected to result in shifts between skotomorphogenesis and photomorphogenesis [39] and, potentially, reduced flowering as illustrated in **Figures 2** and **7**. However, in the underground storage root of cassava,

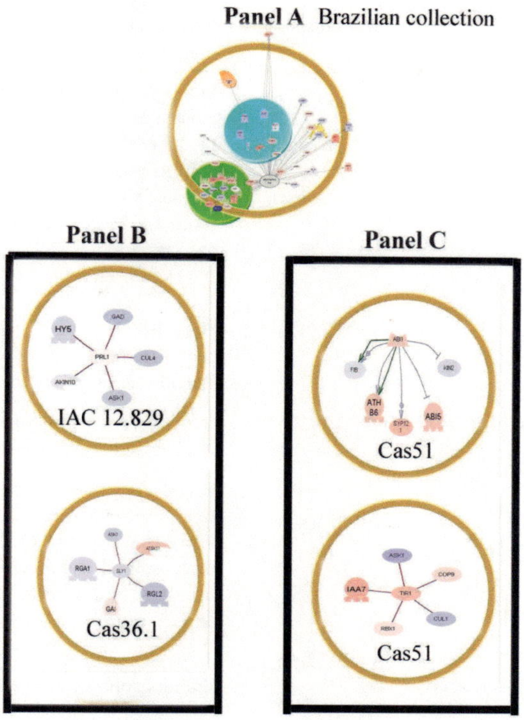

Figure 6. Diagram showing exploratory pathway networks for regulatory genes related to cassava ancestral (*M. esculenta subsp flabellifolia*) domestication storage root trait and its genetic diversity in the cultivated species (*M. esculenta subsp esculenta*). Panel A. Differentially expressed regulatory gene node and regulated genes of ancestral and Brazilian collection representatives. Panel B. Differentially expressed regulatory gene node/hub and regulated genes of ancestral and white color cassava (cv. IAC12.829 and sugary landrace Cas36.1). Panel C. Differentially expressed regulatory gene node/hub and regulated genes of ancestral and pink landrace Cas51. Blue color lines refers to more abundant expression in domesticated species, red color lines refers to less abundant expression in domesticated species, and purple color lines represents protein-protein interactions). Arrow (→) refers to positive regulation of the gene and block line (...⊣...) refers to silencing the gene.

altered regulation of DELLAs may have played some role in the shift observed from a fibrous type to a storage root type, as also illustrated in **Figures 2** and **7**. DELLAs have been reported to impact auxin signaling pathways, and, possibly, DELLA's impact on auxin via jasmonic acid (JA) regulation involving JAZ1 and MYC2, as reviewed by [42], could be involved in this process (**Figure 7**). Further, DELLA and SCARECROW (SCR, see **Figure 5 (Panel A—World Collection)** have known interactions that could also be involved in the domestication trait of cassava storage root formation by impacting GA/auxin/cytokinin cross-talk and signaling and root meristematic development and differentiation [44]. **Figure 5 (Panel A)** and **Figure 6 (Panel C)** highlighted abscisic acid (ABA) signaling as a potential component in the domestication process, which could also have some connection to DELLA and SCR proteins known in integration of GA/ABA cross-talk responses [45]. Finally, as presented in **Figure 5 (Panel B)**, **Figure 5 (Panel C)**, **Figure 6 (Panel B)**, and **Figure 6 (Panel B)**, regulatory pathways

Gene abbreviation	Gene name	ATG number	Function (from www.arabidopsis.org)
14-3-3	14-3-3 family protein		14-3-3 proteins are a family of conserved regulatory molecules that are expressed in all eukaryotic cells. 14–3-3 proteins have the ability to bind a multitude of functionally diverse signaling proteins, including kinases, phosphatases, and transmembrane receptors. More than 200 signaling proteins have been reported as 14–3-3 ligands
ABI1	ABA INSENSITIVE 1	AT4G26080	Involved in abscisic acid (ABA) signal transduction
AP1	APETALA1	AT1G69120	Floral homeotic gene encoding a MADS domain protein homologous to SRF transcription factors. Specifies floral meristem and sepal identity
AXR3	AUXIN RESISTANT 3	AT1G04250	Transcription regulator acting as repressor of auxin-inducible gene expression
BRI1	BRASSINOSTEROID INSENSITIVE 1	AT4G39400	Encodes a plasma membrane-localized leucine-rich repeat receptor kinase involved in brassinosteroid signal transduction. BRI1 appears to be involved in the autonomous pathway that regulates the transition to flowering, primarily through its effects on FLC expression levels
COP1	CONSTITUTIVE PHOTOMORPHOGENIC 1	AT2G32950	Represses photomorphogenesis and induces skotomorphogenesis in the dark
CRY1	CRYPTOCHROME 1	AT4G08920	A flavin-type blue-light photoreceptor with ATP-binding and autophosphorylation activity. Functions in perception of blue/green ratio of light
CTR1	CONSTITUTIVE TRIPLE RESPONSE 1	AT5G03730	Homologous to the RAF family of serine/threonine protein kinases. Negative regulator in the ethylene signal transduction pathway
EIN2	ETHYLENE INSENSITIVE 2	AT5G03280	Involved in ethylene signal transduction. Acts downstream of CTR1
FLC	FLOWERING LOCUS C	AT5G10140	Transcription factor that functions as a repressor of floral transition and contributes to temperature compensation of the circadian clock
LSD1	LESION SIMULATING DISEASE 1	AT4G20380	LSD1 monitors a superoxide-dependent signal and negatively regulates a plant cell death pathway
PHYA	PHYTOCHROME A	AT1G09570	Light-labile cytoplasmic red/far-red light photoreceptor involved in the regulation of photomorphogenesis
PHYB	PHYTOCHROME B	AT2G18790	Red/far-red photoreceptor involved in the regulation of de-etiolation
PRL1	PLEIOTROPIC REGULATORY LOCUS 1	AT4G15900	Mutations confer hypersensitivity to glucose and sucrose and augment sensitivity to cytokinin, ethylene, ABA, and auxin
SCF	Skp, Cullin, F-box containing complex		A multi-protein E3 ubiquitin ligase complex catalyzing the ubiquitination of proteins destined for proteasomal degradation
SLY1	SLEEPY1	AT4G24210	F-box protein that is involved in GA signaling. Component of E3 ubiquitin complex. Interacts with DELLA proteins

Gene abbreviation	Gene name	ATG number	Function (from www.arabidopsis.org)
TIR1	TRANSPORT INHIBITOR RESPONSE 1	AT3G62980	Encodes an auxin receptor that mediates auxin-regulated transcription. It contains leucine-rich repeats and an F-box and interacts with ASK1, ASK2, and AtCUL1 to form SCF-TIR1, an SCF ubiquitin ligase complex
WAVE1	WISKOTT-ALDRICH SYNDROME PROTEIN FAMILY VERPROLIN HOMOLOGOUS PROTEIN 1	AT2G34150	Encodes a member of the SCAR family. These proteins are part of a WAVE complex. The SCAR subunit activates the ARP2/ARP3 complex which in turn acts as a nucleator for actin filaments

Table 4. List of functional node/hub genes with roles in domestication syndrome which play a role in flowering set trait and storage root formation trait in the cultivated plant of relevance for the domestication syndrome traits (storage root trait and flowering set) as reported in **Table 2**

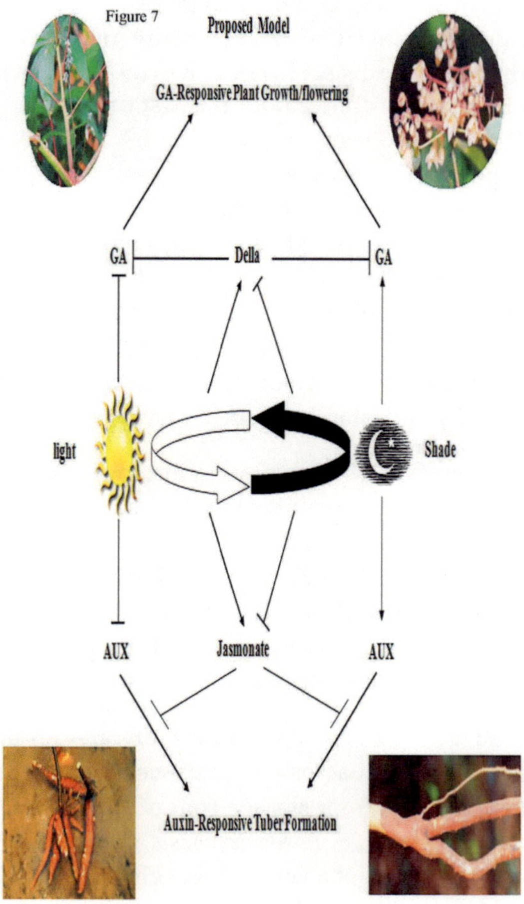

Figure 7. Diagram representing a proposed hormonal gene regulatory model to represent the environmental light alteration when human first removed the forest to cultivate cassava to harvest for edible food. The proposed model is based on results of differentially expressed genes of hormone signaling and biosynthesis pathways identified in storage roots of the cassava ancestor when compared to other domesticated cassava landraces and modern cultivars.

(with node/hubs SLY1, SCF, TIR1, and ubiquitin) also appear to function through proteasome degradation pathways. Collectively, these examples suggest that [39, 40] the domestication syndrome may have evolved through a combination of changes in [39, 40] environmental signaling factors and selection pressures that impacted phytohormone cross-talk, signaling, and protein regulation pathways during the removal of the cassava ancestor from the forest, which, in turn, lead to modern-day domesticated landraces and cultivars of cassava.

4. Synthesis and conclusions

This study highlighted some key factors influencing the fate of gene function in relation to cassava domestication syndrome traits including (1) landscape alterations resulting in sunlight exposure (alteration of light quality and quantity) in early stages of the domestication process of cassava crop due to removal of forest; (2) natural selective pressure, due to high light intensity under open-field cultivation leading to an edible cassava storage root; (3) artificial selective pressure (man selecting edible plant parts), due to harvest of plants showing storage root formation; and (4) artificial selective pressure (man involuntarily selecting plant traits such as plants with low flowering set in relation to the ancestor). Additionally, the data presented in the present study also allowed, for the first time, to propose a hormonal regulating model (**Figure 7**) on the involvement of GA/DELLA and Auxin/Jasmonate in cassava domestication traits. It appears that domestication of cassava originated with the removal of ancestral *M. esculenta subsp. flabellifolia* from the Amazon forest, which, through human trait preference and selection, evolved into current-day cultivated landraces and cultivars of *M. esculenta* due. As a result, ancestral cassava evolved from a vine, prolific flowering, and fibrous root phenotype into a domesticated bushy, reduced flowering, and tuberous storage root phenotype (**Figures 2** and **7**). Selection for specific storage root traits also resulted in domesticated cassava storage root with color diversity [46], storage root phenotypes, and carbon sequestration diversity with starch vs. sugary storage root phenotypes [47–49].

As our understanding of gene expression evolution improves, it should become possible to infer protein function into approaches focused on the use of proteomics technologies [50]. Ancestral protein functions can be estimated using this approach, and efforts to annotate current genes/proteins will benefit from knowledge of the behavior and factors influencing gene expression profiles. Ultimately, gene expression profiles should be equally integrated with structure and sequence to predict and assist in annotating protein function and evolution directly on the genome sequence of the ancestor and cultivated species.

Technological advances have aided our ability to rapidly and affordably obtain and compare transcriptomes from within and across plant species. As presented in this chapter, we compare the global transcriptomes from storage root of various landraces and cultivars of domesticated cassava and ancestral *M. esculenta subsp. flabellifolia* to identify differentially abundance of transcripts, in this case, at a very stringent level ($p < 0.005$). Modern technology also affords advances in bioinformatics approaches for analyzing these large transcriptomic data set using ever-evolving algorithms. In this chapter, we used GSEA and SNEA to identify nodes/hubs and regulatory genes/proteins that provide a snapshot of potential regulatory networks/pathways that differ between ancestral and domesticated cassava storage root and are likely key pathways involved in the domestication syndrome. Based on our results, it appears that as ancestral cassava evolved into domesticated cassava several important regulatory pathways involved in light signaling and regulation, floral signaling and regulation, and hormone signaling and regulation (particularly GA and auxin) were altered. Many of these processes also appear to involve complexes involved in regulating protein degradation through Ubiquitination and Proteasomal trafficking.

Acknowledgements

We would like to acknowledge the financial support provided by the Rockefeller Foundation (RF96010#25 and RF9707#26 in the search for landrace diversity); Conselho Nacional de Desenvolvimento Cientifica e Tecnológico – CNPq (grant #480410.2001-1 for functional genomic work); Programa Nacional de Pesquisa em Biotecnologia – CENARGEN (Project Nº 1 060302058 for carbohydrate analysis); IAEA (contract #13188 for funds supplied to gene expression analysis); National Nature Science Foundation of China (NSFC grant #31271776), and NSFC-CGIAR International (Regional) Cooperation and Exchange Programs (grant #31361140366) for genome proteome sequences, and this publication.

Source of funds

National Nature Science Foundation of China (NSFC grant number 31271776).

Author details

Luiz Joaquim Castelo Branco Carvalho[1*†], James V. Anderson[2†], Songbi Chen[3†], Chikelu Mba[4†] and Münevver Doğramaci[5†]

*Address all correspondence to: luiz.castelo@embrapa.br

1 Embrapa-Cenargen, Brasilia, DF, Brazil

2 USDA-ARS, Sunflower and Plant Biology Research Unit, Fargo, ND, USA

3 Tropical Crop Genetic Resources Institute, CATAS, Danzhou, China

4 FAO-The Plant Production and Protection Division, Rome, Italy

5 Department of Internal Medicine, Sanford School of Medicine, University of South Dakota, Vermillion, SD, USA

†All authors contributed equally to this work

References

[1] Chapin FS III, Autumm K, Pungnaire F. Evolution of suites of traits in response to environmental stress. American Naturalist Supplement. 1993;**142**:S78-S92

[2] Isendahl C. The domestication and early events spread of manioc (*Manihot esculenta* Crantz): A brief synthesis. Latin American Antiquity. 2011;**22**(4):452-468. ISSN: 1045-6635 (Print), 2325-5080

[3] Schaal BA, Olsen KM, Carvalho LJCB. Evolution, domestication, and agrobiodiversity in the tropical crop cassava. In: Timothy J Otley; Nyree Zerega, and Hugh Cross, editors. Darwin's Harvest: New Approaches to New Origins, Evolution, and Conservation of Crops. 1 ed. New York: Columbia University Press; 2005. p. 269-284

[4] Schaal B, Carvalho LJCB, Prinzie T, Olsen K, Hernandez M, Cabral G, Moeller D. Phylogenetic relationships and genetic diversity in *Manihot* species. African Journal of Root and Tuber Crops. 1997;**2**:147-149

[5] Allem AC. The Origins and Taxonomy of Cassava. In: Hillocks RJ, Thresh JM, Bellotti AC, editors, Cassava: biology, production and utilization. CAB International; 2002. p 1-16. DOI: 10.1079/9780851995243.0000

[6] Allem AC. The origin of *Manihot esculenta* Crantz (Euphorbiaceae). Genetic Resources and Crop Evolution. 1994;**41**:133-150

[7] Allem AC. The closest wild relatives of cassava (*Manihot esculenta* Crantz). Euphytica. 1999;**107**:123-133. DOI: 10.1023/A:1026422229054

[8] Allem AC, Roa AC, Mendes RA, Salomão AN, Burle ML, Second G, Carvalho PCL, Cavalcanti J. The primary gene pool of cassava (*Manihot esculenta* Crantz). In: Carvalho LJCB, Thro AM and Vilarinhos AD, editors. Cassava Biotechnology - IV Intl Sci Meet-CBN, EMBRAPA-Genet Res Biotechn, Brazil: Brasilia-DF; 2000. p. 3-14

[9] Rogers DJ, Appan SG. *Manihot* and Manihotoides (Euphorbiaceae), a computer-assisted study. Flora Neotropica, monograph no 13. New York: Hafner Press; 1973; 272 pp

[10] Rogers DJ. Some botanical and ethnological considerations of *Manihot* esculenta. Economic Botany. 1965;**19**:369-377

[11] Olsen KM, Wendel JF. A bountiful harvest: genomic insights into crop domestication phenotypes. Annu Rev Plant Biol. 2013;**64**:47-70. DOI: 10.1146/annurev-arplant-050312-120048

[12] Olsen KM. Population history of *Manihot esculenta* (Euphorbiaceae) inferred from nuclear DNA sequences. Molecular Ecology. 2002;**11**:901-911

[13] Briana LG, Olsen KM. Genetic perspectives on crop domestication. Trends Plant Sci. 2010;**15**(9):529-537. DOI: 10.1016/j.tplants.2010.05.008

[14] Olsen KM, Schaal BA. Evidence on the origin of cassava: phylogeography of Manihot esculenta. Proceedings of the National Academy of Sciences of the USA. 1999;**96**:5586-5591

[15] Olsen K, Schaal B. New evidence on the origins of cassava: Phylogeography of *Manihot esculenta*. Proc. Natl. Acad. Sci. 1999;**96**:5586-5591

[16] Olsen K. Schaal B, Hernandez M. A survey of DNA sequence variation in cassava and other *Manihot* species. In Cassava Biotechnology: IV International Scientific Meeting – *CBN*, 129-134, eds. Carvalho LJCB, Thro AM, Vilarinhos AD. Brasília, DF, Brasil: Embrapa–CENARGEN; 2000. p. 129-134

[17] Olsen K, Schaal B. Microsatellite variation in cassava (Manihot esculenta, Euphorbiaceae) and its wild relatives: Further evidence for a southern Amazonian origin of domestication. Amer. J. Botany. 2001;**88**:131-142

[18] Olsen KM. Minisatellite variation in a single-copy nuclear gene: phylogenetic assessment of repeat length homoplasy and mutational mechanism. Molecular Biology and Evolution, 1999;**16**:1406-1409

[19] Schaal B, Olsen KM. Gene genealogies and variation within plant populations. Proc. Natl. Acad. Sci. 2000;**97**:7024-7029

[20] Beer MA, Tavazoie. Predicting Gene Expression from Sequence. Cell. 2004;**117**:185-198

[21] Gross BL, Olsen KM. Genetic perspectives on crop domestication. Trends in Plant Science. 2010;**15**:529-537. DOI: 10.1016/j.tplants.2010. 05.008

[22] Doebley JF, Gaut BS, Smith BD. The Molecular Genetics of Crop Domestication. Cell. 2006;**127**:1309-1321

[23] Wang W, Feng B, Xiao J, Xia Z, Zhou X, Li P, Zhang W, Wang Y, Møller BL, Zhang P, Luo MC, Xiao G, Liu J, Yang J, Chen S, Rabinowicz PD, Chen X, Zhang HB, Ceballos H, Lou Q, Zou M, Carvalho LJCB, Zeng C, Xia J, Sun S, Fu Y, Wang H, Lu C, Ruan M, Zhou S, Wu Z, Liu H, Kannangara RM, Jørgensen K, Neale RL, Bonde M, Heinz N, Zhu W, Wang S, Zhang Y, Pan K, Wen M, Ma PA, Li Z, Hu M, Liao W, Hu W, Zhang S, Pei J, Guo A, Guo J, Zhang J, Zhang Z, Ye J, Ou W, Ma Y, Liu X, Tallon LJ, Galens K, Ott S, Huang J, Xue J, An F, Yao Q, Lu X, Fregene M, López-Lavalle LA, Wu J, You FM, Chen M, Hu S, Wu G, Zhong S, Ling P, Chen Y, Wang Q, Liu G, Liu B, Li K, Peng M. Cassava genome from a wild ancestor to cultivated varieties. Nat Commun. 2014;**5**:5110-5119

[24] Lokko Y, Anderson JV, Rudd S, Raji A, HorvathD, Mikel MA, Kim R, Liu L, Hernandez A, Dixon AGO, Ingelbrecht IL. Characterization of an 18,166 EST dataset for cassava (Manihot esculenta Crantz) enriched for drought-responsive genes. Plant Cell Report. 2007;**26**:1605-1618

[25] http://www.phytozome.net/search.php?show=blast&targetType=genome&method=Org_Mesculenta

[26] http://cys.bios.niu.edu/plantcazyme/family.php?family=GH13&species=maes

[27] https://www.ncbi.nlm.nih.gov/nucest/?term=cassava%20carvalho

[28] https://www.ncbi.nlm.nih.gov/protein/?term=cassava+carvalho

[29] https://www.ncbi.nlm.nih.gov/genome/?term=cassava+carvalho

[30] http://www.cassavabase.org

[31] www.cenargen.embrapa/genome

[32] http://cassava.psc.riken.jp/ver.1/snp/index.pl?v=1

[33] http://www.R-project.org/

[34] Le S, Josse J, Husson F. Factominer: an R-package for multivariate analysis. Journal of statistical software, 2008;**25**(1):1-18

[35] Chavent M, Kuentz V, Saracco J. Orthogonal Rotation in PCAMIX. Advances in Classification and Data Analysis, 2011;**6**:131-146

[36] Doğramaci M, Horvath DP, Chao WS, Foley ME, Christoffers MJ, Anderson JV. Extended low temperature impacts dormancy status, flowering competence, and transcript profiles in crown buds of leafy spurge. Plant Mol. Biol. 2010;**73**:207-226. DOI: 10.1007/s11103-010-9621-8

[37] Doğramaci M, Foley ME, Chao WS, Christoffers MJ, Anderson JV. Induction of endodormancy in crown buds of leafy spurge (Euphorbia esula L.) implicates a role for ethylene and cross-talk between photoperiod and temperature. Plant Mol. Biol. 2013;**81**:577-593. DOI: 10.1007/s11103-013-0026-3

[38] Mazzella MA, Arana MV, Staneloni RJ, Perelman S, Rodriguez Batiller MJ, Muschietti J. Phytochrome control of the Arabidopsis tran¬scriptome anticipates seedling exposure to light. Plant Cell. 2005;**17**:2507-2516. DOI: 10.1105/tpc.105.034322

[39] Sun TP. Gibberellin metabolism, perception and signaling pathways in Arabidopsis. In: The Arabidopsis Book. American Society of Plant Biologists, Rockville, MD; 2008. doi/10.1199/tab.0103, http://www.aspb. org/publications/arabidopsis/

[40] Harberd NP, Belfield E, Yasumura Y. The angiosperm gibberellin-GID1-DELLA growth regulatory mechanism: How an "inhibitor of an inhibitor" enables flexible response to fluctuating environments. The Plant Cell. 2009;**21**:1328-1339

[41] Kazan K, Manners JM. JAZ repressors and the orchestration of phyhormone crosstalk. Trends in Plant Science, 2012;**17**(1):22-31

[42] Moubayidin L, Salvi E, Giustini L, Terpstra I, Heidstra R, Costantino P, Sabatini SA. SCARECROW-based regulatory circuit controls Arabidopsis thaliana meristem size from the root endodermis. Planta. 2016;**243**:1159-1168

[43] https://www.pathwaystudio.com/

[44] www.arabidopsis.org

[45] Golldack D, Li C, Mohan H, Probst N. Gibberellins and abscisic acid signal crosstalk: living and developing under unfavorable conditions. Plant Cell Rep. 2013;**32**:1007-1016

[46] Carvalho LJCB, Agustini MA, Anderson JV3, Vieira EA, de Souza CR, Schaal BA, Silva JP, Natural variation in expression of genes associated with carotenoid biosynthesis and accumulation in cassava (Manihot esculenta Crantz) storage root. BMC Plant Biol. 2016;**16**(1):133. DOI: 10.1186/s12870-016-0826-0

[47] Carvalho LJCB, de Souza CRB, Cascardo JMC, Bloch Junior C, Campos L. Identification and characterization of a novel cassava (Manihot esculenta Crantz) clone with high free sugar content and novel starch. Plant Mol Biol. 2004;**56**:643-659

[48] Carvalho LJCB, de Souza CRB, Cascardo JCM, Agostini MAV, Vieira EA Anderson JV, Lippolis J. Natural Genetic Variation in Cassava (Manihot esculenta Crantz) Landraces: A Tool for Gene Discovery. In: Shu QY, editor, Induced Plant Mutations in the Genomics Era. Food and Agriculture Organization of the United Nations, Rome; 2009. p. 313-316

[49] Carvalho LJCB, Schaal BA. Assessing genetic diversity in the cassava (Manihot esculenta Crantz) germplasm collection in Brazil using PCR-based markers. Euphytica. 2001; **120**:133-142

[50] An F, Chen T, Djabou MAS, Kaimian L, Qing XL, Carvalho LJCB, Keith T, Li BG, Chen S. Domestication syndrome is investigated by proteomic analysis between cultivated cassava (*Manihot esculenta* Crantz) and its wild relatives. PLoS One. 2016;**11**(3):e0152154. https://doi.org/10.1371/journal.pone.0152154

Ecophysiology and Production Principles of Cassava (*Manihot* species) in Southeastern Nigeria

Martin A.N. Anikwe and Ejike E. Ikenganyia

Additional information is available at the end of the chapter

http://dx.doi.org/10.5772/intechopen.70828

Abstract

Cassava (*Manihot* species) is a crop of the humid tropics that belongs to the family *Euphorbiaceae*. Cultivated forms belong to the species "*Manihot esculenta* Crantz" and "*Manihot utilissima* Pohl." Africa produces about 50–80 million tonnes of cassava annually; this translates into an average of more than 300 calories per day for more than 200 million people. Cassava can grow on relatively marginal soils and erratic rainfall conditions in southeastern, Nigeria. It quickly adapts and integrates into the traditional farming system, is easy to cultivate and process and it is available all year round acting as a buffer against crop failure. These characteristics make this root crop a necessary component of the farming system in many areas of Africa south of the Sahara. Some of the principal recommended cultivated varieties in Nigeria include; TME 419, TMS 90257, TMS 91934, TMS 81/00110, TMS 82/00661, TMS 30001, TMS 30555, TMS 30572 and local cultivars—*Nwugo*, *Nwaiwa*, *Ekpe* and *Okotorowa* that are popular in southeastern Nigeria. Cassava is expected to play increased role in Africa's struggle to attain food and nutrition security through increased production and utilization. This paper examines the ecophysiology, production principles, pest and disease management, uses and constraint hampering cassava production in southeast Nigeria.

Keywords: cassava (*Manihot* species), production, yield, edaphoclimatic requirements, uses, constraints, Nigeria

1. Introduction

Cassava (*Manihot* spp.) belongs to the family *Euphorbiaceae* and is the most important root crop grown in the Tropics. The plant probably originated from South America and was first introduced into Central Africa during the last part of the sixteenth century, into West Africa in the early eighteenth century and into East Africa in the early nineteenth century [1]. Cassava is thus a relatively new crop to African Agriculture. Hahn et al. [2] postulated that farmers rapidly adopted cassava and integrated it into the traditional farming practice because the plant quickly adapts to local farming conditions. The crop is easy to cultivate and process, and it is available all year round thus acting as a buffer against crop failure. Cassava requires minimal input for cultivation and yet produces relatively good yields with little labor input.

Although the crop grows in every country of the sub-continent (sub-Saharan Africa), cultivation is concentrated in the humid tropics [3]. Africa produces about 50–80 million tonnes of cassava annually; this translates into an average of more than 300 calories per day for more than 200 million people [4]. Africa has 7.48×10^6 ha of land under cassava and this account for 52% of the world's area under this crop in 1984 [5]. Four African countries (Mozambique, Nigeria, Tanzania, and Zaire) are among the 10 largest cassava producers in the world.

According to [3], cassava's ability to adapt to biotic and abiotic stress hampering crop production, poor edaphic condition, irregular rain fall, high yield per unit of area of land and labor requirements make cassava a primary component of the farming system in many parts of Africa south of the Sahara. Cassava has the potentials for eliminating food crisis and famine. Cassava is commonly blamed for malnutrition because the roots contain little protein and an anti-nutritive factor (hydro cyanogenic glucoside) which is removed during processing. The absence of cassava in the farming and food systems of Africa would lead to catastrophe. Cassava deserves a primary position as a crop that has saved many lives in Africa. This paper provides detailed information on cassava—growing conditions, production systems, and processing methods in Southeastern Nigeria.

2. Classification

Cassava is a dicotyledonous plant of the Spurge family *Euphorbiaceae*. All cultivated forms belong to the species "*Manihot esculenta* Crantz" and "*Manihot utilissima* Pohl". There are many cultivars of cassava, and these may be distinguished based on many criteria relating to the structural features of the plant. Other features such as tuber shape, earliness of maturity, yield and its content of cyanogenic glycoside (HCN) are characteristics employed in distinguishing the different varieties of the plant. *Manihot utilissima* Phol or sweet and *Manihot aipi* Phol are two edible species reported to have high and low cyanogenic glucoside concentrations respectively. Cassava has 2n = 36 ploidy number. Many closely related species are present in the tropical and subtropical Americas that can be crossed with *M. esculenta* [6].

A current study by IITA revealed that about 4000 local varieties are in 17 major cassava-growing countries in Africa. The research programmes on cassava improvement of the countries surveyed released a total of 206 improved cassava varieties between 1970 and 1998 [7].

Some of the important recommended cultivated varieties in Nigeria include; TME 419, TMS 90257, TMS 91934, TMS 81/00110, TMS 82/00661, TMS 30001, TMS 50395, TMS 84537, NR 87184, NR 41044, NR 8082, NR 8083, NR 8212, NR 83107, TMS 30211, TMS 30555, TMS 30572, U – 41044, U – 7706 and U – 60506 – 2(4) TMS 419, TMS 30001, TMS 300017, TMS 30110, TMS

30337,TMS 4(2)1425, UMUCASS 42, UMUCASS 43 and local cultivars—*Nwugo, Nwaiwa, Ekpe* and *Okotorowa* that are popular in southeastern Nigeria. The best cassava varieties possess short growth cycle, high yields, withstand biotic and abiotic stress, early maturing, give high root yields (fresh and dry) and meet end-users quality characteristics, and store well in the soil for more than 18 months.

3. Edaphoclimatic requirements

Cassava is a crop of the humid tropics. It is warmth – loving plant and performs optimally at a temperature between 25 and 29°C but cannot perform well at temperatures below 10°C. Cassava requires some amount of moisture, and an annual rainfall of 1000–2000 mm is adequate for its growth. Since it is a moderate drought tolerant plant, its cultivation is possible with only 500–750 mm of rainfall. During drought, cassava growth ceases or slows down considerably, the length of internodes shorten, and tuber enlargement stops. The crop, however, cannot tolerate drought a few weeks after planting. The optimal soil for cassava can be described as well drained, light textured, deep and fertile soil. Onwueme and Sinha [8] pointed out that cassava does best on light sandy loam soil with high but not excessive fertility. Excessive fertility, especially in soils with a preponderance of nitrogen content, cause the plant to produce more shoot at the expense of root growth. The crop performs poorly on stony soils, saline/alkaline soils (with pH above 8.0, above 2.5% soil sodium saturation, and above 0.5–0.7 dSm^{-1} electrical conductivity [9] or soils with hardpan and poor drainage that restricts root growth and enhances tuber rot. According to [10], soil physical status often gets less attention although it usually controls the chemical and biological functions of soils concerning crop production. Poor soil structure and acidity are attributes of long term effect of the continuous application of chemical fertilizer. In many heavily weathered soils, subsoil penetration by roots and water percolation are impeded by chemical and physical barriers [11].

Hahn et al. [2] noted that African soils under cassava are over cultivated and have low fertility, high acidity and aluminum levels, low organic matter, are shallow and highly compacted, retain little soil moisture and are high in the soil temperature at certain times of the year. In these soils, with poor fertility, cassava gave 79–80% of the maximum yields after liming while sorghum and maize gave only 9.5 and 52% of their maximum yields respectively. This indicates that cassava performs better than any other food crops on soils of low fertility and high acidity.

4. Crop physiology

Cassava is a woody perennial that can grow up to 5 m in height. The leaves are large, spiral and lobed. Several tubers are produced at growth stage which contains 35% starch and weighs up to 40 kg. The male and female flowers are in clusters, and the plant also produces a non-fleshy fruit capsule [12]. Cassava can be propagated by stem cutting or by seed. Propagation by seed is often slow, and some of the seeds may require to be scarified before germination can take place. However, stem cuttings germinate readily, and this is the usual method of multiplication or establishment.

Germination, shoot expansion, and root spread occurs within the first few weeks after emergence or sprouting. At the early growth stage, the adventitious roots are formed first from the nodes at the base of more than one axillary bud (nodal roots) 5–7 days after planting, which is then followed by the formation of rootlets from a recently formed callus at the base of the cutting (basal roots) [13]. The buds also begin to sprout and enlarge 5–8 days after planting, with the first leaves appearing by 10–13 days. Sprouting is faster at a soil temperature around 28–30°C but ceases at temperatures higher than 37°C and lower than 17°C [14]. The maximum leaf area is reached in 4–5 months after planting. Flowering starts from the first 6 weeks and continues throughout the growth period of the crop. Tuber initiation starts from the 8th week after planting but depends on the variety and environmental conditions. Most of the fibrous roots will develop into tubers but after 6–9 months no fibrous roots will grow into tubers.

The fibrous roots (ranging from about 3 to 20 roots, depending on cultivars and growing conditions) initially penetrate the soil as thin fibrous roots, after that, they undergo secondary thickening and starts swelling from the proximal end where the fibrous root is attached to the stem. The feeder roots grow vertically into the soil to a depth of 1 m, thus the reason for its ability to tolerate drought and low soil fertility [9]. Mature roots which contain 20–30% starch extend 60 cm down into the soil and are around the base of the plant. Fresh root yield at harvest under the most favorable conditions is about 90 t/ha while average world yields from mostly subsistence agricultural systems are 10 t ha^{-1} [15, 16]. The cassava tuber is physiologically inactive and thus cannot be used to propagate the crop. Studies have shown that shoot development takes pre-eminence during the first 3–5 months of the development of the plant whereas root bulking occurs during the subsequent period of the growing season. This may be because the plants mobilize photosynthates to the shoots early in the growing cycle and supply the roots more photosynthates during the later part of the growing cycle. This may, however, depend on critical eco-physiological conditions such as soil and water conditions, temperature regimes and photoperiodism [13, 17–19]. In general, cassava does not have specific water stress sensitive growth stage beyond crop establishment, and the crop tolerates prolonged drought and erratic precipitation [13]. The ability of cassava to tolerate elevated temperature, drought and increased concentration of atmospheric carbon dioxide, places it as a crop that can adapt to climate change impacts.

5. Cultivation of cassava

The planting period, planting density and the position of the cuttings are three primary factors that are important in the growing of cassava [12]. Cassava cultivation starts with the selection of a suitable site. Selection of a good site is a desideratum for production of high yields. Factors such as edaphic condition and land use history, vegetation, topography and specifically soil physical properties should be considered. Although cassava tolerates a broad range of soil conditions, it grows well in well-drained soil of medium texture (loamy soils) with high organic matter content. Soils that are water-logged or imperfectly drained should be avoided as tuber rotting proliferates after more than 10 days of continuous inundation in water. Similarly, cassava will not give good yields in soils that are stony, compacted or soils with hard pans or indurated layers. If it is possible, a site that has been over cropped (used for continuous cultivation for 3–5 years without appropriate fertilization) should be avoided. Land with gentle or rolling slope is preferable to soil on steep slopes for cassava production.

Land preparation for cassava planting usually starts with clearing, stumping and sometimes burning. The cuttings may be planted on unploughed land using minimum tillage technique. Similarly, cassava may be planted on mounds, which are 30–60 cm high and 1–2 m apart, or on large mammoth mounds up to 1 m high in poorly drained soils. The choice of the land preparation method largely depends on soil type and the depth of water table.

In modern agriculture, the land is plowed and harrowed, thereafter cassava cuttings may be planted on the flat or on ridges that are 1–1.5 m apart depending on cultivars. Ridge and mound planting are suitable on heavy soils. This prevents water logging which leads to root rot.

6. Planting material

Cassava is propagated by the use of stem cuttings. The cuttings measuring 15–30 cm long with 4–6 nodes are used for planting. Such cuttings make it easy for mechanization, transportation and handling of large quantities of planting material. However, longer cuttings have been found to produce higher yield because they produce greater number of roots and shoot and contain larger stored food reserves that the plant can utilize before it becomes self-sufficient. The ripe wood (mature cuttings) taken from the middle of the stem is normally preferable to cutting from either the plant tip or base. It is a normal practice to cut cassava stems 2–3 days before they are cut into final planting lengths and planted. This allows the stems to develop callus (auxins), which promote early sprouting and development. Before planting, healthy cassava stems are selected. They are cuttings usually procured from mature plants 8–10 months old and preserved under the shade for a few days before cutting and planting. It is important to place the distal end of the stems on the soil and moisten it regularly making sure that the surroundings are kept weed free [20]. This will make the stems sprout faster than when they are planted freshly cut from the field.

Cassava cuttings are planted in different positions in various countries. They may be planted upright in a vertical position, inclined at an angle of 30–40 or planted horizontally at about 10 cm beneath the soil surface. When cassava sticks are planted vertically, they sprout and acquires healthy foliage slightly more rapidly and produces deeper lying tubers than inclined or horizontally planted ones. Probably, planting in an inclined position, at an angle of 45°, is best. If the cuttings are planted inclined or vertically, the cuttings are buried in the soil with one-third above the ground surface with the buds pointing upwards. This is not necessary for horizontal planting. The planting depth can be 5–15 cm. Vertical or angular planting is recommended in areas of high rainfall whereas horizontal planting is better in dry areas.

Timeliness in planting operations ensures healthy sprouting, good crop establishment, and growth. Dry season planting is not recommended in areas of low rainfall and where water table is low because soil moisture content influences stem sprouting and survival [20, 21].

Cassava is usually planted when there is adequate moisture in the soil. This is important because young plants do not tolerate drought, unlike older plants. The crop is planted one cutting per hole/hill and sprouts within 7–14 days. It can be planted manually or with newly developed mechanical planters. The standard spacing is between 80 and 100 cm apart on ridges, mounds or flat which are 100–150 cm apart depending on cultivars and environmental conditions. When cassava is intercropped with other crops, a much wider spacing is adopted depending on types and number of plants as well as the kind of inter-cropping used. In intensive cultivation, it is recommended that the crop should be alternated with a legume cover crop at rest period. At times intercropping cassava at the end of rotation help conserves soil fertility [12].

7. Weed competition in cassava

Weeds reduce cassava yield if not controlled properly. In cassava fields where weeds are not adequately managed, yield reduction of between 50 and 80% is observed. Weed competition at any stage of growth after rooting reduces yield but the most damaging effects of weeds on yield occur during early canopy formation and tuberization. This starts from planting to 90–120 days. Optimum yield can be achieved if cassava is kept free of weeds during this time. The period, 90–120 days is referred to as the critical duration of weed interference. However, weed competition from 8 to 12 weeks after planting was found to cause the greatest reduction in tuber yield. This period (8–12 weeks after planting) is referred to as the critical period of weed interference in cassava. Thus if there is limited labor for weeding, the farmer must choose to weed at 8–12 weeks to obtain satisfactory yield. Weed control takes up to 60% of the labor and more than 40% of the total cost of growing cassava in sub- Saharan Africa [20].

Two properly spaced weeding at 30 and 60 days after planting could produce a satisfactory yield up to 77% of maximum yield. After that cassava is weeded periodically (every 6 months) especially if the crop is to spend up to 2 years in the field. Cassava can be weeded using hand held hoes (manual method). This method is effective if it is done at the right time but the method may damage plant parts especially the tender superficial roots. Chemical weeding using herbicides is another good option for weed control in cassava farms. Herbicides may be applied as pre-emergence or post–emergence herbicides. Some herbicides used in cassava farms are:

Pre-emergence herbicides

- Atrazine applied at 1.5–3.0 kg ha^{-1}
- Fluometuron applied at 2.0–3.0 kg ha^{-1}
- Premextra applied at 2.5–3.0 kg ha^{-1}
- Diuron applied at 1.5–3.0 kg ha^{-1}

Post-emergence herbicides are:

- Paraquat or Gramoxone applied at 0.5–1.0 kg ha^{-1}

Lower rates of these herbicides are recommended for light soils whereas higher rates are recommended for heavy soils. Most post-emergence herbicides are non-selective and therefore must be directed to the weeds only.

Weeds can also be controlled in cassava using the cultural method, for example, the use of increased plant population and optimal fertilizer use to check weed growth. Similarly, an integrated approach (integrated weed management) can be employed, and this involves the use of a combination of methods to achieve desired results, e.g., use of a pre-emergence herbicide which is followed up by one hand weeding later. Early weeding prevents weeds from competing with the crop for nutrients, water, light, and space. Melifonwu [22] recommended a combination of different cultural practices to control weeds control weeds in small scale farms in Nigeria. These include but not limited to:

a. Removal of weed rhizomes, stolons, and tubers during land preparation.

b. Mulching of cassava seed beds with live or dead mulch materials to reduce weed problems and improve soils.

c. Growing cassava varieties with early, low, and much branching habit; these will suppress weed growth better than varieties with late, high, or no branching habits.

d. Intercropping cassava with appropriate crops to reduce weed problems and improve soils.

e. Using the improved fallow plant as "live mulch" on land for planting cassava.

f. Hand weeding three times within 3 months after planting cassava; this will reduce weed competition with cassava for nutrients.

g. Combining the most appropriate weed control practices for more efficient control of the weeds.

h. Using an appropriate herbicide to control weeds in cassava farm.

8. Fertilization

Cassava grows well in fertile soils. Fertilization can be in the form of organic manure and inorganic fertilizer. Different fertilizer rates and types can be used such as NPK 12:12:17 or NPK 15:15:15 depending on:

a. The nature of the soil, previous cropping history of the field, cultivar of the crop grown and planting density. If a location were previously planted with cowpea or other legumes, that location would need less fertilization than a place that was cropped with other grains continuously for a longer period.

b. The amount of rainfall prevalent in a particular location. It is possible to apply fertilizer at the time of land preparation, especially in low rainfall areas. Under inter-cropping, the fertilizer may be applied at 4 weeks after planting, and in areas with very high rainfall, it is possible to apply fertilizers in split doses at 4 and 8 weeks after planting. This is to prevent most of the fertilizer from being washed away or leached out of the soil solum.

Cassava requires a moderate supply of potassium fertilizer to produce a high yield of tubers. It is usually preferable to apply potassium sulfate instead of muriate of potash (potassium chloride) because potassium sulfate also meets the plant demand for sulfur. Potassium (K) is essential in the absorption of N and P in the soil. One way of adding K to the soil is through bush burning during land preparation. This cultural practice can increase yield if planting is done immediately after burning. Nowadays farmers are not encouraged to use this cultural practice (bush burning) because the disadvantages often outweigh the gains for using it [23].

Cassava needs a moderate supply of nitrogen, as excessive N application will increase shoot to root ratio and also increase the level of hydro cyanogenic glucoside (HCN) in most varieties. The best management practice would be to carry out soil tests before fertilizer application in a particular year. This will help the farmer to know what is lacking and what is to be applied. Soil tests are carried out and interpreted by soil scientists. Cassava can grow in all types of soil but best grown in a well drained sandy loam soil of average fertility. There is every need to adopt the most suitable cultural practices and method that will boost the yield of cassava. Even though cassava is said to have the ability to produce under low soil fertility, the relative yield is higher under high soil fertility, hence the need to improve soils for better yield. Soils in southeastern Nigeria are characterized by kaolinitic clay mineralogy. They are often degraded and are characterized by low fertility and high acidity which may be caused by over exploitation, erosion or leaching. Farmers in attempt to overcome this challenge adopted the strategy of chemical fertilizer application. The approach of farmers toward chemical fertilizer usage has posed a threat to soil physical quality status [10]. Various soil amendments are used by farmers to improve the productivity of soils. For example, the use surface applied gypsum, lime, and organic manures especially poultry droppings, cow dung, etc. in ameliorating chemical and physical barriers associated with improving soil productivity are promising technologies. The effect of these improvements on soil physical condition is attributed to the flocculation and aggregation effect of these amendments on the soil properties [11, 24].

Anikwe et al. [25] found that soil application of a combination of 5000 kg ha^{-1} of lime and 2500 kg $ha^{-1}$of gypsum increased soil exchangeable Ca^{2+}, soil percent base saturation, soil total porosity, soil water transmissivity and decreased soil dry bulk density in the unamended control plots. The changes in soil conditions increased mean plant height of cassava and fresh tuber yield by 30–36%.They concluded that lime and gypsum influenced the soil physical and chemical properties through the addition Ca^{2+} that helped to flocculate soil particles, promoted nutrient uptake, proper moisture infiltration, aeration, increased exchangeable P and optimum pH for growth of cassava. Odedina et al. [26] postulated that cost, availability and technical reasons constrain the use of chemical fertilizer input in Nigeria. Animal manure, an alternative to mineral fertilizer is cheap, but the disadvantage in it usage is that there are not available in the required quantities needed for bountiful crop growth. Essential nutrients such as N, P and K are often low. They studied the effect of an integrated use of organic manures and inorganic fertilizer on cassava production and soil parameters and found that the highest crop responses were observed in plots amended with organic manures with no productivity gaps when compared with scenarios where the recommended fertilizer and manure rates were used. They found more plant-available nutrients in plots where a combination of organic and inorganic manures was used. The study also found that whereas root yield was higher in plots where organic and inorganic fertilizers were used together, stem yield was greater in inorganic fertilizer managed plots. The lowest crop yield responses were found in unamended plots. The productive capacity of soils could be maintained by the use of integrated nutrient management techniques.

Howeler [27] observed that a combination of good agricultural practices that do not harm the environment with the right mix of inorganic fertilizers is recommended for sustainable crop production. They noted that cassava could endure acid soils and in symbiotic association with soil fungi intertwined with cassava roots, the crop can absorb nutrients especially phosphorus which they use in the production of photosynthates. Most nutrients absorbed by the cassava plant are mobilized to the leaves and stems, and this gives the plant the advantage of recycling nutrients within the plant-soil system. Although cassava has been reputed to produce well in soils of marginal fertility, many trials as shown by FAO imply that cassava responds favorably to mineral fertilizers especially in soils of low fertility.

9. Harvesting

Cassava tubers should be harvested when the tuber has not become fibrous or woody (when the starch content is at its peak). It should be harvested at 7–15 months after planting depending on the cultivar. The sweet types are harvested around 7 months while the bitter varieties are harvested at about 12–15 months after planting. It is typically recommended that mature cassava should be harvested before the dry season starts to reduce the loss of tubers during the dry season when the soil is hard and dry. Cassava is usually harvested piece-meal over a period after maturity. This means that you harvest, as you need since the keeping quality of cassava after harvesting is poor.

The crop can be harvested manually or mechanically using mechanical harvesters. For manual harvesting, cassava is mostly harvested by hand. In this process, the stem of the plant is cut off usually by using machetes, then the remaining lower part of the stem is lifted out of the ground by hand. Levers and ropes can be used to assist harvesting. A mechanical harvester can also be used. Mechanical harvesters, like those developed in Brazil, would grab onto the stem and lift the roots from the ground [16, 28]. Low productivity is caused by an inappropriate or near absence of mechanization, farming tools and limited market opportunities in southeastern Nigeria. This scenario encourages drudgery and delay in farm operations. Thus, with mechanical harvesters, cassava harvesting time frame is reduced from approximately 8–10 days per hectare to about 6 h [28].

Mechanical harvesters are less cumbersome but can damage the tubers, leave some of them buried in the soil and make separation of the tubers from soil and plant residues difficult. Thus, developing a labor-saving technology for cassava harvesting is the most critical challenge for cassava farmers and a limiting factor to cassava production in Africa. This challenge is more urgent than further increases in cassava yield [29]. This is because current harvesting methods are cumbersome, i.e., labor intensive. There is a need to develop seamless mechanical harvesting methods that can enhance harvesting of vast hectares of land. The bane of commercial production of cassava is mostly caused by manual harvesting reflecting drudgery and time consumption, especially in dry season. As enunciated by [30]. Africa is the highest producer of cassava in the world, but lack of access to appropriate mechanization to support production and processing of cassava is impeding the development of the cassava market in Africa. This challenge has limited the capacity of the farmers to increase production. Africa utilizes more than 90% of the cassava it produces for food whereas in Asia and Latin America, only about less than half of its cassava production is consumed.

The unavailability of equipment for cassava mechanization is a major problem militating against the expansion of output and utilization of cassava in Africa.

Frimpong [31] reported that the Agricultural Engineering Department of the Kwame Nkrumah University of Science & Technology (KNUST) in Ghana recently developed a unique cassava harvester designed to enhance the mechanization of root and tuber crops cultivation, mainly cassava and yam. The device 'Tek Mechanical Cassava Harvester' (TEK-MCH) has been engineered to address the difficulty in commercially harvesting root and tuber crops. The development and use of this machine is expected to revolutionize the technology for cassava production for use as food and industrial raw material. The TEK-MCH can harvest a hectare within a maximum of 2 h. This technology will increase efficiency and reduce drudgery thus allowing many farmers especially youths to embrace cassava production. The tek mechanical harvester worked best on fields with minimal trash or weeds and relatively dry soils with moisture content from 12 to 16% d.b. and requires drafts of up to 10.33 kN with penetration depth from 23 to 29 cm. Best harvesting performance was achieved at a tractor speed of 5 km/h giving a field capacity of 1.9–2.5 h/ha. After mechanical harvesting, the field is left plowed with savings on fuel, time and cost. However, it is recommended to field evaluate the harvester in all agro-ecological zones and through a wide range of soil moisture regimes in Ghana to determine suitable areas for mechanical harvesting and to promote nationwide adoption [32]. Amponsah [16] reviewed available manual, semi-manual and mechanical cassava harvesting methods and equipment and postulated that in the traditional method of cassava harvesting traditional farm implements including machete or cutlass is used to severe the stem from the base. The hoe or mattock is used to dig a hole around the tuber and subsequently the farmer pulls the tuber out of the soil. This method is laborious especially during the dry season when soil moisture is at lower levels [33]. According to [34], manual harvesting requires about 22–62 man-days per hectare.

The International Institute of Tropical Agriculture (IITA) and the National Centre for Agricultural Mechanization in Nigeria also invented a manually operated cassava root tuber lifter and a semi-mechanized cassava harvester respectively. These implements are to be used by small scale cassava farmers [35].

The implements use mainly human efforts to operate them efficiently, and they have been tested to harvest up to 200 plants per man-hour [35].

Mechanical harvesting of cassava relies on the deployment of tractor mounted equipment to harvest cassava roots. However, human efforts are still required to collect and separate the cassava tubers from the stump [36]. Other existing mechanical cassava harvesters include the Leipzig Mechanical Cassava Harvester that is capable of digging, lifting and transporting of cassava root cluster into a windrow. This has been demonstrated under a Ghanaian condition using a prototype cassava harvester developed at the Leipzig University, Germany [36]. The equipment reduces to the minimum the heavy physical work involved in manual cassava harvesting using the hoe and cutlass, especially in the dry season.

According to [36], tests to date show that several factors are critical for the successful mechanized harvesting of fully matured cassava crop. These key factors include soil penetration resistance, maintenance of tractor speed, soil moisture content, ridge height and depth of penetration of share in the soil. Other factors include configuration of roots, stem height and diameter, and planting density, etc. Other mechanical cassava harvesters include, CLAYUCA Cassava Harvester Model P600, semi-mechanized cassava harvester prototypes developed in Brazil and the NCAM Tractor-drawn Tuber Harvester developed by the National Centre for Agricultural Mechanization (NCAM) in Nigeria. Agbetoye et al. [37] reported that most of the experimental cassava harvesters in literature are based on the elevator digger principle whereby the share cuts through the soil 0.3–0.4 m deep and 0.7–0.8 m wide and handling about 0.23 m^3 or about 500 kg of soil to harvest a plant. All these unique characteristics must be appropriately considered to design an efficient harvester [16, 38]. Mechanization of cassava production using tractor drawn implements successfully improved efficiency and reduced time used for land preparation from 240 to 3 man-hours; planting from 64 to 1 man-hour and harvesting from 320 to 8 man-hours in 5500 households in Nigeria, Zambia and Uganda [29].

10. Yield of cassava

Fresh root yield of cassava under farmer's condition is about 2–10 tons per hectare, but cassava may yield up to 40–60 tons per hectare where growing conditions are best. In sub-Saharan Africa, Nweke [39] found the mean root yield per hectare to be 11.89 t, with the range from 0.4 to 67.3 tonnes in 196 cassava growing communities in Cote 'd'ivore, Nigeria, Tanzania, and Uganda. Root yield in cassava is influenced by cultivar, cultural operations like weeding, fertilization, field spacing, climate, etc.

As elucidated by [40] when food and nutrition security are considered, crop yield and nutritional quality become important factors. These prompted the efforts to develop cassava as a major cash crop in sub-Saharan Africa. Cassava varieties fortified with Vitamin A have already been developed by IITA and National Root Crops Research Institute in Nigeria and are already popular in Southeastern Nigeria.

The cassava value chain in south-east Asia already produces cassava products like industrial starch, ethanol and cassava chips for livestock feed on a large scale. From that perspective, yield still matters, but so too do others such as dry matter content for processing into starch or high-quality cassava flour. Today we plant 10,000 plants per hectare, and each plant produces 1.5–3 kg of the root or 15–30 t/ha (tonnes per hectare) of fresh root.

11. Pests and diseases

Like other crops, cassava is susceptible to pests and diseases. These pests and diseases reduce yields drastically. In some regions, pests and diseases proliferate as a result of the expansion of land under cassava cultivation and intensification of the cultivation processes [41]. More than 30 diseases of cassava have been identified. Lozano and Booth [42] listed the important bacterial diseases as Bacterial blight (*Xanthomonas campestris* pv. *Manihotis*), Bacterial angular leaf spot (*X. campestris* pv. *Cassava*), Bacterial stem gall (*Agrobacterium tumefaciens*), Bacterial stem rot (*Erwinia carotovora* subsp. *Carotovora*), Bacterial wilt (*Erwinia herbicola*). Fungal Diseases include Anthracnose (*Colletotrichum gloeosporioides, Colletotrichum graminicola*), Armillaria root rot (*Armillaria mellea*), Black root and stem rot (*Scytalidium* sp.), Blight leaf spot (*Cercospora vicosae*), Brown leaf spot (*Cercosporidium henningsii*), Cassava ash (*Oidium manihotis*), Concentric ring leaf spot (*Phyllosticta manihotae* Viegas: *Phyllosticta manihoticola* Syd.), Dematophora root rot (*Dematophora necatrix*), Rosellinia root rot (*Rosellinia nec*), Diplodia root and stem rot (*Diplodia manihotis* Sacc.), Fusarium root rot (*Fusarium oxysporum; Fusarium solani*), Phytophthora root rot (*Phytophthora cryptogea*), Pythium root rot (*Pythium* spp.), Rigidopurus root rot (*Rigidoporus microporus*), Rust (*Uromyces* spp.), Sclerotium root rot (southern blight; *Sclerotium rolfsii* Sacc.), Superelongation (*Sphaceloma manihoticola*), Verticillium root and stem rot (*Verticillium dahliae* Kleb.), White leaf spot (*Phaeoramularia manihotis*). They also listed diseases caused by viruses as African cassava mosaic caused by African cassava mosaic virus, Antholysis, Cassava common mosaic caused by Cassava common mosaic virus, Cassava frogskin, Phytoreo-like virus, Cassava green mottle disease caused by Cassava green mottle virus, Cassava symptomless infections by Cassava American latent virus and Cassava Ivorian bacilliform virus. Others include Cassava vein mosaic caused by Cassava vein mosaic virus, Indian cassava mosaic by Indian cassava mosaic virus. Other miscellaneous diseases or disorders are Post-harvest root rot, Physiologic and pathogenic deteriorations, Root smallpox disease, Microbial rotting after feeding by *Cyrtomenus bergi*.

The most dangerous of them is the African mosaic disease (AMD, a virus), and cassava bacterial blight (CBB), which is caused by *Xanthomonas* spp. The mosaic virus is spread by *Bermisia* spp.; other diseases are brown streak (virus), brown leaf spot (Fungus) and Anthracnose disease. The most important control measures for AMD and CBB are to use tolerant varieties and use of disease free planting materials. Similarly, insecticides like *vetox 85* could be used to control *Bermisia* spp., which transmit AMB.

The most important insect pests are green spider mite, red spider mite, web mite, scale insects, white flies, termites and mealy bugs. They can be controlled using one or more of the following methods:

- Early planting of high yielding and early maturing varieties
- Planting of stakes by burying at a shallow depth to obtain a clean crop.

- Fertilization.
- By use of biological agents called parasitoids e.g. *Epidinocarsis lopezi* which eat up mealy bugs and mites.
- By application of insecticides, e.g., Dimethoate and methidathion.
- By removing and burning infected plant materials
- By dipping stem cuttings in Nuvacron 40 EC for about 5 min before planting.

Other pests of cassava are mammals especially rodents, goats, and monkeys.

12. Uses

Cassava is regarded as one of the principal plants of use to man because of the important part it plays as food. It is the tuberous roots that are most important as a foodstuff, but it should be noted that the leaves may also be eaten and that they possess nutritional characteristics, which are complementary to those of the roots. However, it must be noted that cassava contains a glucoside, which must be removed by processing before the product is consumed [43].

The major cassava products include:

a. *Dried cassava*. Tuberous roots either peeled or unpeeled are cut into chips and dried. This can be used as animal feed. When they are crushed, sieved and reduced to flour, they serve as food for man. The flour is popular in Brazil and Indonesia and can be used to make porridges, dough or bread.

b. *Cassava steeped in water*. Cassava peeled, or unpeeled are immersed in water, preferably slow running water for 3–5 days after which the starch is sieved out and used to prepare 'Fufu.' This is a pleasant food called *Farinha de agua* in Brazil and *Farine* in the Central African Republic. Cassava can also be made into "cassava bread" and "cassava stick."

c. *Grated cassava product*. Fresh tubers are peeled and grated, then left to ferment for 2–3 days, it is then dried, broken – up, sieved and fried or cooked in hot plates. This produces 'garri' (West Africa), "Farinha de Mesa" (Brazil) or 'Melange flours' (Angola). This product is attractive because it keeps for a longer period.

d. *Pellets*. The industrial product, which is currently most important regarding the volume of trade, is cassava pellets. The tubers, peeled or unpeeled are cut – up, using root cutters, into chips which are dried in the sun, the chips are then pelleted in the factory by compressing chips against grids with mesh to produce uniformly sized pellets.

e. *Starch*. This is virtually pure carbohydrate used for various purposes in food industries (sweetened products, thickeners etc.), in textile, paper, and other industries.

Cassava is also used to produce tapioca, industrial sweeteners, and alcohol.

13. Constraints to improved production of cassava

- Lack of seed material for those high-yielding varieties.
- Price changeability of fresh tubers because of the absence of a stable market—when production is low, prices are high on the other hand when production is high, prices are too low that cost of production cannot be covered.
- Budgetary constraints—Governments are unable to mobilize extension workers and to purchase processing equipment to process cassava into chips and garri.
- Lack of resources to conduct permanent research on cassava.
- Low multiplication ratio of cassava.
- Poor storability of cuttings—They do not usually keep for more than 14 days.
- Bulky planting material.
- The rapid deterioration of tubers.

14. Solutions

- Adaptation of improved methods for increased stake production.
- Development of farmer efficiency by encouraging farmers to plant high yielding and disease resistant varieties.
- Finding new uses of cassava and its products.
- Provision of better facilities for processing cassava.
- Increased extension and research work on better methods of production, utilization, multiplication, and distribution of improved planting material.

15. Conclusion

Finally, the tolerance of cassava to extreme conditions, its biological efficiency in the production of food energy, its low production resource requirements, its availability throughout the year and its suitability for farming systems will make cassava more popular to African farmers. With improved varieties, cultural practices and processing, cassava yield and product quality (gari is one of the best-processed food) could be equaled or bettered with less land and labor. For this reason, cassava has great potential as a crop of the future in Africa's struggle to attain household food sufficiency and security through increased production and utilization [44].

Author details

Martin A.N. Anikwe* and Ejike E. Ikenganyia

*Address all correspondence to: anikwema@yahoo.co.uk

Department of Agronomy and Ecological Management, Faculty of Agriculture and Natural Resources Management, Enugu State University of Science and Technology, Enugu, Nigeria

References

[1] Jones WO. Manioc in Africa. Sanford, USA: Sanford University Press; 1959. pp. 31

[2] Hahn SK, Mahungu NM, Otoo JA, Msabaha MAM, Lutaladio NB, Dahniya MT. Cassava and the African food crisis. In: Terry ER et al., editors. Tropical Root Crops—Root and the African Food Crisis. Proceedings of the Third Triennial Symposium of the International Society for Tropical Root Crops—Africa Branch, held in Owerri, Nigeria, 17-23 August 1986; 1987

[3] Asadu CLA, Nweke FI. Soils of Arable Crop Fields in the Sub-Saharan Africa: Focus on Cassava-growing Areas. Collaborative study of cassava in Africa (COSCA) working paper no. 18; Ibadan, Nigeria: IITA; 1999. pp. 182.

[4] Dorosh P. The Economics of Root and Tuber Crops in Africa. Resource and Crop Management Programme; RCMD, Resource Monograph No. 1. Ibadan Nigeria: IITA; 1988. pp. 68

[5] Okigbo BN. Roots and tubers in the African food crisis. In: Tropical Root Crops: Root Crops and the African Food Crisis: Proceedings of the Third Triennial Symposium of the International Society for Tropical Root Crops—Africa Branch held in Owerri, Nigeria, 17-23 August 1986. Ottawa, Ont., Canada: International Development Research Centre; c1987

[6] O'Hair SK. Tropical root and tuber crops. Horticultural Reviews. 1990;**12**:157-196

[7] Manyong VM, Houndékon VA. Land tenurial systems and the adoption of *Mucuna* planted fallow in the derived savannas of West Africa. CGIAR System-wide Program on Property Rights and Collective Action. International Food Policy Research Institute; 2000

[8] Onwueme IC, Sinha IT. Field Crop Production in Tropical Africa. Wageningen, Netherlands: CTA; 1991

[9] Cock JH, Howeler RH. The Ability of Cassava to Grow on Poor Soils [Includes Fertilizer Requirements]. USA: ASA Special Publication-American Society of Agronomy; 1978

[10] Anikwe MAN, Emmanuel OP, Eze JC, Ibudialo AN, Edeh VN. Identifying fertilizer management strategies to maximize soil nutrient acquisition by cocoyam (*Colocasia esculenta*) in a degraded Ultisol in Agbani, Enugu area, Southeastern Nigeria. American Journal of Plant Nutrition and Fertilization Technology. 2015;**5**(2):61-70

[11] Sumner ME, Radcliffe DE, McCray M, Carter E, Clark, RL. Gypsum as an ameliorant for subsoil hardpans. Soil Technology. 1990:**3**:253-258

[12] Kouakou J, Nanga Nanga S, Plagne-Ismail C, Pali AM, Ognakossan KE. Cassava, Production and Processing. The Pro-Agro Collection is a joint publication by Engineers Without Borders, Cameroon (ISF Cameroun) and The Technical Centre for Agricultural and Rural Cooperation (CTA); 2016

[13] El-Sharkawy MA. Global warming: Causes and impacts on agroecosystems productivity and food security with emphasis on cassava comparative advantage in the tropics/subtropics. Photosynthetica. 2014;**52**(2):161-178

[14] Keating BA, Evenson JP. Effect of soil temperature on sprouting and sprout elongation of stem cuttings of cassava (*Manihot esculenta* Crantz.). Field Crops Research. 1979;**2**:241-251

[15] Rawls WJ, Pachepsky YA, Ritchie JC, Sobecki TM, Bloodworth H. Effect of soil organic carbon on soil water retention. Geoderma. 2003;**116**(1):61-76

[16] Amponsah SK. Performance evaluation of the TEK mechanical cassava harvester in three selected locations of Ghana [Doctoral dissertation, MS thesis]. Kumasi: Agricultural Engineering Dept., Kwame Nkrumah University of Science and Technology; 2011. Available at: http://dspace.knust.edu.gh:8080/jspui/bitstream/123456789/3960/1/Final.pdf [Accessed: 20 October 2016]

[17] Pellet DM, El-Sharkawy MA. Cassava varietal response to fertilization: Growth dynamics and implications for cropping sustainability. Experimental Agriculture. 1997;**33**(3):353-365

[18] Alves AAA. Cassava botany and physiology. In: Hillocks RJ, Thresh MJ, Bellotti AC, editors. Cassava: Biology, Production, and Utilization. Oxford: CABI International; 2002. p. 67-89

[19] El-Sharkawy MA, Cadavid LF. Response of cassava to prolonged water stress imposed at different stages of growth. Experimental Agriculture. 2002;**38**(3):333-350

[20] Abass AB, Towo E, Mukuka I, Okechukwu R, Ranaivoson R, Tarawali G, Kanju E. Growing Cassava: A Training Manual from Production to Postharvest. Ibadan, Nigeria: IITA; 2014

[21] Anikwe MAN, Onyia VN, Ngwu OE, Mbah CN. Ecophysiology and Cultivation Practices of Arable Crops. Enugu, Nigeria: New Generation Books; 2005

[22] Melifonwu AA. Weeds and their control in cassava. African Crop Science Journal. 1994;**2**(4):519-530

[23] Ngwu OE. Evaluation of cassava varieties on sandy loam soil of the rain forest zone of Nigeria under two fertility regimes. African Journal of Environmental Studies. 2003;**2**:106-109

[24] Anikwe MAN. Amelioration of a heavy clay loam soil with rice husk dust and its effect on soil physical properties and maize yield. Bioresource Technology. 2000;**74**(2):169-173

[25] Anikwe MAN, Eze JC, Ibudialo AN. Influence of lime and gypsum application on soil properties and yield of cassava (*Manihot esculenta* Crantz.) in a degraded Ultisol in Agbani, Enugu Southeastern Nigeria. Soil and Tillage Research. 2016;**158**:32-38

[26] Odedina SA, Odedina JN, Ogunkoya MO, Ojeniyi SO. Agronomic evaluation of new cassava varieties introduced to farmers in Nigeria. Journal of Life Sciences. 2012;**6**(8):939

[27] Howeler RH. Cassava mineral nutrition and fertilization. In: Hillocks RJ, Thresh JM, Bellotti AC, editors. Cassava: Biology, production, and utilization. CAB International, Oxon, GB.2002. p. 115-147

[28] Kuiper L, Ekmekci B, Hamelinck C, Hettinga W, Meyer S, Koop K. Bio-ethanol from Cassava. Netherlands: Ecofys Netherlands BV; 2007. pp. 1-38. 104. Available online: http://www.probos.net/biomassa-upstream/pdf/FinalmeetingEcofys.pdf [Accessed: 24 June 2017]

[29] Fagbemi A. How to Raise Africa's Cassava Yields; 2016. https://guardian.ng/news/how-to-raise-africas-cassava-yields/ [Accessed: 21 August 2017]

[30] James D, Faleye T. Cassava mechanization prospects and future in Nigeria. International Research Journal of Agricultural Science and Soil Science. 2015;**5**(3):98-102

[31] Frimpong EA. KNUST Develops Cassava Harvester; Feb 8, 2016. thebftonline.com/business/agribusiness/17264/knust-develops-cassava-harvester-.html

[32] Amponsah SK, Bobobee EY, Agyare WA, Okyere JB, Aveyire J, King SR, Sarkodie-Addo J. Mechanical cassava harvesting as influenced by seedbed preparation and cassava variety. Applied Engineering in Agriculture. 2014;**30**(3):391-403

[33] IITA (International Institute for Tropical Agriculture). Sustainable Production in Sub-Saharan Africa. 1. IITA's Contributions. Ibadan, Nigeria: IITA; 1992

[34] Nweke FI, Spencer DS, Lynam JK. The Cassava Transformation: Africa's Best-kept Secret. Michigan State University Press. East Lansing, Michigan USA; 2002

[35] Oni KC, Eneh MCC. Strategies for reducing the cost of production/processing of cassava. In: Facilitation and Sensitization Workshop for Cassava Producing States under the Presidential Initiatives on Cassava Production and Export. Vol. 8; 2004

[36] Bobobee EYH, Okyere JB, Twum A, Neumann RH, Knechtges H. Performance evaluation of a mechanical cassava harvester. In: Proc. Intl. Commission of Agric. Eng. (CIGR) XII World Congress; 1994, August

[37] Agbetoye LAS, Dyson J, Kilgour J. Prediction of the lifting forces for cassava harvesting. Journal of Agricultural Engineering Research. 2000;**75**(1):39-48

[38] Odigboh EU, Moreira CA. Development of a complete cassava harvester: II—Design and development of the uprooter/lifter system. Agricultural Mechanization in Asia, Africa and Latin America. 2002;**33**(4):50-58

[39] Nweke FI. Cassava Production Prospects in Africa (No. 13). IITA. Ibadan, Nigeria; 1996

[40] Dashiell K. Cassava Can Become Africa's New Cash Crop. Global Development Professionals Network. (Ed) van Vark C, Guardian Professional. London UK. https://www.theguardian.com/global-development-professionals-network/2013/mar/28/cassava-food-security-sub-saharan-africa [Accessed: 22 August 2017]

[41] Howeler RL, Thomas N, Calderón Cedeño C, Araujo QR, RKP Ney JRM, Bonsu O, FM Oppong FK. Save and Grow: Cassava. A Guide to Sustainable Production Intensification Produire plus avec moins Ahorrar para crecer (No. FAO 633.6828 S266). Roma (Italia): FAO. CTA, Wageningen,The Netherlands; 2013

[42] Lozano JC, Booth RH. Diseases of cassava (*Manihot esculenta* Crantz). PANS Pest Articles & News Summaries. 1974;**20**(1):30-54

[43] Silvestre P. Cassava: The Tropical Agriculturist. Hong Kong: Macmillan Publishers Ltd; 1989. 1-6

[44] Ikpi AE, Gebremeskel T, Hahn ND, Ezumah H, Ekpere JA, UNICEF. Cassava—A Crop for Household Food Security. A 1986 situation analysis for Oyo local government area-Nigeria. IITA-UNICEF Collaborative Program on Promotion of Household Food Production and Nutrition, Socio-economic unit, IITA, Ibadan, Nigeria. 1986

Chapter 8

Microbial Fermentation as Means of Improving Cassava Production in Indonesia

Andri Frediansyah

Additional information is available at the end of the chapter

http://dx.doi.org/10.5772/intechopen.71966

Abstract

Cassava is one of the pivotal carbohydrate sources for millions of people in Indonesia. Its production up to 20 million ton a year made this country to become the third most prominent producer of cassava. However, cassava is often considered as food sources for marginal people. The majority of individuals still depend on rice and wheat flour for carbohydrate intake. Unfortunately, the elevating consumption of those sources is an imbalance with its products nationally. Both use of rice product and wheat flour is more than 8.5 kg/capita/year. The critical fact is that Indonesia is one of the biggest countries in rice production globally. However, it is also one of the largest, rice importers. Another hand, the existence of wheat flour is the result of imports from other countries and always increasing every year. The Indonesian government has contributed actively to resuscitate local foods including the cassava. There are numerous strategies that have been applied to substitute the wheat flour, however, the characteristic was always far different from its flour. *Mocaf* is the recent trend for Indonesian food industry. It is free of gluten and can easily substitute with wheat flour to produce several types of wheat-dependent-products.

Keywords: cassava, *mocaf*, free gluten, wheat flour, Indonesia

1. Introduction

Cassava (*Manihot esculenta*) is one of the vital carbohydrate sources for millions of people in Indonesia. Moreover, it is categorized as the sixth most essential food crop regarding annual production globally [1]. This crop species belong to the order of Malpighiales and family of Euphorbiaceae. Based on the study by Gibbons, cassava originated from Amazon region in Brazil and domesticated since more than 5000 years BC [2]. This root species spread to other places between sixteenth and nineteenth centuries by Western people [3].

Indonesia is the third biggest producer of cassava after Nigeria and Thailand, in which the production is up to 23.4 million ton in 2014 [1]. Cassava categorized as a friendly crop, since tolerant to drought, can grow on soil with limited nutrient, and resistant to the pest. However, cassava often considered as inferior food sources for middle- to low-income people. It also belongs to one of marginal food crops which is almost never mentioned in the colonial literature. Moreover, cassava is also considered as a crop which has the low amount of protein, minerals, and vitamin substances. Another limitation is that cassava root has a very short shelf life in fresh form up to 2 days [4], and some eatable parts of cassava contain toxic substances called cyanogenic glycosides including linamarin and lotaustralin. If the fresh form is digested without enough pre-treatment, some people may develop intoxication.

The consumption of cassava in Indonesia is higher in rural areas, especially in Java and Sumatra Island. The majority of individuals still depend on rice and wheat flour for the carbohydrate intake. Unfortunately, the increasing consumption of those sources is an imbalance with its products nationally. Both use of rice and wheat flour is more than 8.5 kg/capita/year. The new fact is that Indonesia is one of the most enormous countries in rice production globally. However, it is also one of the largest rice importers. In addition, the presence of wheat flour in Indonesia is the result of imports and always increasing from year to year (**Figure 2**).

2. The use of cassava

In Indonesia, cassava is used as a food product up to 53% and the rest as animal feed, in food industry, and as sources of bio-ethanol. Cassava roots are consumed variously, mostly as a side-dish or snack. In some areas, cassava roots are consumed as the fresh form, which is directly eaten after boiling or frying. Fried cassava could be served by giving a different type of spices such as cheese, BBQ, seaweed, chilli, and salty taste. In some urban areas, especially in Central Java and the Special Region of Yogyakarta, peeled fresh cassava is used as a raw material of solid fermented cassava called *tape singkong* (**Figure 1b**). In West Java, the solid fermented cassava called *peyeum* (**Figure 1a**). This product is made by unpeel cassava as raw material which make this food is different with *tape singkong*. Those fermented by microbial mixed contain a large number of *Saccharomyces cerevisiae* called *ragi* or *usar*.

In some parts of Java Island, especially in the Special Region of Yogyakarta and Central Java, cassava prepared as a dried form called *gaplek* (**Figure 1c**), a chip of roots which was drawn up by peeled, is sliced then and dried in the sun for up to 3 days. When needed, *gaplek* is pounded into tiwul, a small granule made by mixing its flour and water which is quite similar to rice grains in shape and size or *gathot* (**Figure 1e**), a steam of slice *gaplek* with brown sugar and grated coconut. Also, root cassava chip can be converted into cassava flour (*tepung singkong*) (**Figure 1j**) which has a rough texture. Then, it can be used to make several types of snacks such as *timus, getuk, gemblong, keripik/opak* (**Figure 1f, g, h, i**). In advance, cassava roots are extracted to provide starch called tapioca (**Figure 1k**). In short, peeled form is grated and washed with the amount of water using sieves and decantation. In large scale, tapioca is usually dried by using flash driers or wholly automated machine. However, many households still use the sun as a natural dryer. Tapioca mixed with other flours is used to make some products such as gluten-free bread, flatbread, desserts, binding agent, and thickener. The most popular cassava product is a chip. In short, fresh cassava root is piled up and chipped into the diesel-powered chipping machines, and let it dry by the sun or dry machine. After the moisture content of the dried-chips under 15%, then its chip can be fried or heated in the oven. The use of various types of spices will increase consumer appetite (**Figure 1**).

Figure 1. Indonesian cassava-based product available in market: (a) *peyeum*, (b) *tape singkong*, (c) *gaplek*, (d) *tiwul* (e) *gathot*, (f) *timus*, (g) *getuk*, (h) *gemblong*, (i) *opak/keripik singkong*, (j) *tepung singkong/cassava flour*, (k) *tepung tapioca/tapioca*, and (l) *mocaf*. Fermented based products are shown in a, b, and l.

3. Fermented based cassava product: Modified cassava flour (*mocaf*) and its production

Nowadays, cassava roots become one of the trends in Indonesian food industry since its modification form, called *mocaf* (**Figure 1g**), can provide quite a similar characteristic to wheat flour. This fact made the economic value of cassava increasing. The Indonesian government has contributed actively to resuscitate local foods including the cassava. There are numerous strategies that have been applied to substitute the wheat flour; however, the characteristic was always far different from its powder. *Mocaf* is free of gluten and can easily replace with wheat flour to produce several types of wheat-dependent-products. *Mocaf* is a product derived from cassava flour which uses the principle of modifying cassava flour during fermentation. *Mocaf* has better physical characteristic compared to cassava flour on viscosity, gelatinized ability, rehydration capacity, and the solubility. Besides, *mocaf* has a preferable aroma and sensory as resulted from the fermentation. Its native aroma has a cover by the volatile organic compounds such as lactic acid, acetic acid or alcohol. Another advantage is that the product can quickly digest when ingested due to its simple structures that are formed as a result of microbial fermentation. There are numerous ways to produce *mocaf*, however not all methods give similar characteristic. In general, *mocaf* is produced by the step as follows.

3.1. Preparation

Preparation of raw material is one of essential steps in providing excellent quality of *mocaf*. In general, cassava is ready to harvest after 8–12 months after planting. There are two types of cassava: sweat cassava that contains hydrogen cyanide (HCN) content which is less than 40 mg/kg of root cassava and bitter cassava with more than 40 mg/kg of HCN content. HCN is a chemical compound which could release from cyanogenic glycosides from cassava root. The presence of linamarase which present in the cell wall of cassava root will break down the cyanogenic glycosides which placed in vacuoles resulting on releasing acetone cyanohydrin and 2-butanone cyanohydrins, subsequently chemically convert into HCN by alpha hydroxynitrilelyase. Linamarin, a beta-glucosidase, accounts for 95% of total cyanoglycosides and the rest contains lotaustralin [5]. The resulting compound is highly toxic for both animal and human and grouped as the systemic poison. The toxicity is due to the inhibition of cytochrome oxidase together with the presence of a ferric ion in a mitochondrial system. The presence of HCN can affect both acute and chronic onset to human. Moreover, the lack of cobalamin may predispose a human into the higher risk of cyanide-associated neuropathies [6]. World Health Organization recommended that the maximum safe intake of cyanide-containing food for human be 10 mg HCN/kg as described in Codex standard 176-1989 [7], which is much lower than the acceptable limit in Indonesia which is 40 mg HCN/kg [8]. There are several factors that influence the amount of HCN in cassava including harvesting time [9, 10], cultivar [11, 12], and environmental condition [13]. There is much variety of cassava in Indonesia, for low cyanide content including *Krentil, Mentega, Adira 1, Malang 1, Malang 2, Darul Hidayah, Telo Ketan* and *Markonah*. The high content of cyanide is present in *Adira 4, Malang 4, Malang 6, UJ-3*, and *UJ-5*.

3.2. Peeling, washing, chipping, and soaking

The outer layer (skin) cassava was removed using a sharp knife. The peeled cassava is subsequently washed with water until no slime or dirt is found. The resulting violet color after peeling indicates that the cassava contains the amount of cyanide. The peeled cassava is then cut into small round pieces called chip or *sawut* (Javanese language) with the thickness about 0.5 cm. This process could be done using a sharp knife or chipper machine. The process is contributed to the small reduction of cyanide glycoside [14]. The cell wall of cassava root will be damaged and will release endogenous linamarase which is important in converting linamarin from vacuole part into glucose and cyanohydrins [5]. In the last chemical form, HCN will quickly evaporate at 30°C [15]. Additional soaking can be performed before fermentation to reduce the rest of cyanogenic compounds [15–17] (**Figure 2**).

3.3. Fermentation

Peeled cassava roots are soaked in water for 18–72 h with bio-starter. This step is an essential step in producing the excellent quality of *mocaf*. Bio-starter is contained of large number microorganism to accelerate the fermentation process. It can be in liquid or solid form such as powder. The dose for cassava fermentation can be different for various starter products

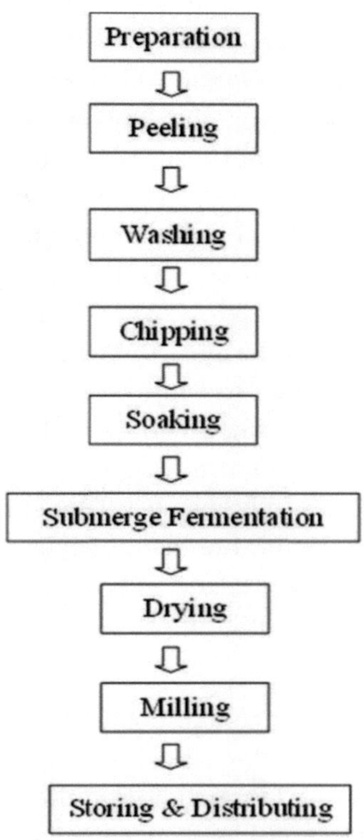

Figure 2. *Mocaf* production step.

in the market. As an example, 1 kg of *Bimo CF* solid starter, and an example of commercial culture starter, can be used for 10 ton of peeled cassava for *mocaf* production. In general, the function of bio-starter is to convert the chemical substance of cassava root during fermentation, which modifies the natural structure, enrich the nutritional value and contribute on other miscellaneous [35]. There are several bio-starters that had been applied to produce good quality of *mocaf*. However, not all of them are sold into the market. The starter can be a single culture, co-culture, and mixed culture. However, spontaneous or natural fermentation not preferably use on large scale of *mocaf* production due to the competitive activities of various micro-flora resulting in unfocus in converting target substance, which is not fit with the requirement of the *mocaf* standard. The repetition of natural fermentation challenging to monitor primarily on a large scale with different batches since the community of microorganism easily changed.

Lactic acid bacteria (LAB) are one of the groups which dominate during natural fermentation of cassava [19–21]. *Lactobacillus plantarum* is one of the LABs which dominated during natural cassava fermentation [19, 20, 22–24]. *L. fermentum* and *L. brevis* are also present during cassava fermentation with the frequency up to 16% [24]. Those bacteria can be potentially used as bio-starter for *mocaf* production. The use of lactic acid bacterial culture starter during cassava fermentation can improve the proximate compositions, fiber content and structure, cyanogenic glycoside reduction, and other miscellaneous properties such as whiteness and viscosity. During fermentation, excreted natural enzymes by that organism will breakdown the cell wall of cassava root and hydrolyze polysaccharide into subtle sugar [28]. Thus, the soluble fiber is elevating while cassava tissue is the breakdown, resulting in soft texture. This process can also improve swelling power and the viscosity of *mocaf* paste [25]. The resulting uncomplex sugar is subsequently converted into another volatile organic substance, which can cover the native aroma of cassava [26]. There are several enzymes which involve in cell wall breakdown including cellulose, hemi-cellulose, amylase, and pectinase. Previous studies showed that *L. plantarum* has amylolytic activity up to 20 U/ml and cellulolytic activity up to 12 U/ml after 18 h of fermentation at a steady temperature of 37°C during cassava fermentation [27]. Enzyme amylase activity from *L. plantarum* also has been observed by several researchers [29–31]. Cellulolytic property of microorganism is essential for breaking down the cell wall of cassava roots, resulting in improving physical properties of a *mocaf* product. Also, the presence of amylase is vital on hydrolyzing starchy substances into subtle sugar. That sugar is necessary for cell growth of bacterial culture. On the other hand, the rest of sugar will be converted into the trace of volatile fatty acids (acetic, propionate, and butyrate) and alcohol by other indigenous activities that happen on the last fermentation to cover unpreferable cassava sensory and aroma [30].

Another advantage of using bio-starter containing LAB is that *L. plantarum* could eliminate the cyanogenic glycoside up to 80% during single fermentation [18, 30, 32–34]. So, the addition of exogenous linamarase during fermentation could be excluded. Gunawan et al. also reported that the use of *L. plantarum* could significantly increase the protein content of *mocaf* products. The presence of carbon and nitrogen will use on developing protein during fermentation. *L. plantarum* improved protein considerably, and cyanide acid content during cassava fermentation compared to another type of microorganisms such as *Saccharomyces cerevisiae* and *Rhizopus oryzae* [34]. The use of glucose and cellobiose as additional nutrients during fermentation by *L. plantarum* also could increase the activity of linamarase and amylase [30]. The temperature setting and salinity also have a significant effect on survival rate

Criteria	Requirement
Morphology	
Form	Fine particles
Smell	Normal
Color	White
Foreign bodies	Not detected
Insects and their stadia	Not detected
Fineness	
Passes sieve 100 mesh	Min 90%
Passes sieve 80 mesh	100%
Moisture	Max 13%
Ash	Max 1.5%
Crude fiber	Max 2%
Degree of whiteness (MgO = 100)	Min 87
Sulfur dioxide (SO_2)	Negative
Acid degree (ml NaOH 1 N/100 g)	Max 4
HCN (mg/kg)	Max 10
Metal contamination (mg/kg)	
Cadmium (Cd)	Max 0.2
Lead (Pb)	Max 0.3
Tin (Sn)	Max 40
Mercury (Hg)	Max 0.05
Arsenic (AS)	Max 0.5
Microbial contamination (colony/g)	
Total plate count (35°C, 48 h)	Max 10^6
Escherichia coli	Max 10
Bacillus cereus	Max 10^4
Fungi	Max 10^6

Table 1. The Indonesian national standard for *mocaf*.

and either acetic or lactic acid production during fermentation [27]. The use of single starter does not mean that one microorganism does the whole fermentation process. The addition of only culture starters such as *L. plantarum* inhibited the natural development of hetero-lactic microorganism. Kresnowati et al. reported that the combination between L. *plantarum* and *Bacillus subtilis, L. plantarum* and *Aspergillus oryzae, B. subtilis* and *A. oryzae* as co-culture starter improved *mocaf* production [36]. The presence of *A. oryzae* elevated the protein content. The proximity of *L. plantarum* and *B. subtilis* on co-culture fermentation gives a better effect on reduction of cyanogenic glycosides and sugar hydrolysis. Verachtert et al. explained that the use of mixed culture could elevate the growth rate, improved biotransformation, and higher yield in the products [37]. The successful stories of diverse culture are on *tempeh* fermentation [38], beer production [37], and wine production [38–40]. The commercial culture starter for *mocaf* output in Indonesia called *Bimo CF,* a biologically modified cassava flour, used a different type of lactic acid bacteria as the mixed-culture starter. The application of *Bimo CF* is conducted in several studies [41, 42]. An experiment using beta-carotene-producer - cassava cultivar called *Adira 1* using *Bimo CF* as a culture starter has been conducted in small scale [43].

3.4. Drying, milling, storing

After fermentation, chips were dehydrated using drying machine or sun drying for maximum 1 day. It depends on the heat transfer in relation with water evaporation on the surface area of chips. Low moisture content (less than 13%) of chips can store into a plastic bag for the long-term storage. So it cannot quickly absorb the water from outside. Dried fermented chip was then milled to produce a grayish-white flour and then was sieved by 80 mesh and 100 mesh filter to achieve the standard size of commercial flour as setup by Indonesian government authorities. *Mocaf* also can be stored using a transparent plastic bag to barrier the product from the air, water or animal such as bugs.

3.5. Quality control

To evaluate the quality of the output and to protect the people, Indonesian authority called *Badan Standardisasi Nasional,* the Republic of Indonesia has setup *mocaf* standard for commercial use as mentioned in *Standard Nasional Indonesia* (SNI) 7622: 2011. The details of information are projected in **Table 1**.

4. The use of *mocaf* as wheat flour replacement and its implication

Currently, Indonesia is facing the limitation of wheat flour, which is the primary alternative to other staple foods including rice and corn. Moreover, it has been categorized as the essential core of food stabilization program that has been setup by Indonesian government since a long time ago. But, wheat flour is also one of the national burdens, since the plant itself cannot grow well in this land. It has been imported since the 1950s, and the number of import is increasing dramatically every year [44]. Wheat flour is the primary material for most of the Indonesian food products, such as noodle, bread, biscuits, cake, various fried-food products called *gorengan*, etc. In the past, the use of wheat flour was likely only distributed to the rural area, but from time to time, it has also been spread to the urban area. The presence of wheat in the Indonesian market is mostly an import from Australia. It accounts about 65% of the total wheat import, followed by Canada, India, and United States. Australian wheat meets the requirement of noodle industry in Indonesia. The Australian wheat shows proper milling extraction, less foreign material, and low moisture. The resulting flour produces bright noodle color and consistent dough properties. Moreover, those are cheaper, and the country is relatively close to Indonesia, so the shipping is much easier [45]. Based on the data in 2015, wheat in Indonesia is mostly used as noodle materials. It calculates about 58% which consists of 18% of the wet noodle, 36% for instant noodle, and 4% of dried noodle. The rest is used as bread (16%) and biscuit (26%) materials [46]. Instant noodle is the dominant choice for Indonesian people who do not have enough time to cook. Its price is also cheaper than rice. Bread and biscuits are likely used as an alternative to breakfast option especially for urban people.

As mentioned earlier, the characteristic of *mocaf* is relatively same as wheat flour. In comparison to native cassava flour, *mocaf* shows better aroma, flavor, and other physical and chemical properties. The characteristic of cassava flour improved during microbial fermentation and additional physical treatment. On the other hand, *mocaf* shows lower price when compared to wheat flour and is safe to people who have gluten intolerance and gluten allergy. These are the first onset on developing the autoimmune disorder called celiac disease. Furthermore, the presence of gluten for people who had celiac disease can destroy their villi of the digestive tract. This protein is found in several grains such as wheat, rye, and barley. The absence of gluten in *mocaf* could be used for health campaign to attract people in replacing wheat flour to *mocaf*. Also, *mocaf* easily digests in the body due to the fermentation process. Fresh cassava also rich of hydrocoumarin such as scopoletin and scopolin which have pharmacological activities such as anti-cancer, anti-inflammatory, anticoagulant, and anti-microbial [47]. However, the study about the presence and its effect on these secondary metabolites in *mocaf* has never been conducted. So, *mocaf* provides healthier impact in people, and this can be the primary advantage to compete with the wheat flour.

As *mocaf* has similar characteristic to wheat flour, it can be used as important materials for making noodle including instant noodle [48, 49], bread [50–53], and other products [54]. One of commercial *mocaf*-based instant noodle called *Mie Ayo*, has successfully launched in the public market (**Figure 3**). This product development involves university, government institution, and micro, a small and medium enterprise called *Putri 21*. *Mocaf* is a potential source for Indonesian industry to diver and replace the use of wheat flour in the future. The availability of *mocaf* in the market could pull down the dependency of wheat flour time to time. To substitute or even replace wheat flour with mocaf, the Indonesian government encourages people to use local materials including cassava for *mocaf* production. The government launched commercially *mocaf* standard SNI 7622 in 2011. There are several funding sources from the government to the

Figure 3. *Mie Ayo,* one of *mocaf*-based noodle product in the market.

institution that actively involved in public service, especially in giving knowledge to people to build community capacity on *mocaf* development. The funding by Indonesian government is variety including *mocaf* production, *mocaf* usage, and the marketing strategies. Java is the most focus island on *mocaf* development. Several districts have been setup to be a center for *mocaf* production including Trenggalek, Pacitan, Ponorogo, Blitar, Malang, Tulungagung, and Kediri. Those belong to East Java. However, *mocaf* also has been developed in Central Java and the Special Region of Yogyakarta. The biggest *mocaf* mill is in Trenggalek. The production of mocaf on industrial scale also spread to outside Indonesia including Malaysia. That country has factory called Malaysian *Mocaf* Sdn Bhd which has been operated since 2015 [55] (**Figure 3**).

5. Conclusion

Mocaf transform from cassava flour which is less useful to substitute or even replace the presence of wheat flour. The use of microbial fermentation and additional physical treatment can develop new transform product, one step after cassava flour, called *mocaf*. This powder is a new hope for Indonesian people to reduce wheat grain import that has been conducted since the 1950s and dramatically increased from time to time. Because the wheat plant itself difficult to grow in every part of Indonesia. On the contrary, wheat-based food product seems gradually stapled in replacing rice, corn, and cassava mainly in the form of noodle and bakery products. *Mocaf* has similar characteristic to wheat flour. It can be use to substitute or replace the use of wheat flour. However, this can go slowly due to political issue of big industries or the dependency of people to wheat flour as staple product for more than 50 years. Indonesian government encourages people to shift wheat flour with *mocaf* by giving various funding, education, machine, and workshop to people especially for micro, small and medium enterprises (*Usaha Mikro Kecil Menengah*, UMKM). On the other hand, different types of cassava products still exist and are accepted by people such as *peyeum, tape singkong, gaplek, tiwul, cempong, gemblong, gethuk, opak*, etc. Those belonging to food heritage needs to be preserved.

Author details

Andri Frediansyah

Address all correspondence to: andri.frediansyah@lipi.go.id

Research Unit for Natural Product Technology (BPTBA), Indonesian Institute of Sciences (LIPI), Yogyakarta, Indonesia

References

[1] Food and Agricultural Organization of the United Nations. Crops [Internet]. 2017. Available from: http://www.fao.org/faostat/en/#data/QC [Accessed: 31-08-2017]

[2] Gibbons A. New view of early Amazonia. Science. 1990;**244**(4962):1488-1491

[3] Allem AC. The origin and taxonomy of cassava. In: Hillock RJ, Thresh JM, Bellotti AC, editors. Cassava: Biology, Production and Utilization. Oxon, UK: CABI Publishing; 2002. pp. 1-16

[4] Wetsby A. Cassava utilitation, storage, and small scale processing. In: Hillocks RJ, Thresh JM, Bellotti AC, editors. Cassava: Biology, Production and Utilization. Oxon, UK: CABI Publishing; 2002. pp. 281-300

[5] Conn EE. Cyanogenesis—A Personal Perspective. In: International Workshop on Cassava Safety; 1-4 March 1994; Ibadan, Nigeria. Wageningen: WOCAN in Cooperation with ISHS and ISTRC; 1994. p. 31-44

[6] Briani C, Torre CD, Citton V, Manara R, Pompanin S, Binotto G, Adami F. Nutrients. Cobalamin deficiency: Clinical picture and radiological findings. 2013;**5**(11):4521-4539. DOI: 10.3390/nu5114521

[7] Food and Agricultural Organization of the United Nations/World Health Organization. Codex standard for edible cassava flour. In: Joint FAO/WHO Food Standard Program. Rome, Italy; Codex Legal Alimentarius, Canadian Legal Info. Institute. 1995;**7**:133-136

[8] Damardjati DS, Widowati S, Rachim A. Cassava flour production and consumers acceptance at village level in Indonesia. Indonesian Agricultural Research and Development Journal. 1993;**15**:16-25

[9] Hue KT, Do Thi TV, Inger L, Wredle E, Sporndly E. Effect of harvesting frequency, variety and leaf maturity on nutrient composition, hydrogen cyanide content and cassava foliage yield. Asian-Australasian Journal of Animal Sciences. 2012;**25**(12):1691

[10] Kayode GO. Effects of various planning and harvesting times on the yield, HCN, dry matter accumulation and starch content of four cassava varieties in a tropical rainforest region. The Journal of Agricultural Science. 1983;**101**(3):633-636

[11] Frediansyah A, Kurniadi M, Nurhikmat A, Susanto A. Improving quality of mocaf (modified cassava flour) by bioprocess using *Lactobacillus plantarum* and its utility for foodstuff. In: International seminar on enhancing grassroots innovation competitiveness for poverty alleviation (EGICPA); 16-18-10-2012; Yogyakarta, Indonesia. Indonesian Institute of Sciences (LIPI)

[12] Santana MA, Vásquez V, Matehus J, Aldao RR. Linamarase expression in cassava cultivars with roots of low-and high-cyanide content. Plant Physiology. 2002;**129**(4):1686-1694

[13] Mburu FW, Swaleh S, Njue W. Potential toxic levels of cyanide in cassava (*Manihot esculenta Crantz*) grown in Kenya. African Journal of Food Science. 2012;**6**(16):416-420

[14] Cardoso AP, Mirione E, Ernesto M, Massaza F, Cliff J, Haque MR, Bradbury JH. Processing of cassava roots to remove cyanogens. Journal of Food Composition and Analysis. 2005;**18**(5):451-460

[15] Bradbury JH. Simple wetting method to reduce cyanogens content of cassava flour. Journal of Food Composition and Analysis. 1994;**19**(4):388-393

[16] Padmaja G, Steinkraus KH. Cyanide detoxification in cassava for food and feed use. Critical Reviews in Food Science and Nutrition. 1995;**35**(4):299-339

[17] Montagnac JA, Davis CR, Tanumihardjo SA. Processing technique to reduce toxicity and antinutrients of cassava for use as a staple food. Comprehensive Reviews in Food Science and Food Safety. 2009;**8**(1):17-27

[18] Iwuoha GN, Ubeng GG, Onwuachu UI. Detoxification effect of fermentation on cyanide content of cassava tuber. Journal of Applied Science & Environmental Management. 2013;**17**(4):567

[19] Kostinek M, Specht I, Edward VA, Schilinger U, Hertel C, Holzapfel WH, Franz CM. Diversity and technological properties of predominant lactic acid bacteria from fermented cassava used for the preparation of Gari, a traditional African food. Systematic and Applied Microbiology. 2005;**28**(6):527-540

[20] Oyewole OB, Odunfa SA. Characterization and distribution of lactic acid bacteria in cassava fermentation during fufu production. Journal of Applied Microbiology. 1990;**68**(2):145-152

[21] Odunfa SA, Oyewole OB. African fermented food. In: Wood BJB, editor. Microbiology of Fermented Foods. Boston, MA: Springer; 1987. pp. 713-752

[22] Oguntoyinbo FA. Identification and functional properties of dominant lactic acid bacteria isolated at different stages of solid state fermentation of cassava during traditional gari production. World Journal of Microbiology and Biotechnology. 2007;**23**(10):1425-1432

[23] Mante ES, Sakyi-Dawson E, Amoa-Awua WK. Antimicrobial interactions of microbial species involved in the fermentation of cassava dough into agbelima with particular reference to the inhibitory effect of lactic acid bacteria on enteric pathogens. International Journal of Food Microbiology. 2003;**89**(1):41-50

[24] Amowa-Awua WKA, Appoh FE, Jakobsen M. Lactic acid fermentation of cassava dough into agbelima. International Journal of Food Microbiology. 1996;**31**(1-3):87-98

[25] Misgiyarta, Suismono, Suyanti. Bimo cassava flour prospective increasingly. Warta Penelitian dan Pengembangan Pertanian. 2009;**31**:1-4 [Indonesian]

[26] Subagiyo A. Industrialization of modified cassava flour (Mocaf) as a raw material for food industry to support diversification of national staple food [Dissertation]. Indonesia: Faculty of Agricultural Technology, Universitas Jember; [Indonesian]. 2007

[27] Frediansyah A, Kurniadi M. Comparative influence of salinity and temperature on cassava flour product by *Lactobacillus plantarum* and *Lactobacillus acidophilus* during single culture fermentation. Nusantara Bioscience. 2016;**8**(2):207-214

[28] Frediansyah A, Kurniadi M. *Michaelis* kinetic analysis of extracellular cellulase and amylase excreted by Lactobacillus plantarum during cassava fermentation. AIP Conference Proceedings. 2017;**1**(1788):0301111-0301116

[29] Sanni AI, Morlon-Guyot J, Guyot JP. New efficient amylase-producing strains of Lactobacillus plantarum and L. fermentum isolated from different Nigerian traditional fermented foods. International Journal of Food Microbiology. 2002;**72**(1):53-62

[30] Giraud E, Gosselin L, Raimbault M. Production of a *Lactobacillus plantarum* starter with linamarase and amylase activities for cassava fermentation. Journal of the Science of Food and Agriculture. 1993;**62**(1):77-82

[31] Jeon HY, Kim NR, Lee HW, Choi HJ, Choung WJ, Koo YS, Shim JH. Characterization of a novel maltose-forming α-amylase from Lactobacillus plantarum subsp. plantarum ST-III. Journal of Agricultural and Food Chemistry. 2016;**64**(11):2307-2314

[32] Lei V, Amoa-Awua WKA, Brimer L. Degradation of cyanogenic glycosides by Lactobacillus plantarum strains from spontaneous cassava fermentation and other microorganisms. International Journal of Food Microbiology. 1999;**53**(2):169-184

[33] Obilie EM, Tano-Debrah K, Amoa-Awua WK. Souring and breakdown of cyanogenic glucosides during the processing of cassava into akyeke. International Journal of Food Microbiology. 2004;**93**(1):115-121

[34] Gunawan S, Widjaja T, Zullaikah S, Ernawati L, Istianah N, Aparamarta HW, Prasetyoko D. Effect of fermenting cassava with *Lactobacillus plantarum*, *Saccharomyces cereviseae*, and *Rhizopus oryzae* on the chemical composition of their flour. International Food Research Journal. 2015;**22**(3):1280-1287

[35] Kresnowati MTAP, Listianingrum, Zaenudin A, Trihatmoko K. The effect of microbial starter composition on cassava chips fermentation for the production of fermented cassava flour. AIP Conference Proceedings. 2016;**1**(1699):030002

[36] Verachtert H, Shanta Kumara HMC, Dawoud E. Yeasts in mixed cultures. In: Verachtert H, De Mot R, editors. Yeast in Biotechnology and Biocatalysis. New York, UK: Marcel Dekker Inc; 1990. pp. 429-449

[37] Buckle KA. Reduction of phytic acid levels in soybeans during tempeh production, storage and frying. Journal of Food Science. 1985;**50**(1):260-263

[38] Ciani M, Comitini F, Mannazzu I, Domizio P. Controlled mixed culture fermentation: A new perspective on the use of non-Saccharomyces yeast in winemaking. FEMS Yeast Research. 2010;**10**(2):123-133

[39] Garcia A, Carcel C, Dulau L, Samson A, Aguera E, Agosin E, Gunata Z. Influence of mixed culture with Debaryomyces vanriji and Saccharomyces cerevisiae on the volatile of Muscat wine. Journal of Food Science. 2002;**67**(3):1138-1143

[40] Zironi RP, Romano G, Suzzi F, Battistutta CG. Volatile metabolites produced in wine by mixed and sequential cultures of Hanseniaspora guilliermondii or Kloeckera apiculata and Saccharomyces cerevisiae. Biotechnology Letters. 1993;**15**(3):235-238

[41] Suismono S, Misgiarta M. Tepung cassava termodifikasi pengembangan agroindustri (tepung Bimo-CF). Jurnal Pangan. 2016;**21**(1):44-54

[42] Zulaidah A. Modifikasi ubi kayu secara biologi menggunakan starter Bimo-CF menjadi tepung termodifikasi pengganti gandum [Dissertation]. Indonesia: Universitas Dipenegoro

[43] Hartati H, Fathoni A, Kurniawati S, Hartati NS, Sudarmonowati E. Technological innovation in the protection of beta carotene on MOCAF production which is rich in beta carotene. Nusantara Bioscience. 2017;**9**(1):6-11

[44] Magiera SL. The role of wheat in the Indonesian food sector. Bulletin of Indonesian Economic Studies. 1981;**17**(3):48-73

[45] USDA Foreign Agricultural Service. Indonesia Grain and Feed Annual Report 2014 [Internet]. 2014. Available from: https://gain.fas.usda.gov/Recent%20GAIN%20Publications/Grain%20and%20Feed%20Annual_Jakarta_Indonesia_4-23-2014.pdf [Accessed: 30-08-2017]

[46] Grain Growers. The Indonesian market for Australian grains: An overview. 2016. Available from:www.aph.gov.au/DocumentStore.ashx?id=f27387b1-b111-4878-a225-2ce837945bb9&subId=463774 [Accessed: 30-08-2017]

[47] Bayoumi SAL, Rowan MG, Blagbrough IS, Beeching JR. Biosynthesis of scopoletin and scopolin in cassava roots during post-harvest physiological deterioration: The E-Z-isomerisation stage. Phytochemistry. 2008;**69**(17):2928-2936

[48] Vinsensia IR, Bella Nina M, Catarina SB. Pemanfaatan tepung umbi gadung (Dioscorea hispida Dennst) dan tepung mocaf (modified cassava flour) sebagai bahan substitusi dalam pembuatan mie basah, mie kering dan mie instant. Jurnal Teknologi Kimia dan Insustri. 2013;**2**(2):246-256

[49] Wahdini AI, Susilo B, Yulianingsih R. Uji karakteristik mi instant berbahan dasar tepung terigu dengan substitusi mocaf dan pati jagung. Jurnal Keteknikan Pertanian Tropis dan Biosistem. 2014;**2**(3):8-9

[50] Pato U, Restuhadi F, Ali A, Ulfah R. Evaluasi mutu dan daya simpan roti manis yang dibuat melalui susbtitusi tepung terigu dengan pati sagu dan mocaf. Jurnal Sagu. 2013; **11**(1):10

[51] Raharja S, Udin F, Suparno O, Febrianti FH, Nuraisyah A. Mocaf cros-linking with gluten to improve the quality of mocaf dough. AIP Conference Proceedings. 2017;**1**(1823):020050

[52] Saloko S, Handito D, Cicilia S, Dwiani A. Physicochemical and sensory characteristics of patisseries made from mocaf. Rekapangan. 2017;**10**(1):12

[53] Yenrina R, Murtius WS, Putri NN. Mocaf bread enriched with mung bean (Vigna radiate L.) as source of protein. Asia Pasific Journal of Sustainable Agriculture, Food and Energy. 2013;**1**(1):10-13

[54] Subagio A, Windrati WS. Effect of composition mocaf (modified cassava flour) and rice flour on characteristic of beras cerdas. Jurnal Pangan. 2012;**21**(1):29-38

[55] Shakir BBM. Production of modified cassava flour (Mocaf) [Thesis]. Malaysia: Faculty of Chemical and Natural Resources Engineering, Universiti Malaysia Pahang; 2013

Chapter 9

Recent Biotechnological Advances in the Improvement of Cassava

Vincent N. Fondong and Chrissie Rey

Additional information is available at the end of the chapter

http://dx.doi.org/10.5772/intechopen.70758

Abstract

Cassava (*Manihot esculenta* Crantz) is the fourth most important source of carbohydrates for human consumption in the tropics and thus occupies a uniquely important position as a food security crop for smallholder farmers. Consequently, cassava improvement is of high priority to most national agricultural research institutions in the tropics. With advances in functional genomics and genome editing approaches in this post genomics era, there are unprecedented opportunities and potential to accelerate the improvement of this important crop. These new technologies will need to be directed toward addressing major cassava production constraints, notably virus resistance, protein content, tolerance to drought and reduction of hydrogen cyanide content. Here, we discuss the important role novel functional genomics and genome editing technologies have and will continue to play in cassava improvement efforts. These approaches, including artificial miRNA (amiRNA), trans-acting small interfering RNA (tasiRNA), clustered regularly interspaced short palindromic repeat (CRISPR)-associated protein 9 (Cas9), and Targeting Induced Local Lesions IN Genomes (TILLING), have been shown to be effective in addressing major crop production constraints. In addition to reviewing specific applications of these technologies in cassava improvement, this chapter discusses specific examples being deployed in the amelioration of cassava or of other crops that could be applied to cassava in future.

Keywords: amiRNA, cassava, CRISPR, genetic engineering, tasiRNA, TILLING

1. Introduction

Cassava, *Manihot esculenta* Crantz, was transported to Africa by the Portuguese in the sixteenth century, and was initially grown in and around trading posts in the Gulf of Guinea in West Africa; it was subsequently introduced into East Africa from Madagascar in the later part of the eighteenth century [1]. Today, cassava is a staple food to an estimated 800 million people worldwide [2] and is grown almost exclusively by smallholder farmers (**Figure 1**) and in isolated areas where soils are poor and rainfall is low or unpredictable. Additional attributes of this crop include low-cost and readily planting material, tolerance to acid soils, forms a symbiotic association with soil fungi to help its roots absorb phosphorus and micronutrients. Thus, cassava production requires very low input and gives reasonable harvests where other crops would fail [2]. Cassava is also increasingly being adopted as a source of family income following the fall of coffee and cocoa in the world market. Consequently, improvement of this crop is of high priority to most national agricultural research institutions in Africa. Moreover, the recognition that cassava industrial starch-based products, especially in renewable energy, could enhance food security and livelihoods, makes this crop a potentially valuable source of economic growth on the African continent.

Figure 1. Africa produces more cassava than any other crop. (A) Cassava is a woody shrub that grows well in marginal lands; (B) a family in Southern Cameroons transporting cassava tuberous root harvest; (C) cassava tuberous roots are processed into many food types, here a family in Southern Cameroons preparing "garri," a flour produced from cassava tuberous roots.

Cassava is cultivated principally for its tuberous roots, which are a good source of energy; additionally, in some parts of central Africa, leaves are also consumed as a source of protein, vitamins, and minerals. Cassava roots and leaves are deficient in sulfur-containing amino acids (methionine and cysteine) and some nutrients are not optimally distributed within the plant [3], leading to a deficiency in protein content, especially in roots. Cassava also contains antinutrients, most notably cyanogens, that can interfere with nutrient absorption and utilization and may have toxic side effects [4]. There are efforts to add nutritional value to cassava (biofortification) by increasing the contents of protein, minerals, starch, and β-carotene through biotechnological approaches [3]. Thus, nutritional content and production of cassava will benefit greatly from advances in genomics and biotechnological approaches.

Cassava improvement, either through conventional breeding or through genetic engineering, is challenging. The most reliable regeneration system cassava so far is through somatic embryogenesis (**Figure 2**) [5]. In the case of conventional breeding, which so far is the most routinely used approach to improve this crop is challenging due to several factors associated with several factors, include: (1) Lack of useful genes in the core cassava germplasm collections; (2) Heterozygosity and allopolyploidy of the cassava genomes; (3) Irregular flowering;

Figure 2. Regeneration of cassava cultivars from Cameroon [5]. Callus with proembryogenic masses (A); clusters of organized embryogenic structures consisting of globular, heart and torpedo structures, early cotyledonary stage, asynchronous development of somatic embryos (B); organogenic callus with green cotyledons developed clusters of shoot buds (C); shoot buds rooted and developed into whole plantlets in vitro (D).

and (4) Low fertility, seed set, and germination rates. As for genetic engineering and gene transfer, over the past few decades, this approach has been used to complement conventional breeding [6]. Undoubtedly, advances in modern technologies such as transcriptomics, proteomics, and metabolomics are likely to benefit breeding and genetic engineering strategies from an understanding of plant metabolic pathways and the role of key genes associated with their regulation. In this chapter, we identify some of the most important nutritional characteristics of cassava and production constraints that can benefit from advances in genome editing and functional genomics approaches are discussed in this section.

2. Important characteristics of cassava requiring improvement

2.1. Protein content

Cassava tuberous roots have relatively low protein content, which on the average ranges from 2 to 3% dry weight [7], compared with 9–11% for maize grain [8]. Indeed, a 500-g cassava meal provides only 30% of the daily protein requirement. Added to the low protein content, is the fact that roots are processed and the processed product is essentially protein-free. Consequently, individuals consuming exclusively or predominantly cassava usually suffer from protein-deficiency symptoms [9]. There is evidence suggesting that protein content in the roots can be considerably higher (6–8%) in some landraces [7] and such an important attribute can be introgressed into cassava through classical breeding. However, as indicated above, cassava breeding is rather challenging. Thus, fortification via genetic engineering is a more feasible option in efforts aimed at improving cassava protein content. For example, cyanide derived from linamarin is a major cause of reduced nitrogen for cassava root protein synthesis, thus disruption of linamarin transport from leaves to the roots through gene silencing-mediated inhibition of the two cytochrome P450s genes, CYP79D1/D2, resulted in an increase in nitrogen levels in cassava roots and higher levels of root protein content [10]. There have also been attempts to increase protein root content by producing transgenic cassava expressing genes that enhance protein root accumulation, including an artificial storage protein gene, *ASP1* [11]. Thus, advances in genomics and transcriptomics will undoubtedly identify genes and pathways that will provide new opportunities to increase protein content using genetic engineering approaches.

2.2. Hydrogen cyanide content

Consumption of residual cyanogens (linamarin and lotaustralin) in incompletely processed cassava roots can cause various health disorders that render a person unsteady and uncoordinated [12]. Hydroxynitrile lyase (HNL) catalyzes the conversion of acetone cyanohydrin to cyanide and is expressed predominantly in the cell walls and laticifers of leaves, compared with tuberous roots, which exhibit very low [10]. Transgenic cassava over-expressing HNL was shown to display significantly reduced acetone cyanohydrin levels and exhibited increased cyanide volatilization in processed or homogenized roots [12]. It has been shown that the genomic region surrounding the cytochrome P450, CYP79D3, contains all genes required for cyanogenic glucoside biosynthesis in cassava [13]. As indicated above, this provides an additional opportunity to reduce cyanogen content in cassava by for example tissue specific suppression of two P450 genes, CYP79D1/D2, that catalyze the first-dedicated step in cyanogen synthesis [14]. Thus, at the molecular level, cyanogen detoxification can either be achieved by gene overexpression or through gene suppression, either of which can be achieved through genome editing techniques.

2.3. Starch quality

Due to its high starch content, cassava provides a source of dietary carbohydrate to an estimated 800 million people worldwide [2]. Insight in cassava development and starch biosynthesis is necessary to improve cassava starch quality and quantity [15]. Isolation and characterization of cassava gene homologs implicated in processes affecting the conversion of assimilated carbon to sucrose in photosynthetic cells, the phloem transport of sucrose to storage organs, the transition of sucrose to starch, and the degradation of starch into simple sugars, could be exploited to improve starch quality. Molecular and functional characterization of the genes involved in these processes will greatly enhance cassava varietal improvement by altering the gene activities either via genetic manipulation or through gene editing. Also, application of advanced systemic-based computational techniques to understand the physiological regulation and control of starch metabolism in plant plastids would be the basis for understanding these processes in cassava.

Cassava is also a good source of industrial starch and bioethanol [16, 17]; in both situations, the quality of starch is important. Starch consists of two glucan polymers: amylose and amylopectin. Amylopectin is extremely soluble in water whereas amylose has a strong tendency to recrystallize after dispersion in water, a property referred to as retrogradation. Retrogradation is undesirable for many applications of starch in which a defined and stable viscosity is required. Therefore, for industrial purposes, starch is often treated with chemicals in order to make the amylose less sensitive to crystallization [18]. As retrogradation is caused mainly by the amylose fraction in starch, amylose-free starches do not have to be treated with chemicals [19]. There are therefore efforts to generate amylose-free cassava through genetic engineering; for example, starch-free cassava was obtained by silencing *GBSSI*, the granule-bound starch synthase gene, which is required for the synthesis of amylose [20].

2.4. Postharvest physiological deterioration and storage

Harvested cassava tuberous roots undergo rapid postharvest physiological deterioration (PPD) [21, 22]. PPD is initiated by mechanical damage, which typically occurs during tuberous roots harvesting and progresses from the proximal site of damage to the distal end, making the roots unpalatable within 72 h [22, 23]. Reactive oxygen species (ROS) production has been identified as one of the earliest events in PPD [21, 22, 24]. Under conditions of stress, the equilibrium between the production and scavenging of ROS is disturbed, resulting in a rapid increase in the buildup of ROS known as an oxidative burst [25]. In cassava roots, an oxidative burst occurs within 15 min of harvest [21], resulting therefore in an early PPD. Other early events that result in rapid PPD include, increased activity of enzymes that modulate ROS levels, such as catalase, peroxidase, and superoxide dismutase [22]. Further evidence in support of a role of oxidative stress in PPD comes from the observation that cassava cultivars that have high levels of b-carotene (which quenches ROS) are less susceptible to PPD [26]. Reduction of ROS and PPD was also shown to be induced by cyanogenesis, suggesting that a possible solution to cassava PPD is to reduce the cyanide-induced accumulation of ROS.

2.5. Cassava pathogens

Cassava viruses constitute a major challenge to cassava production; of particular importance are cassava mosaic geminiviruses (CMGs) (Family, *Geminiviridae*: Genus, *Begomovirus*), which cause the cassava mosaic disease (CMD) in all cassava growing regions of Africa and the Indian subcontinent. CMGs are transmitted by the whitefly vector, *Bemisia tabaci* (Gennadius) and through cuttings used routinely for vegetative propagation. Tuberous root losses due to CMD range from 20 to 100% [27]. With the emergence of new molecular and sequencing capabilities, CMGs have been shown to exhibit considerable sequence and biological differences and so far, 11 species have been described in the cassava growing regions of African and the Indian subcontinent [28] and some of these viruses co-infect the same plant resulting in a synergistic interaction, characterized by severe symptoms (**Figure 3**). Interestingly, cassava was introduced in Africa from South America [29], yet CMGs are not found in South America and therefore these viruses are likely recent descendants of geminiviruses adapted to indigenous uncultivated African plant species [30]. The problem of CMGs has been compounded by the emergence, in eastern Africa, of cassava brown streak disease (CBSD), which is caused by cassava brown streak viruses (CBSVs (Family, *Potyviridae*: Genus, *Ipomovirus*). Like CMGs, CBSVs are transmitted by the whitefly vector [31] and through infected stem propagules. For a long time, CBSD was considered to be limited to lowland coastal regions of Tanzania, and to a limited extent in lowland areas of Uganda [32], northwestern Tanzania, southern Uganda [32–34]. Since 2004, however, the CBSD epidemic has spread around the Great Lakes Region to affect eastern Uganda, western Kenya, the Lake Zone of Tanzania, Rwanda, Burundi and the DRC [35, 36]. The most damaging symptoms of CBSD are found in tuberous roots, including brown, corky necrosis of the starchy tissue, occasional radial constrictions and a reduction in the content of starch and cyanide [32, 34]. Yield losses are estimated to be up to 70% in highly susceptible cultivars [37]. In additional to viral pathogens, is cassava bacterial blight (CBB), caused by *Xanthomonas axonopodis* pv. *manihotis (Xam)*. CBB is considered to be one of the most relevant plant pathogenic bacteria because of the yield losses, estimated to be 70%, it causes in cassava [38, 39].

Use of resistant varieties has been the most effective in controlling CMD in Africa thanks to the discovery in the 1930s that some of the cassava varieties being grown were less affected by CMD than others. Thus, resistance breeding began in Ghana, Madagascar, Tanzania and

Figure 3. Cassava plant mixed infected by two cassava mosaic geminiviruses displaying severe mosaic and leaf distortion and size reduction, resulting in plant stunting.

elsewhere in Africa [40, 41]. In the last 2 decades, use of genetic engineering to produce virus resistant cassava has gained considerable attention, especially with the discovery of RNA interference pathways [42].

2.6. Tolerance to drought

In most cassava growing regions of the world, the cassava growth cycle is typically interrupted by months of drought, influencing various plant physiological processes and resulting in depressed growth, development and yield [43, 44]. Although cassava is a drought tolerant crop, there is a range of drought-tolerance levels in available germplasm. Thus, growth and productivity of genotypes with a low threshold of drought tolerance in marginal areas are constrained by severe drought stress, especially during the earlier stages of growth [45]. Indeed, experimental data suggest that root production is positively correlated with the life span of individual leaves [46] and increased leaf retention was found to increase root yield under irrigated and stressed conditions [47]. With continuous advances in genome science, there will be opportunities to enhance drought tolerance in the cassava crop. For example, Zhang et al. [46] have shown that transgenic cassava expressing *isopentenyltransferase (IPT) gene* under the control of senescence-activated promoter (SAG12), delayed leaf senescence under both greenhouse and field conditions, leading an increase in drought resistance. Also, identification of miRNA gene targets involved in post-transcriptional abiotic stress regulation could prove useful in engineering cassava for drought resistance [48].

3. RNA-based functional genomics technologies in cassava improvement

3.1. Hairpin dsRNA, co-suppression and antisense RNA silencing

The hairpin dsRNA (hpRNA), anti-sense silencing and co-suppression strategies have been extensively employed in crop improvement [49–51]. In cassava improvement, hpRNA and antisense silencing procedures have been employed mostly in virus control [52–56]. An indirect approach where the hpRNA is used to knockdown the expression of V-ATPase A, an enzyme that provides force for many transport processes, has been used to control whitefly vectors of CMGs and CBSVs [57, 58].

The antisense strategy has also been used to inactivate allergens and toxins in cassava, especially in the inhibition of hydrogen cyanide (HCN), which is the product of linamarase-mediated hydrolysis of linamarin. The presence of residual linamarin and its breakdown product (acetone cyanohydrin) in cassava-based food products has been a cause for concern because of their possible effects on human health. As discussed above, the first committed steps in linamarin biosynthesis is catalyzed by cytochrome P450 genes (CYP79D1 and CYP79D2) and therefore efforts have been made to knockdown CYP79D1 and CYP79D2 using hpRNA-mediated silencing so as to reduce HCN toxicity in cassava. Thus, transgenic cassava lines containing antisense copies of both genes exhibited almost complete absence of linamarin in tuberous roots [10]. Unfortunately, this approach could not be applied extensively as transgenic cassava lines exhibited poor tuberous root development.

In spite of the encouraging early results obtained from the use of hpRNA, co-suppression and antisense RNA silencing in crop improvement, this approaches have been tempered by several disadvantages associated with these approaches, these include poor stability of the transgene in transformed plants, dependence on the expression levels of the antisense strand, and limited penetration of the silencing signal to the appropriate target cells due to target-sequence folding (reviewed in Fondong et al. [59].

3.2. Small RNA (sRNA)-mediated silencing

The limitations of hpRNA, co-suppression and antisense RNA silencing strategies are, to a large extent overcome in sRNA strategies, including especially artificial microRNAs (amiRNAs) and trans-acting siRNA (tasiRNA). microRNAs constitute a well-studied class of sRNAs; their biogenesis starts with the transcription of long primary RNAs (pri-miRNAs) [60, 61]. miRNAs function in a homology-dependent manner against target mRNAs to typically either directly cleave at highly specific sites or to suppress translation. The amiRNA silencing technique exploits the biogenesis and function of endogenous miRNAs to silence genes in plants. In this approach, the endogenous miRNA-miRNA duplex in a native miRNA precursor is replaced with a customized sequence designed from the target gene. Upon processing, the amiRNA redirects the miRNA-induced silencing complex to silence the targeted mRNA, thereby generating a loss-of-function phenotype for the gene of interest [62–66]. The amiRNA strategy has especially been used in targeting plant viruses (reviewed in Fondong et al. [59]. However, there has been little application in cassava improvement. Indeed, to our knowledge, the only report of use of amiRNA in cassava improvement is the replacement of miR159 precursor with amiRNAs from cassava brown streak viruses in miR159 precursor; transgenic *Nicotiana benthamiana* lines thus produced were virus resistant [67].

It is important to note that the amiRNA platform has several advantages over the hpRNA strategy, including the fact that amiRNAs are small and thus have a reduced likelihood of off-targeting and the approach can easily be multiplexed via use of polycistronic miRNA backbone. In addition, processing of miRNA is not affected by changes in temperature compared with hpRNA-derived siRNAs whose levels decrease at low temperatures [68]. Thus, it is likely that this platform will prove useful in studying cassava gene function. A major limitation of the amiRNA strategy is that the small size of the amiRNA (21nt) increases opportunities for loss of complementarity between the amiRNA and the target gene, and genes from the same family with variations may not be silenced using a single amiRNA. To reduce these risks, a multimeric amiRNA approach in which multiple amiRNAs targeting different conserved regions of the gene can be adopted as has been reported in plant virus control [69, 70].

A second class of sRNAs used in crop improvement is tasiRNAs, which are produced from noncoding *TAS* genes, which have been identified in all examined land plants. *TAS* genes differ from most other genes in that they do not code for a protein, but rather produce long noncoding RNA transcripts, which are subsequently processed into 21nt tasiRNAs. Synthesis of tasiRNA is initiated by miRNA-directed and Argonaute (AGO) protein-mediated cleavage of *TAS* transcripts, of which four (TAS1, 2, 3, 4) have been extensively studied in Arabidopsis (see reviews Allen and Howell [71] and Yoshikawa [72]. Two models of tasiRNA biogenesis, referred to as "one-hit" and "two-hit", have been described in Arabidopsis [73]. The tasiRNA strategy is very efficient, highly predictable in processing siRNAs and can easily be multiplexed to target multiple genes, especially genes from the same family; yet it remains an underutilized strategy in plant improvement. It has been used to successfully engineer resistance to plant viruses [74, 75]. In cassava virus control, transgenic *N. benthamiana* containing Arabidopsis *TAS1a* gene modified with tasiRNA from the cassava geminivirus, *East African cassava mosaic Cameroon virus* (EACMCV) exhibits strong resistance to the virus (Fondong et al., unpublished). Fifty-four tasiRNAs and fifteen possible *cis*-nat- siRNAs were identified in cassava infected with cassava bacterial blight, and many of these loci were induced or repressed in response to *Xam* infection [76]. A similar transgenic strategy using a *TAS* gene modified with tasiRNAs from *Xam* could be promising. This finding emphasizes the potential potency of this strategy in plant virus control.

4. Role of reverse genetics and gene editing techniques in cassava improvement

4.1. TILLING and EcoTILLING

TILLING is a non-recombinant reverse genetics approach used to identify novel sequence variation in genomes, with the aims of investigating gene function and/or developing useful alleles for breeding. TILLING involves induction of mutations in the plant genome using classical mutagenesis approaches followed by traditional or high throughput deep sequencing to identify the mutations in the gene of interest [77–79]. This technique has been used in allele discovery in different plant species [80–83]. EcoTILLING, which is an adaptation of the TILLING, is used in detecting rare single nucleotide polymorphism (SNPs) or small INDELs in target genes in natural populations [84]. In EcoTILLING, mismatches formed by hybridization of different genotypes in a test panel are cleaved with CEL I, which is a mismatch-specific endonuclease from celery. A valuable application of EcoTILLING in plants is in the search for variation in disease resistance genes. There are only a few reports of the use of TILLING or EcoTILLING in cassava improvement. Of these, is the recent report of irradiation of seeds of elite cassava lines and wild *Manihot* species in an effort to broaden the genetic base of the germplasm pool so as to expand the industrial uses of cassava [85]. The study led to the discovery of small granules, which are abnormal amylose starch molecules resulting from a mutation. These small granules are ideal for industrial ethanol production due to the fact that they facilitate the activity of starch-degrading enzymes [86]. Because of the promise of the technique in cassava improvement, the International Institute of Tropical Agriculture (IITA) in Ibadan, Nigeria, is developing a TILLING protocol for discovery of important cassava traits [87].

TILLING has several advantages over other crop improvement techniques: (1) it produces a spectrum of allelic mutations that are useful for genetic analysis, (2) it is applicable to any organism, (3) mutations that are difficult to be detected by forward genetics can be revealed via TILLING since it can focus at on the gene of interest, and (4) it is a non-transgenic method, hence there are no biosafety or environmental concerns [88]. The main disadvantages of TILLING are the requirement of locus-specific polymerase chain reaction (PCR) products (difficult for gene families with very similar sequences and in polyploids) and the inability to detect mutations near simple sequence repeats (SSRs) (because of the flare caused by polymerase slippage-induced deletions) [89].

4.2. Clustered regularly interspaced short palindromic repeat (CRISPR)

As indicated above, cassava transformation and crossing are challenging and thus gene editing is potentially a method that can be used to improve the crop. The clustered regularly interspaced short palindromic repeats (CRISPRs) and associated protein (Cas) approach has recently gained wide application in gene editing. In bacteria and especially archaea, CRISPRs/Cas is a nucleic acid-based adaptive immune system, which confers molecular immunity to foreign nucleic acids, including plasmids and viruses (see review Barrangou [90]. CRISPR genomic loci consist of repeat sequences, typically 20–50 bp in length, separated by variable spacer sequences (or protospacers) of similar length that match a segment of invading nucleic acids. These protospacers serve as a molecular memory of prior infections and together with repeat sequences, constitute CRISPR RNAs in the CRISPR locus [90, 91]. CRISPR RNAs are used as guides by Cas proteins for base-pairing with and degradation of complementary sequences in invading DNAs [90, 91]. The CRISPR/Cas system is functional in eukaryotic systems, for which the *Streptococcus pyogenes* endonuclease Cas9 (Cas9) has been harnessed for efficient eukaryotic genome editing and gene regulation [92, 93]. The ease of deployment of the CRISPR/Cas9 system is due to its dependence on RNA as the moiety that directs the Cas9 nuclease to a desired DNA sequence [94, 95].

The functionality of CRISPR/Cas9 system in eukaryotes has revolutionized genome editing and in a very short time since its discovery, has become a very useful tool in crop improvement. Successful examples have been reported for several crops with complex genomes (reviewed in Paul and Qi [96]). However, only a few reports of use of CRISPR/Cas9 system in cassava improvement exist and are still in the preliminary stage, these include CRISPR/Cas9-mediated modification of cassava flowering genes to induce flowering in this predominantly clonally propagated crop [97]. Because of the successful development of a modified geminivirus vector based on *Cabbage leaf curl virus* for a virus-guided delivery of CRISPR/Cas9 [98], it is likely that a similar vector system can be developed for cassava using the cassava geminivirus, *African cassava mosaic virus*.

There are drawbacks of the CRISPR/Cas system, including: (1) imbalance in stoichiometry between Cas9 and sgRNA ratio that may lead to off-target cleavage [99, 100]. (2) Many protospacer adjacent motif (PAM) sites may lead to undesired cleavage of DNA regions [101]; to resolve this problem, bioinformatics tools are being developed at whole genome sequence level to improve specificity [102]. (3) Codon usage varies across species and may affect Cas9 translation; several codon-optimized versions of Cas9 genes have therefore been harnessed for several individual crops [102] and there may be need for a cassava codon optimized Cas9. (4) CRISPR/Cas9 systems use exogenous promoters for Cas9 and sgRNA expression; for cassava, *Cassava vein mosaic virus* promoter [103] has been shown to be very efficient. (5) Homology between gene family members may complicate sequence targeting and directing sgRNAs to the 5′ region of the targeted gene has been proposed improve target specificity [102].

5. Natural host resistance to pathogen in cassava

5.1. Natural pathogen resistance and cassava virus control

It is now clear that the cassava geminiviruses and cassava brown streak viruses are the most important constraints to cassava production in the African [104]. Correspondingly, in Latin America, a diverse set of virus species that cause the cassava frog skin disease syndrome has a serious impact on cassava production [105]. Thus, considerable effort will be required to expand sources of resistance to cassava viral diseases and advances in genomics have provided new opportunities to explore sources of natural resistance. An important source of resistance that may be useful in cassava is non-host resistance. Mechanistically, non-host resistance is likely due to an intrinsic lack of susceptibility, which is a multigenic trait. It is now known that natural compounds, such as melatonin that modulate immune responses, such as ROS metabolism, calcium signaling and mitogen-activated protein kinase (MAPK) cascades, can be used to enhance natural resistance [106]. Notably, the recent identification and functional analysis of melatonin synthesis genes in cassava has provided a direct link between melatonin and immune responses [107]. Furthermore, the importance of resistance targets that function as host susceptibility factors, such as translation initiation factors 4E and 4G in RNA viruses, have been studied in model systems and can potentially be exploited for CBSV resistance in cassava [108].

Viruses that successfully infect the host induce changes in host cells by manipulating the host molecular pathways and host responses can provide clues for functional manipulation of resistance traits. It has been shown that CMGs [109] and CBSVs [110] induce global transcriptome reprogramming of cassava. In the case of CBSD, of the 700 overexpressed genes in a resistant cassava variety, none of the genes was identified as a resistance gene, instead most belonged to hormone signaling and metabolic pathway gene classes [110]. Interestingly, three functional genomic studies with *South African cassava mosaic virus* in three hosts, Arabidopsis [111], cassava [109] and *N. benthamiana* [112] revealed a small number of common differentially expressed genes at the early infection stage of full systemic symptoms. However, a common theme in all three hosts was virus-induced changes in hormone signaling, and primary and secondary metabolisms. Understanding the roles of host reprogramming and RNA silencing during cassava-virus interactions could be exploited to improve natural immunity in cassava.

5.2. Identification of cassava immunity-related or resistance (R) genes

Dominant and recessive genes have been associated with natural plant virus resistance [113]. Using a combination of genotype-by-sequencing (GBS)-based SNPs and physical mapping of scaffolds from cassava whole genome sequencing (WGS), 1061 cassava immunity-related genes were mapped [114]. Notably, from 105 putative *CMD2* genes identified from the CMD2 locus on chromosome 8 [115], 35 were identical to those identified in a RNA-seq study of SACMV-infected cassava genotype TME3 [109]. These genes could be strong candidates contributing to resistance in cassava. Proteins encoded by *R* gene usually occur as large families of proteins with nucleotide binding-leucine rich repeat (NLR) domains and function as indirect perception sensors of pathogen avirulence (avr) proteins. The determinants of apparent virus R gene-wide specificity lies in the leucine-rich repeat (LRR) domains and sequencing of wild cassava varieties may provide a source for discovery of new cassava virus resistance genes. Recently, 228 NLR and 99 partial NBS genes were mapped to the cassava reference genome (http://phytozome.jgi.doe.gov) and these genes show high sequence similarity to genes found in other plant species [116]. However, involvement of these genes in CMD or CBSD resistance is not known. Furthermore, microRNAs are master regulators that trigger processing of genes coding for NLR into phased small interfering RNAs (phasiRNAs) [117] and are therefore regulators of genes that are the first line of defense. Unveiling the role of miRNAs in cassava virus resistance would provide new tools in the combat against these viruses.

Based on a holistic approach, combining high-throughput transcriptome sequence data, public genomic data from cassava and Arabidopsis, Leal et al. [118] identified predicted immunity related gene (IRG) pathways, which showed that several cellular pathways are strongly related to immune response pathways. We will need to exploit these genomics data to identify evolutionarily diverse resistance or immunity genes in different cassava genotypes for development of durable resistance to cassava viruses.

5.3. Resistance against whitefly, *B. tabaci* (Gennadius)

Another approach to generate resistance to CMGs and CBSVs is through use of functional genomics to control the whitefly (*B. tabaci*), vector of both virus groups. Until recently, little was known about the molecular mechanisms of insect defense. Development of *B. tabaci* type B on Arabidopsis was shown to rely on the concomitant increase of salicylic acid and decline or unchanged levels of jasmonic acid and ethylene defense pathways [119]. Transgenic mediated overexpression or down-regulation of genes involved in lignin or other defenses against insect pests could be exploited to develop insect resistant cassava [120]. Application of functional genomics in insect resistance was recently elucidated by expressing an insecticidal fern protein in cotton, which exhibited resistance to whitefly [121]. Efforts in editing genes that play a role in whitefly resistance in cassava will thus play a role in developing cassava with resistance to the whitefly, vector of many cassava viruses.

6. Conclusion

The future challenge in cassava is the ability to combine desirable traits with different agronomic requirements using molecular breeding, gene editing and RNAi technologies. This is critically important, given that cassava is fundamental to food security in many parts of the world. In this chapter, we have discussed advances in the improvement of this crop, especially with regards to nutrient quality and biotic as well as abiotic constraints. We have also proposed novel genome editing technologies that will likely address some of the challenges faced by this crop. These include technologies such as amiRNA, tasiRNA, TILLING/EcoTILLING and CRISPR-Cas9, which provide enormous potentials in cassava improvement. Also, the increasing reduction in the cost of high-throughput sequencing and lessons from ongoing and past work will continue to provide new insights into additional new genome-editing and functional genomic approaches for the improvement of the crop.

Acknowledgements

The disease resistance work in Vincent N. Fondong's laboratory is supported by NSF-IOS-1212576.

Author details

Vincent N. Fondong[1]* and Chrissie Rey[2]

*Address all correspondence to: vfondong@desu.edu

1 Department of Biological Sciences, Delaware State University, Dover, USA

2 School of Molecular and Cell Biology, University of the Witwatersrand, Johannesburg, South Africa

References

[1] Legg JP, Thresh JM. Cassava mosaic virus disease in East Africa: A dynamic disease in a changing environment. Virus Research. 2000;**71**(1-2):135-149

[2] FAO. FAOSTAT Statistical Database 2013 Available from: http://faostat.fao.org [Accessed: August 2017]

[3] Montagnac JA, Davis CR, Tanumihardjo SA. Nutritional value of cassava for use as a staple food and recent advances for improvement. Comprehensive Reviews in Food Science and Food Safety. 2009;**8**:181-194

[4] Montagnac JA, Davis CR, Tanumihardjo SA. Processing techniques to reduce toxicity and antinutrients of cassava for use as a staple food. Comprehensive Reviews in Food Science and Food Safety. 2009;**8**:17-27

[5] Mongomake K, Doungous O, Khatabi B, Fondong VN. Somatic embryogenesis and plant regeneration of cassava (*Manihot esculenta* Crantz) landraces from Cameroon. SpringerPlus. 2015;**4**:477. DOI: 10.1186/s40064-015-1272-4

[6] Sanghera GS, Wani SH, Gill MS, Kashyap PL, Gosal SS. RNA interference: Its concept and application in crop plants. In: Malik CP, Verma A, editors. Biotechnology Cracking New Pastures. New Delhi: MD Publishers; 2010. p. 33-78

[7] Ceballos H, Iglesias CA, Pérez JC, Dixon AG. Cassava breeding: Opportunities and challenges. Plant Molecular Biology. 2004;**56**(4):503-516

[8] Ensminger ME, Oldfield JE, Heinemann WW. Feeds and Nutrition. 2nd ed. Clovis, CA: Ensminger Publishing Co; 1990 505 p

[9] Sayre R, Beeching JR, Cahoon EB, Egesi C, Fauquet C, Fellman J, Fregene M, Gruissem W, Mallowa S, Manary M, et al. The BioCassava plus program: Biofortification of cassava for sub-Saharan Africa. Annual Review of Plant Biology. 2011;**62**:251-272. DOI: 10.1146/annurev-arplant-042110-103751

[10] Siritunga D, Arias-Garzon D, White W, Sayre RT. Over-expression of hydroxynitrile lyase in transgenic cassava roots accelerates cyanogenesis and food detoxification. Plant Biotechnology Journal. 2004;**2**(1):37-43

[11] Zhang P, Jaynes J, Potrykus I, Gruissem W. Transfer and expression of an artificial storage protein (ASP1) gene in cassava (*Manihot esculenta* Crantz). Transgenic Research. 2003;**12**(2):243-250

[12] Narayanan NN, Ihemere U, Ellery C, Sayre RT. Overexpression of hydroxynitrile lyase in cassava roots elevates protein and free amino acids while reducing residual cyanogen levels. PLoS One. 2011;**6**(7):e21996. DOI: 10.1371/journal.pone.0021996

[13] Takos AM, Knudsen C, Lai D, Kannangara R, Mikkelsen L, Motawia MS, Olsen CE, Sato S, Tabata S, Jørgensen K, Møller BL, Rook F. Genomic clustering of cyanogenic glucoside

biosynthetic genes aids their identification in *Lotus japonicus* and suggests the repeated evolution of this chemical defence pathway. The Plant Journal. 2011;**68**(2):273-286. DOI: 10.1111/j.1365-313X.2011.04685.x

[14] Andersen MD, Busk PK, Svendsen I, Møller BL. Cytochromes P-450 from cassava (*Manihot esculenta* Crantz) catalyzing the first steps in the biosynthesis of the cyanogenic glucosides linamarin and lotaustralin. Cloning, functional expression in *Pichia pastoris*, and substrate specificity of the isolated recombinant enzymes. The Journal of Biological Chemistry. 2000;**275**(3):1966-1975

[15] Karlström A, Calle F, Salazar S, Morante N, Dufour D, Ceballos H. Biological implications in cassava for the production of amylose-free starch: Impact on root yield and related traits. Frontiers in Plant Science. 2016;**20**(7):604. DOI: 10.3389/fpls.2016.00604

[16] Shetty JK, Chotani G, Duan G, Bates D. Cassava as an alternative feedstock in the production of transportation fuel. International Sugar Journal. 2007;**109**:3-11

[17] Pandey A, Soccol CR, Nigam P, Soccol VT, Vanderberghe LPS, Mohan R. Biotechnological potential of agro-industrial residue. II: Cassava bagasse. Bioresource Technology. 2000;**74**: 81-87

[18] Bruinenberg PM, Jacobsen E, Visser RGF. Starch from genetically engineered crops. Chemistry and Industry. 1995;**95**:881-884

[19] Koehorst-van Putten HJ, Sudarmonowati E, Herman M, Pereira-Bertram IJ, Wolters AM, Meima H, de Vetten N, Raemakers CJ, Visser RG. Field testing and exploitation of genetically modified cassava with low-amylose or amylose-free starch in Indonesia. Transgenic Research. 2012;**21**(1):39-50. DOI: 10.1007/s11248-011-9507-9

[20] Raemakers K, Schreuder M, Suurs L, Furrer-Verhorst H, Vincken JP, de Vetten N, Jacobsen E, Visser RGF. Improved cassava starch by antisense inhibition of granule-bound starch synthase I. Molecular Breeding. 2005;**16**:163-172

[21] Reilly K, Gómez-Vásquez R, Buschmann H, Tohme J, Beeching JR. Oxidative stress responses during cassava post-harvest physiological deterioration. Plant Molecular Biology. 2004;**56**(4):625-641

[22] Iyer S, Mattinson DS, Fellman JK. Study of the early events leading to cassava root post-harvest deterioration. Tropical Plant Biology. 2010;**3**:151-165

[23] Buschmann H, Reilly K, Rodriguez MX, Tohme J, Beeching JR. Hydrogen peroxide and flavan-3-ols in storage roots of cassava (*Manihot esculenta* Crantz) during postharvest deterioration. Journal of Agricultural and Food Chemistry. 2000;**48**(11):5522-5529

[24] Zidenga T, Leyva-Guerrero E, Moon H, Siritunga D, Sayre R. Extending cassava root shelf life via reduction of reactive oxygen species production. Plant Physiology. 2012;**159**(4):1396-1407. DOI: 10.1104/pp.112.200345

[25] Apostol I, Heinstein PF, Low PS. Rapid stimulation of an oxidative burst during elicitation of cultured plant cells: Role in defense and signal transduction. Plant Physiology. 1989;**90**(1):109-116

[26] Sanchez T, Dufour D, Moreno IX, Ceballos H. Comparison of pasting and gel stabilities of waxy and normal starches from potato, maize, and rice with those of a novel waxy cassava starch under thermal, chemical, and mechanical stress. Journal of Agricultural and Food Chemistry. 2010;**58**(8):5093-5099. DOI: 10.1021/jf1001606

[27] Masinde EA, Ogendo JO, Maruthi MN, Hillocks R, Mulwa RMS, Arama PF. Occurrence and estimated losses caused by cassava viruses in Migori county, Kenya. African Journal of Agricultural Research. 2016;**11**:2064-2074

[28] Patil BL, Fauquet CM. Cassava mosaic geminiviruses: Actual knowledge and perspectives. Molecular Plant Pathology. 2009;**10**(5):685-701. DOI: 10.1111/j.1364-3703.2009.00559.x

[29] Olsen K, Schaal B. Evidence on the origin of cassava: Phylogeography of *Manihot esculenta*. Proceedings of the National Academy of Sciences of the United States of America. 1999;**96**(10): 5586-5591

[30] De Bruyn A, Villemot J, Lefeuvre P, Villar E, Hoareau M, Harimalala M, Abdoul-Karime AL, Abdou-Chakour C, Reynaud B, Harkins GW, et al. East African cassava mosaic-like viruses from Africa to Indian ocean islands: Molecular diversity, evolutionary history and geographical dissemination of a bipartite begomovirus. BMC Evolutionary Biology. 2012;**12**:228. DOI: 10.1186/1471-2148-12-228

[31] Storey HH. 1939 Report of the Plant Pathologist. East African Agricultural Research Station; 1939. p. 9

[32] Nichols RFW. The brown streak disease of cassava: Distribution, climatic effects and diagnostic symptoms. The East African Agricultural Journal. 1950;**15**:154-160

[33] Storey HH. Virus diseases of East African plants. VI. A progress report on studies of the disease of cassava. The East African Agricultural Journal. 1936;**2**:34-39

[34] Hillocks RJ, Jennings DL. Cassava brown streak disease: A review of present knowledge and research needs. International Journal of Pest Management. 2003;**49**:225-234

[35] Patil BL, Ogwok E, Wagaba H, Mohammed IU, Yadav JS, Bagewadi B, Taylor NJ, Kreuze JF, Maruthi MN, Alicai T, Fauquet CM. RNAi-mediated resistance to diverse isolates belonging to two virus species involved in Cassava brown streak disease. Molecular Plant Pathology. 2011;**12**(1):31-41. DOI: 10.1111/j.1364-3703.2010.00650.x

[36] Legg JP, Jeremiah SC, Obiero HM, Maruthi MN, Ndyetabula I, Okao-Okuja G, Bouwmeester H, Bigirimana S, Tata-Hangy W, et al. Comparing the regional epidemiology of the cassava mosaic and cassava brown streak virus pandemics in Africa. Virus Research. 2011;**159**(2):161-170. DOI: 10.1016/j.virusres.2011.04.018

[37] Hillocks RJ, Raya MD, Mtunda K, Kiozia H. Effects of brown streak disease on yield and quality of cassava in Tanzania. Journal of Phytopathology. 2001;**149**:389-394

[38] Mansfield J, Genin S, Magori S, Citovsky V, Sriariyanum M, Ronald P, Dow M, Verdier V, Beer SV, Machado MA, Toth I, Salmond G, Foster GD. Top 10 plant pathogenic bacteria in molecular plant pathology. Molecular Plant Pathology. 2012;**13**(6):614-629. DOI: 10.1111/j.1364-3703.2012.00804.x

[39] Wydra K, Verdier V. Occurrence of cassava diseases in relation to environmental, agronomic and plant characteristics. Agriculture, Ecosystems and Environment. 2002;**93**:211-226

[40] Thresh JM, Cooter RJ. Strategies for controlling cassava mosaic virus disease in Africa. Review article. Plant Pathology. 2005;**54**:587-614

[41] Fondong VN. The search for resistance to cassava mosaic geminiviruses: How much we have accomplished, and what lies ahead. Frontiers in Plant Science. 2017;**8**:408. DOI: 10.3389/fpls.2017.00408

[42] Vanderschuren H, Akbergenov R, Pooggin M, Hohn T, Gruissem W, Zhang P. Transgenic cassava resistance to *African cassava mosaic virus* is enhanced by viral DNA-A bidirectional promoterderived siRNAs. Plant Molecular Biology. 2007;**64**(5):549-557

[43] Turyagyenda LF, Kizito EB, Ferguson M, Baguma Y, Agaba M, Harvey JJ, Osiru DS. Physiological and molecular characterization of drought responses and identification of candidate tolerance genes in cassava. AoB Plants. 2013;**5**:plt007. DOI: 10.1093/aobpla/plt007

[44] Bakayoko S, Tschannen A, Nindjin C, Dao D, Girardin O, Assa A. Impact of water stress on fresh tuber yield and dry matter content of cassava (*Manihot esculenta* Crantz) in Côte d'Ivoire. African Journal of Agricultural Research. 2009;**4**(1):021-027

[45] Perez JC, Lenis JI, Calle F, Morante N, Sanchez T, Debouck D, Ceballos H. Genetic variability of root peel thickness and its influence in extractable starch from cassava (*Manihot esculenta* Crantz) roots. Plant Breeding. 2011;**130**(6):688-693. DOI: 10.1111/j.1439-0523.2011.01873.x

[46] Zhang P, Wang WQ, Zhang GL, Kaminek M, Dobrev P, Xu J, Gruissem W. Senescence-inducible expression of isopentenyl transferase extends leaf life, increases drought stress resistance and alters cytokinin metabolism in cassava. Journal of Integrative Plant Biology. 2010;**52**(7):653-669. DOI: 10.1111/j.1744-7909.2010.00956.x

[47] Lenis JI, Calle F, Jaramillo G, Perez JC, Ceballos H, Cock JH. Leaf retention and cassava productivity. Field Crops Research. 2006;**95**:126-134

[48] Ballén-Taborda C, Plata G, Ayling S, Rodríguez-Zapata F, Becerra Lopez-Lavalle LA, Duitama J, et al. Identification of cassava microRNAs under abiotic stress. International Journal of Genomics. 2013;**53**:257-269. DOI: 10.1155/2013/857986

[49] Wesley SV, Helliwell CA, Smith NA, Wang MB, Rouse DT, Liu Q, Gooding PS, Singh SP, Abbott D, Stoutjesdijk PA, et al. Construct design for efficient, effective and high-throughput gene silencing in plants. The Plant Journal. 2001;**27**(6):581-590

[50] Helliwell C, Waterhouse P. Constructs and methods of high-throughput gene silencing in plants. Methods. 2003;**30**(4):289-295

[51] Serio FD, Schob H, Iglesias A, Tarina C, Bouldoires E, Meins FJ. Sense- and antisense-mediated gene silencing in tobacco is inhibited by the same viral suppressors and is associated with accumulation of small RNAs. Proceedings of the National Academy of Sciences of the United States of America. 2001;**98**(11):6506-6510

[52] Dogar AM. RNAi dependent epigenetic marks on a geminivirus promoter. Virology Journal. 2006;30;**3**:5

[53] Vanderschuren H, Alder A, Zhang P, Gruissem W. Dose-dependent RNAi-mediated geminivirus resistance in the tropical root crop cassava. Plant Molecular Biology. 2009;**70**(3):265-272. DOI: 10.1007/s11103-009-9472-3

[54] Ntui VO, Kong K, Khan RS, Igawa T, Janavi GJ, Rabindran R, Nakamura I, Mii M. Resistance to Sri Lankan cassava mosaic virus (SLCMV) in genetically engineered cassava cv. KU50 through RNA silencing. PLoS One. 2015;**10**(4):e0120551. DOI: 10.1371/journal.pone.0120551

[55] Yadav JS, Ogwok E, Wagaba H, Patil BL, Bagewadi B, Alicai T, Gaitan-Solis E, Taylor NJ, Fauquet CM. RNAi-mediated resistance to Cassava brown streak Uganda virus in transgenic cassava. Molecular Plant Pathology. 2011;**12**(7):677-687. DOI: 10.1111/j.1364-3703.2010.00700.x

[56] Odipio J, Ogwok E, Taylor NJ, Halsey M, Bua A, Fauquet CM, Alicai T. RNAi-derived field resistance to Cassava brown streak disease persists across the vegetative cropping cycle. GM Crops & Food. 2014;**5**(1):16-19. DOI: 10.4161/gmcr.26408

[57] Upadhyay SK, Chandrashekar K, Thakur N, Verma PC, Borgio JF, Singh PK, Tuli R. RNA interference for the control of whiteflies (*Bemisia tabaci*) by oral route. Journal of Biosciences. 2011;**36**(1):153-161

[58] Malik HJ, Raza A, Amin I, Scheffler JA, Scheffler BE, Brown JK, Mansoor S. RNAi-mediated mortality of the whitefly through transgenic expression of double stranded RNA homologous to acetylcholinesterase and ecdysone receptor in tobacco plants. Scientific Reports. 2016;**6**:38469. DOI: 10.1038/srep38469

[59] Fondong VN, Nagalakshmi U, Dinesh-Kumar SP. Novel functional genomics approaches: A promising future in the combat against plant viruses. Phytopathology. 2016;**106**(10):1231-1239

[60] Voinnet O. Origin, biogenesis, and activity of plant microRNAs. Cell. 2009;**136**(4):669-687. DOI: 10.1016/j.cell.2009.01.046

[61] Xie M, Zhang S, Yu B. microRNA biogenesis, degradation and activity in plants. Cellular and Molecular Life Sciences. 2015;**72**(1):87-99. DOI: 10.1007/s00018-014-1728-7

[62] Li JF, Chung HS, Niu Y, Bush J, McCormack M, Sheen J. Comprehensive protein-based artificial microRNA screens for effective gene silencing in plants. The Plant Cell. 2013;**25**(5):1507-1522. DOI: 10.1105/tpc.113.112235

[63] Fahim M, Larkin J. Designing effective amiRNA and multimeric amiRNA against plant viruses. Methods in Molecular Biology. 2013;**942**:357-377. DOI: 10.1007/978-1-62703-119-6_19

[64] Khraiwesh B, Ossowski S, Weigel D, Reski R, Frank W. Specific gene silencing by artificial MicroRNAs in *Physcomitrella patens*: An alternative to targeted gene knockouts. Plant Physiology. 2008;**148**(2):684-693. DOI: 10.1104/pp.108.128025

[65] Ossowski S, Schneeberger K, Clark RM, Lanz C, Warthmann N, Weigel D. Sequencing of natural strains of *Arabidopsis thaliana* with short reads. Genome Research. 2008;**18**(12):2024-2033. DOI: 10.1101/gr.080200.108

[66] Schwab R, Ossowski S, Riester M, Warthmann N, Weigel D. Highly specific gene silencing by artificial microRNAs in Arabidopsis. The Plant Cell. 2006;**18**(5):1121-1133

[67] Wagaba H, Patil BL, Mukasa S, Alicai T, Fauquet CM, Taylor NJ. Artificial microRNA-derived resistance to Cassava brown streak disease. Journal of Virological Methods. 2016;**231**:38-43. DOI: 10.1016/j.jviromet.2016.02.004

[68] Szittya G, Silhavy D, Molnar A, Havelda Z, Lovas A, Lakatos L, Banfalvi Z, Burgyan J. Low temperature inhibits RNA silencing-mediated defence by the control of siRNA generation. The EMBO Journal. 2003;**22**(3):633-640

[69] Fahim M, Millar AA, Wood CC, Larkin PJ. Resistance to Wheat streak mosaic virus generated by expression of an artificial polycistronic microRNA in wheat. Plant Biotechnology Journal. 2012;**10**(2):150-163. DOI: 10.1111/j.1467-7652.2011.00647.x

[70] Kung YJ, Lin SS, Huang YL, Chen TC, Harish SS, Chua NH, Yeh SD. Multiple artificial microRNAs targeting conserved motifs of the replicase gene confer robust transgenic resistance to negative-sense single-stranded RNA plant virus. Molecular Plant Pathology. 2012;**13**(3):303-317. DOI: 10.1111/j.1364-3703.2011.00747.x

[71] Allen E, Howell MD. miRNAs in the biogenesis of trans-acting siRNAs in higher plants. Seminars in Cell & Developmental Biology. 2010;**21**(8):798-804. DOI: 10.1016/j.semcdb.2010.03.008

[72] Yoshikawa M. Biogenesis of trans-acting siRNAs, endogenous secondary siRNAs in plants. Genes & Genetic Systems. 2013;**88**(2):77-84

[73] Fei Q, Xia R, Meyers BC. Phased, secondary, small interfering RNAs in posttranscriptional regulatory networks. The Plant Cell. 2013;**25**(7):2400-2415. DOI: 10.1105/tpc.113.114652

[74] Singh A, Taneja J, Dasgupta I, Mukherjee SK. Development of plants resistant to tomato geminiviruses using artificial trans-acting small interfering RNA. Molecular Plant Pathology. 2015;**16**(7):724-734. DOI: 10.1111/mpp.12229

[75] Chen L, Cheng X, Cai J, Zhan L, Wu X, Liu Q, Wu X. Multiple virus resistance using artificial trans-acting siRNAs. Molecular Plant Pathology. 2015;**16**(7):724-734. DOI: 10.1111/mpp.12229

[76] Quintero A, Pérez-Quintero AL, López C. Identification of ta-siRNAs and cis-nat-siRNAs in cassava and their roles in response to cassava bacterial blight. Genomics, Proteomics & Bioinformatics. 2013;**11**(3):172-181. DOI: 10.1016/j.gpb.2013.03.001

[77] Henikoff S, Till BJ, Comai L. TILLING. Traditional mutagenesis meets functional genomics. Plant Physiology. 2004;**135**(2):630-636

[78] Kurowska M, Daszkowska-Golec A, Gruszka D, Marzec M, Szurman M, Szarejko I, Maluszynski M. TILLING: A shortcut in functional genomics. Journal of Applied Genetics. 2011;**52**(4):371-390. DOI: 10.1007/s13353-011-0061-1

[79] McCallum CM, Comai L, Greene EA, Henikoff S. Targeting induced local lesions IN genomes (TILLING) for plant functional genomics. Plant Physiology. 2000;**123**:439-442

[80] Colasuonno P, Incerti O, Lozito ML, Simeone R, Gadaleta A, Blanco A. DHPLC technology for high-throughput detection of mutations in a durum wheat TILLING population. BMC Genetics. 2016;**17**:43. DOI: 10.1186/s12863-016-0350-0

[81] Harloff H, Lemcke S, Mittasch J, Frolov A, Wu J, Dreyer F, et al. A mutation screening platform for rapeseed (*Brassica napus* L.) and the detection of sinapine biosynthesis mutants. Theoretical and Applied Genetics. 2012;**124**(5):957-969. DOI: 10.1007/s00122-011-1760-z

[82] Wang N, Wang Y, Tian F, King G, Zhang C, Long Y, et al. A functional genomics resource for *Brassica napus*: Development of an EMS mutagenized population and discovery of *FAE1* point mutations by TILLING. The New Phytologist. 2008;**180**(4):751-765. DOI: 10.1111/j.1469-8137.2008.02619.x

[83] Stephenson P, Baker D, Girin T, Perez A, Amoah S, King J, et al. A rich TILLING resource for studying gene function in *Brassica rapa*. BMC Plant Biology. 2010;**10**:62. DOI: 10.1186/1471-2229-10-62

[84] Comai L, Young K, Till BJ, Reynolds SH, Greene EA, Codomo CA, Enns LC, Johnson J, Burtner C, Oden AR, et al. Efficient discovery of DNA polymorphisms in natural populations by Ecotilling. The Plant Journal. 2004;**37**(5):778-786

[85] Ceballos H, Okogbenin E, Pérez JC, Becerra LA, Debouck D. Cassava. In: Bradshaw JE, editor. Root and Tuber Crops. New York: Springer Publishers; 2010. p. 53-96

[86] Lehman U, Robin F. Slowly digestible starch – Its structure and health implications: A review. Trends in Food Science and Technology. 2007;**18**:346-355

[87] Esfeld K, Uauy C, Tadele Z. Application of TILLING for Orphan Crop Improvement. In: Jain SM, Gupta SD, editors. Biotechnology of Neglected and Underutilized Crops. Dordrecht: The Netherlands Springer; 2013. pp. 83-113

[88] Tadele Z, Esfeld K, Plaza S. Applications of high-throughput techniques to the understudied crops of Africa. Aspects of Applied Biology. 2010;**96**:233-240

[89] Ülker B, Weisshaar B. Resources for reverse genetics approaches in *Arabidopsis thaliana*. In: Schmidt R, Bancroft I, editors. Genetics and Genomics of the Brassicaceae. Plant Genetics and Genomics: Crops and Models, New York, NY: Springer; 2011. pp. 527-560

[90] Barrangou R. The roles of CRISPR-Cas systems in adaptive immunity and beyond. Current Opinion in Immunology. 2015;**32**:36-41. DOI: 10.1016/j.coi.2014.12.008

[91] Barrangou R, Marraffini LA. CRISPR-Cas systems: Prokaryotes upgrade to adaptive immunity. Molecular Cell. 2014;**54**(2):234-244. DOI: 10.1016/j.molcel.2014.03.011

[92] Cong L, Ran FA, Cox D, Lin S, Barretto R, Habib N, Hsu PD, Wu X, Jiang W, Marraffini LA, et al. Multiplex genome engineering using CRISPR/Cas systems. Science. 2013;**339**(6121):819-823. DOI: 10.1126/science.1231143

[93] Jinek M, East A, Cheng A, Lin S, Ma E, Doudna J. RNA-programmed genome editing in human cells. eLife. 2013;**2**:e00471. DOI: 10.7554/eLife.00471

[94] Sternberg SH, Doudna JA. Expanding the biologist's toolkit with CRISPR-Cas9. Molecular Cell. 2015;**58**(4):568-574. DOI: 10.1016/j.molcel.2015.02.032

[95] Wright AV, Nunez JK, Doudna JA. Biology and applications of CRISPR systems: Harnessing nature's toolbox for genome engineering. Cell. 2016;**164**(1-2):29-44. DOI: 10.1016/j.cell.2015.12.035

[96] Paul JW 3rd, Qi Y. CRISPR/Cas9 for plant genome editing: Accomplishments, problems and prospects. Plant Cell Reports. 2016;**35**(7):1417-1427. DOI: 10.1007/s00299-016-1985-z

[97] Odipio J, Alicai T, Nusinow D, Bart R, Ingelbrecht I, Taylor N. CRISPR/CAS9 genome editing of putative cassava flowering genes. In: Third International Conference of the Global Cassava Partnership for the 21st Century; January 18-23, 2016; Nanning, China; 2016

[98] Yin K, Han T, Liu G, Chen T, Wang Y, Yu AY, et al. A geminivirus-based guide RNA delivery system for CRISPR/Cas9 mediated plant genome editing. Scientific Reports. 2015;**5**:14926. DOI: 10.1038/srep14926

[99] Hsu PD, Scott DA, Weinstein JA, Ran FA, Konermann S, Agarwala V, Li Y, Fine EJ, Wu X, Shalem O, Cradick TJ, Marraffini LA, Bao G, Zhang F. DNA targeting specificity of RNA-guided Cas9 nucleases. Nature Biotechnology. 2013;**31**(9):827-832. DOI: 10.1038/nbt.2647

[100] Pattanayak V, Lin S, Guilinger JP, Ma E, Doudna JA, Liu DR. High-throughput profiling of off-target DNA cleavage reveals RNA-programmed Cas9 nuclease specificity. Nature Biotechnology. 2013;**31**(9):839-843. DOI: 10.1038/nbt.2673

[101] Jiang Y, Chen B, Duan C, Sun B, Yang J, Yang S. Multigene editing in the *Escherichia coli* genome via the CRISPR-Cas9 system. Applied and Environmental Microbiology. 2015;**81**(7):2506-2514. DOI: 10.1128/AEM.04023-14

[102] Samanta MK, Dey A, Gayen S. CRISPR/Cas9: An advanced tool for editing plant genomes. Transgenic Research. 2016;**25**(5):561-73. DOI: 10.1007/s11248-016-9953-5

[103] Verdaguer B, de Kochko A, Beachy RN, Fauquet C. Isolation and expression in transgenic tobacco and rice plants, of the cassava vein mosaic virus (CVMV) promoter. Plant Molecular Biology. 1996;**31**(6):1129-1139

[104] Legg JP, Lava Kumar P, Makeshkumar T, Tripathi L, Ferguson M, Kanju E, Ntawuruhunga P, Cuellar W. Cassava virus diseases: Biology, epidemiology, and management. Advances in Virus Research. 2015;**91**:85-142. DOI: 10.1016/bs.aivir.2014.10.001

[105] Legg JP, Lava Kumar P, Makeshkumar T, Tripathi L, Ferguson M, Kanju E, Ntawuruhunga P, Cuellar W. Cassava virus diseases: Biology, epidemiology, and management. Advances in Virus Research. 2015;**91**:85-142. DOI: 10.1016/bs.aivir.2014.10.001

[106] Lee W-S, Fu S-F, Li Z, Murphy AM, Dobson EA, Garland L, Chaluvadi SR, Lewsey MG, Nelson RS, Carr JP. Salicylic acid treatment and expression of an RNA-dependent RNA polymerase 1 transgene inhibit lethal symptoms and meristem invasion during tobacco

mosaic virus infection in *Nicotiana benthamiana*. BMC Plant Biology. 2016;**16**:15. DOI: 10.1186/s12870-016-0705-8

[107] Wei Y, Hu W, Wang Q, Liu W, Wu C, Zeng H, Yan Y, Li X, He C, Shi H. Comprehensive transcriptional and functional analyses of melatonin synthesis genes in cassava reveal their novel role in hypersensitive-like cell death. Scientific Reports. 2016;**6**:35029. DOI: 10.1038/srep35029

[108] Bastet A, Robaglia C, Gallois JL. eIF4E resistance: Natural variation should guide gene editing. Trends in Plant Science. 2017;**22**(5):411-419. DOI: 10.1016/j.tplants.2017.01.008

[109] Allie F, Pierce EJ, Okoniewski MJ, Rey C. Transcriptional analysis of *South African cassava mosaic virus*-infected susceptible and tolerant landraces of cassava highlights differences in resistance, basal defense and cell wall associated genes during infection. BMC Genomics. 2014;**15**:1006. DOI: 10.1186/1471-2164-15-1006

[110] Maruthi M, Bouvaine S, Tufan HA, Mohammed IU, Hillocks RJ. Transcriptional response of virus-infected cassava and identification of putative sources of resistance for cassava brown streak disease. PLoS One. 2014;**9**(5):e96642. DOI: 10.1371/journal.pone.0096642

[111] Pierce EJ, Rey MEC. Assessing global transcriptome changes in response to *South African cassava mosaic virus* [ZA-99] infection in susceptible *Arabidopsis thaliana*. PLoS One. 2013;**8**(6):e67534. DOI: 10.1371/journal.pone.0067534

[112] Allie F, Rey MEC. Transcriptional alterations in model host, *Nicotiana benthamiana*, in response to infection by *South African cassava mosaic virus*. European Journal of Plant Pathology. 2013;**137**:765-785

[113] Maule AJ, Caranta C, Boulton MI. Sources of natural resistance to plant viruses: Status and prospects. Molecular Plant Pathology. 2007 Mar;**8**(2):223-231. DOI: 10.1111/j.1364-3703.2007.00386.x

[114] Soto JC, Ortiz JF, Perlaza-Jiménez L, Vásquez AX, Lopez-Lavalle LA, Mathew B, Léon J, Bernal AJ, Ballvora A, López CE. A genetic map of cassava (*Manihot esculenta* Crantz) with integrated physical mapping of immunity-related genes. BMC Genomics. 2015;**16**:190. DOI: 10.1186/s12864-015-1397-4

[115] Wolfe MD, Kulakow P, Rabbi IY, Jannink J-L. Marker-based estimates reveal significant non-additive effects in clonally propagated cassava (*Manihot esculenta*): Implications for the prediction of total genetic value and the selection of varieties. G3 (Bethesda). 2016;**6**(11):3497-3506. DOI: 10.1534/g3.116.033332

[116] Lozano R, Hamblin MT, Prochnik S, Jannink JL. Identification and distribution of the NBS-LRR gene family in the Cassava genome. BMC Genomics. 2015;**16**:360. DOI: 10.1186/s12864-015-1554-9

[117] Zhai J, Jeong DH, De Paoli E, Park S, Rosen BD, Li Y, Gonzalez AJ, Yan Z, Kitto SL, Grusak MA, et al. MicroRNAs as master regulators of the plant *NB-LRR* defense gene family via the production of phased, *trans*-acting siRNAs. Genes & Development. 2011;**25**(23):2540-2553. DOI: 10.1101/gad.177527.111

[118] Leal LG, Perez Á, Quintero A, Bayona Á, Ortiz JF, Gangadharan A, Mackey D, López C, López-Kleine L. Identification of immunity-related genes in Arabidopsis and cassava using genomic data. Genomics, Proteomics & Bioinformatics. 2013;**11**(6):345-353. DOI: 10.1016/j.gpb.2013.09.010

[119] Walling LL. Avoiding effective defenses: Strategies employed by phloem-feeding insects. Plant Physiology. 2008;**146**(3):859-866. DOI: 10.1104/pp.107.113142

[120] Wu J, Baldwin IT. New insights into plant responses to the attack from insect herbivores. Scientific Reports. 2017;**7**:44729. DOI: 10.1038/srep44729

[121] Shukla AK, Upadhyay SK, Mishra M, Saurabh S, Singh R, Singh H, Thakur N, Rai P, Pandey P, Hans AL, et al. Expression of an insecticidal fern protein in cotton protects against whitefly. Nature Biotechnology. 2016;**34**(10):1046-1051. DOI: 10.1038/nbt.3665

Chapter 10

Soil-Water-Crop Relationship: A Case Study of Cassava in the Tropics

Saurau O. Oshunsanya and Nkem J. Nwosu

Additional information is available at the end of the chapter

http://dx.doi.org/10.5772/intechopen.71968

Abstract

Cassava is the most important food crop in Africa occupying about 6 million hectares (ha). Several factors have limited the continuous and sustainable production of cassava in tropical Africa. Some of these factors include (but not necessarily limited to) soil and water, which are the two basic fundamental resources for cassava production. The demand on soil and water resources is increasing, especially for new and conflicting soil functions like enhancing crop production, improving water quality and mitigating climate change. Soil–plant-water relations relate to the physical properties of soil and plants that affect the movement, retention and use of water. This chapter reviewed the soil, water and plant relationship for cassava production in tropical Nigeria. The study observed that understanding the effects of soil quality and water characteristics on cassava production and its management as well as the relationship between soil, water and crop for sustainable optimum cassava production is highly imperative now than ever before, especially in developing countries of Africa (like Nigeria) that are characterised by high risks of soil degradation, rising populations and pressure on agricultural lands juxtaposed with predominant resource—poor and small landholders.

Keywords: cassava, soil quality, water characteristics, sustainable production, tropics

1. Introduction

Soil is the Earth's fragile skin that anchors all life comprising countless species that create a dynamic and complex ecosystem. Soil is a major component of the environmental system. It is a major resource of the earth with a lot of potentials. Generally, soil has been described as the basis of human civilisation. This is because soil supports plants, which provide nutrition for man and his livestock [1]. The ability of the soil to continue providing essential services in the face of disturbance, whether natural or human induced, is essential to maintain or improve crop production overtime. Crop production is a function of the soil, management and the environment (of which water availability and quality are important indices). The quality of tropical soils, water quality and the synergy between them is highly imperative for optimum crop production. The availability of both water and plant nutrients is largely controlled by the physico-chemical properties, micro-environment of soils, the success and failure of any species of a particular area are, therefore, governed by the quality of soil characteristics [2].

Soil quality is the ability of a soil to perform functions that are essential to people and the environment [3]. It is the capacity of soil to perform specific functions of interest to human [4]. Soil quality has hitherto been equated with agricultural productivity. Soil conservation practices geared to maintain soil productivity are as old as agriculture itself. Soil quality is implied in many decisions farmers make about land purchases and management, and in the economic value rural assessors place on agricultural land for purposes of taxation. The concern of soil quality has challenged human kind for over 10,000 years; the definition and the basic concept remain a work in progress and keep evolving with every generation [3].

Similarly, the importance of water in any crop production cannot be overemphasised. Cassava requires a certain amount of water daily to meet its crop water requirements [5]. Water is regarded as an intrinsic factor in any crop production as their function in food production, nutrient mineralisation and distribution, as well as plant turgidity in every crop production cannot be extremely emphasised. Water is a major constituent of living plant tissues, which consist of about 90% water. Water is involved in most of the physical, chemical and biological processes that occur in soils. In addition, all biological processes within the living plants depend on water. The optimal moisture conditions for any crop vary depending on many factors, such as soil type, climate conditions, growth rate and habit [5]. The favourable soil moisture tension should be maintained throughout the entire growth period of plants due to the relationship between evapotranspiration and biomass production.

Cassava (Manihot esculenta Crantz) has been identified as one of the most important food crops in Africa. It derives its importance from the fact that it is starchy and thickened, with tuberous roots that serve as valuable sources of cheap calories, especially in developing countries with widespread calorie deficiency and malnutrition. In many parts of Africa, the leaves and tender shoots of cassava are also consumed as vegetables [6]. Over two-thirds of the total production of cassava is consumed in various forms by humans. The International Food Policy Research Institute (IFPRI) in 2014 noted that cassava is an insurance crop that increases food security because they can be left in the ground until needed [7]; and their usage as a source of ethanol for fuel, energy in animal feed and starch for industry is increasing. The crop is amenable to agronomic as well as genetic improvement juxtaposed with a high yield potential under good conditions and performs better than other crops under suboptimal conditions. It is grown widely in several countries in sub-Saharan Africa and Madagascar. Cassava was introduced into Africa in the latter half of the sixteenth century from South America and, perhaps, also from Central America, where it is believed to have originated. Globally, there has been widespread production of cassava across continents. Thailand, Vietnam, Indonesia and Costa Rica have been reported as the world leading exporters of cassava [8].

Although Nigeria is the world's largest producer of cassava (**Figure 1**), however, its exports have been progressively reduced by the brunt population increase in the nation, implying

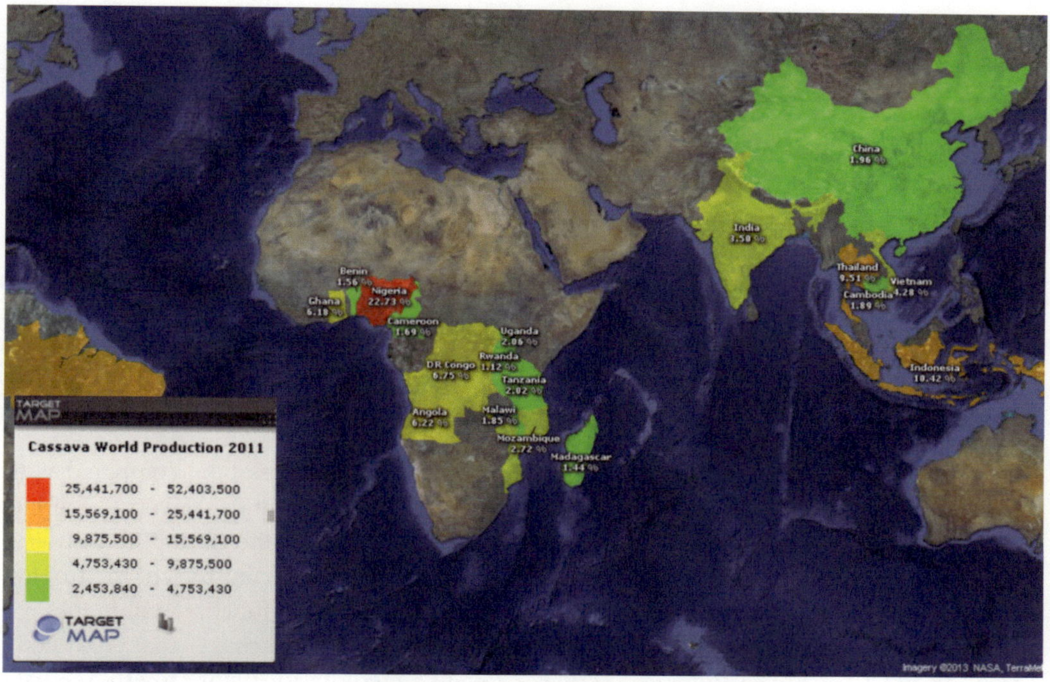

Figure 1. World cassava production in 2011 (source: www.targetmap.com).

that production is not enough to feed the current population in the country. The African Development Bank (AfDB) reported in 2015 that although Nigeria produces 20% of the world's cassava, the country exports less than 1% of its produce. It was concluded that cassava production in the tropical Nigeria needs to increase with the rising population and this has been significantly influenced by the changing climate and the poor soil quality conditions arising due to soil degradation [8]. This invariably laid credence to the pertinence of galvanising strategies for boosting cassava production and export in the tropical Africa. This review will however consider soil quality, water characteristics and their synergy for suitable and optimum production of cassava in tropical soils.

2. Soil quality and cassava production

The demand on soil resources in increasing, especially for new and conflicting soil functions like enhancing food security, improving water quality, disposing urban and industrial wastes and mitigating climate change. Thus, soil quality and its management are more important now than ever before, especially in developing countries that are characterised by high risks of soil degradation, predominantly resource—poor and small landholders [4].

Soil quality is the capacity of soil to perform specific functions of interest to human. Soil quality has historically been equated with agricultural productivity. Soil conservation practices to maintain soil productivity are as old as agriculture itself. Soil quality is implied in many decisions farmers make about land purchases and management and in the economic value rural assessors place on agricultural land for purposes of taxation. The concern of soil quality has challenged human kind for at least 10,000 years; the definition and the basic concept remain a work in progress and keep evolving with every generation.

Furthermore, soil quality is defined as the ability of a soil to perform functions that are essential to people and the environment [9]. Soil quality is not limited to agricultural soils. The first step in science of agriculture is the recognition of soils and of how to distinguish that which is of good quality and that which is of inferior quality. However, in spite of numerous definitions of soil quality, reviewed reports suggest that the widely accepted definition of the concept of soil quality was laid down by the Soil Science Society of America (SSSA) in 1996 which states that soil quality is the capacity of a specific kind of soil to function within natural or managed ecosystem boundaries, sustain plant and animal productivity, maintain or enhance water and air quality and support human health and habitation.

Soil functions keep changing with time and are different with developing compared with developed countries [4]. Although the definition of soil quality may be universal, the fact that its application is soil/society specific needs to be recognised for the concept to be useful in addressing the problems of resource-poor farmers in developing countries of the tropics. Larson and Pierce [10] defined soil quality as the capacity of a soil to function within the ecosystem boundaries and interact positively with the environment external to that ecosystem. Three soil functions are considered essential: provide a medium for plant growth, regulate and partition water flow through the environment and serve as an effective environmental filter. However, no soil is likely to successfully provide all these functions, some of which occur in natural ecosystems and some of which are the result of human modification. Hence, soil quality depends on the extent to which soil functions to benefit humans.

The qualities of tropical soils are imperative indices towards the sustainable production of cassava. Cassava is known to be a heavy feeder, and literatures have opined that more than average output is obtainable from cassava grown on marginal lands. However, in view of the ever-increasing population of the tropical Africa (most notably in Nigeria), and with the production rates seldom meeting the increasing market demands of the produce, production of cassava on quality soils is therefore an imperative factor which when juxtaposed with good management and adequate climatic conditions, the production of cassava can be improved.

More so, speaking in an interview in 2016, the Provost of the Federal College of Agriculture, Akure (FECA), Dr. Samson Odedina, while demonstrating the profitability of cassava production enterprise to young people and emerging farmers noted that farmers obtain an average yield of 8–10 tonnes of cassava per hectare, adding that the yield is far below the potentials of the crop. He further stressed that if the soil conditions are well managed, farmers can get up to 50–60 tonnes per hectare if they follow the recommended soil management practices. This increasingly justifies the pertinence of the quality of soils used for cassava production in tropical Nigeria.

In addition, the FAO in 2017 reported that cassava has the reputation of causing serious erosion when grown on sloppy soils [11]. Researchers have also argued that this reputation is undeserved, since cassava is often grown on already-eroded soils where few other crops can survive and be productive. Nevertheless, concise reviews of related literatures generally maintained that cassava production on slopes causes increased erosion on an annual basis than

http://dx.doi.org/10.5772/intechopen.71968

other crops grown under the same circumstances. Cassava, in conjunction with common bean (Phaseolus vulgaris), upland rice and cotton, tends to cause considerably more erosion than cereals (like maize), peanut, sugarcane, pineapple or sweet potato. This was predominantly attributed to the fact that cassava needs to be cultivated at a relatively wide spacing. The canopy formation and the initial growth phase are slow, leaving the soil exposed to the direct impact of rainfall during the first 3–4 months of the cropping season. Contrarily, once the crop canopy is closed, erosion is usually minimal during the remainder of the crop cycle (**Figure 2**).

The soil condition used for the production of cassava in Nigeria is of utmost importance if the demands for the produce are to be met before the year 2030 (**Figure 3**). For good growth of cassava, the soil used for production must have adequate room for water and air movement

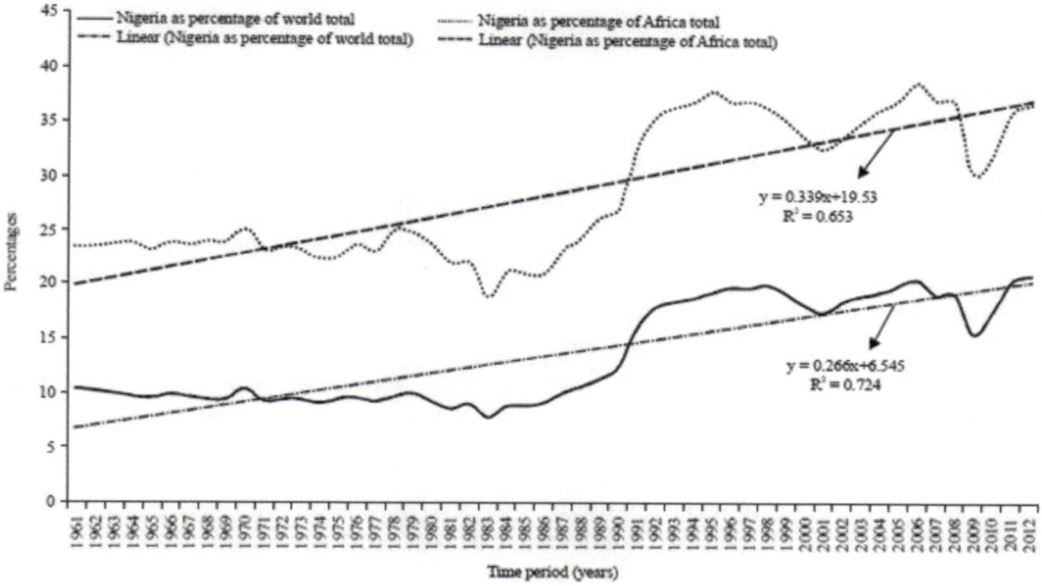

Figure 2. Nigeria's cassava production in the world and Africa totals, 1961–2012 [12].

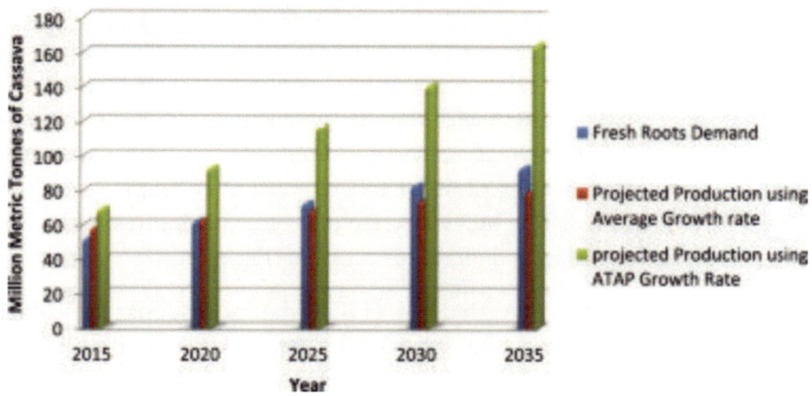

Figure 3. Cassava demand and supply projections in Nigeria [13].

and for root growth. Also, the rising pressure on agricultural lands has made it difficult to obtain high-quality lands for sustainable production of deep-rooted crops like cassava. The insurgence of climate change and its effects on tropical soils has also increased this malady. These, therefore, lay emphasis on the true need of establishing soil management techniques aimed at boosting soil physical, chemical and biological conditions—which are main indices for soil quality towards the optimum production of cassava in this high-demand region of West Africa. Hence, based on reviewed statistics, cassava production in Nigeria will increase greatly with optimum soil, environmental and management conditions.

3. Soil quality assessment for cassava production

Soil is likely to show great variability in their physical, chemical and biological properties because the soil is a heterogeneous unit. Knowledge of variability of soil properties is highly indispensable as this can affect crop yield. A study of the variability trends of soils is essential in order to highlight the soil potentials and enhance their management and productivity [14].

Anikwe et al. [15] noted that the spatial variability of soil properties has effects on crop production across an agricultural field. They emphasised that it is important to be aware of the effect of spatial variability of soil properties when choosing indicator variables of soil quality for crop production. Although when, how and where to collect soil samples for soil quality determination may differ according to the objective of the assessment being made, management history and current inputs should also be considered to ensure valid interpretation of the information.

Soil quality assessment for agricultural production is an important operation towards sustainable crop and livestock production in tropical Africa. Owing to the high degree of variability that is characteristic of tropical soils, there is a need for the assessment of soil condition and capability to offer suitable crop outputs. Fundamentally, soil productivity for cassava production is a function of soil quality and management. Soil quality assessment is the process of measuring the management-induced changes in soil as we attempt to get soil to do what we want it to do. The ultimate purpose of assessing soil quality is to provide the information necessary to protect and improve long-term agricultural productivity, water quality and habitats of all organisms including people [16].

Basically, Soil Science Society of America [3] reported that soil quality is an inherent attribute of a soil that is inferred from soil characteristics or indirect observations. Papendick et al. [17] suggested that a minimum data set (MDS) of soil characteristics representing soil quality must be selected and quantified. The MDS may include biological, physical or chemical soil characteristics otherwise known as soil quality indicators (**Figure 4**).

Furthermore, the US Department of Agriculture [19] defined soil quality indicators as physical, chemical or biological properties, processes and characteristics that can be measured to monitor changes in the soil. The types of indicators that are the most useful depend on the function of soil for which soil quality is being evaluated. Sojka and Upchurch [20] highlighted that while recognising some controversies about the basic concept of soil quality, considerable

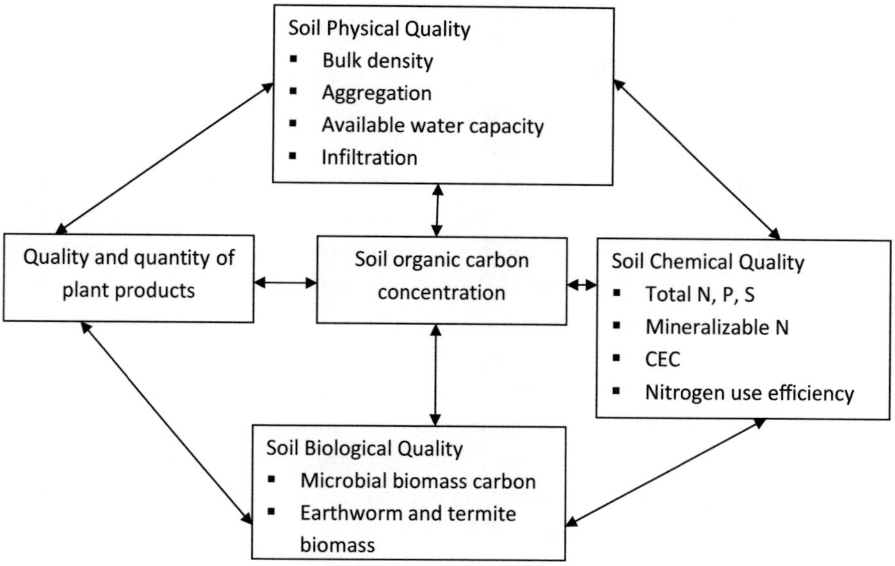

Figure 4. Showing key indicators of soil quality [18].

progress had been made in the 1990s in identifying the indicators of soil quality. However, indicators of soil quality can be generally categorised into four groups: visual, physical, chemical and biological quality indicators [19].

3.1. Visual indicators

This may be obtained from observation or photographic interpretation. Exposure of subsoil, change in colour, ephemeral gullies, ponding, run off, plant response, weed species, blowing soil and deposition are only a few examples of potential locally determined indicators. Visual evidence can be a clear indication that soil quality is threatened or changing [21]. Adeoye and Agboola [22] maintained that for sustainable cassava (or any other crop) production, the presence of spear grass on the field to be cultivated is a good indication of a soil with good fertility conditions. In addition, the presence of Chromolaena odorata suggests that the soil possesses good hydrological properties, which is a necessity for cassava production in the early 3–4 months of the growth period.

3.2. Physical indicators

These are related to the arrangement of solid particles and pores. The soil physical characteristics are necessary part of soil quality assessment for cassava production because they often cannot be easily improved [23] during the course of the cropping season. Lal [18] reported that important soil physical parameters to be assessed include soil aggregation, available water capacity, texture, saturated hydraulic conductivity, bulk density, infiltration rate and rooting depth. Researchers have further stressed the need for establishing a quantitative assessment of these soil physical parameters in order to predict biomass productivity, soil organic carbon dynamics, transport processes of water and solutes, etc.

3.3. Chemical indicators

Assessment of soil quality based on soil chemistry, whether the property is a contaminant or part of a healthy system requires a sampling protocol, a method of chemical analysis and an understanding of how its chemistry affects biological systems and interacts with mineral forms and standards for soil characterisation and suitability classification for cassava production in tropical soils. In light of these, Larson and Pierce [10] laid emphasis on those chemical properties that either inhibit the root growth or affect nutrient supply due to the quantity present or the availability. Also, Reganold and Palmer [24] suggested chemical parameters related to nutrient availability as measures of soil quality, including cation exchange capacity (CEC), total nitrogen and phosphorus, soil pH and extractable phosphorus, sulphur, exchangeable calcium, magnesium and potassium, while Karlen et al. [25] opined that total and available plant nutrients and nutrient cycling rates should be included in soil quality assessments for optimum crop productivity. Nevertheless, Abua [26] highlighted the importance of maintaining high levels of nitrogen and phosphorus in the soils as chemical indices for quality soils to be used for cassava production in southern Nigeria.

3.4. Biological indicators

Basically, microorganisms and microbial communities are dynamic and diverse, making them sensitive to changes in soil conditions [27]. Their populations include fungi, bacteria including actinomycetes, protozoa and algae. However, some soil organisms such as nematodes and bacterial and fungal pathogens reduce plant productivity. Visser and Parkinson [28] reported that diverse soil microbiological criteria may be used to indicate deteriorating or improving soil quality, and measurement of one or more components of the nitrogen cycle including ammonification, nitrification and nitrogen fixation may be used to assess soil fertility and soil quality.

Nevertheless, USDA [19] devised biological indicators of soil quality to include measurement of micro- and macro-organisms, their activity, or by-products; and also suggested measurement of decomposition rates of plant residue in bags or measurement of weed seed numbers, or pathogen population can also serve as biological indicators of soil quality.

4. Soil-water characteristics for cassava production

Water is a major constituent of living plant tissues, which consist of about 90% water, and all biological processes within the living plants depend on it [5]. Water is regarded as the most important of the four soil physical factors that affect plant growth (mechanical impedance, water, aeration and temperature) [29]. The optimal moisture conditions for any crop vary depending on many factors such as soil type, climate conditions, growth rate and habit, etc. [30]. The water movement in soils for any given crop production (a case study of cassava) is defined by the soil water characteristic curve.

The soil-water characteristic curve (SWCC) defines the relationship between (pore water suction) matric suctions (ψ) and water content [gravimetric (w) or volumetric (θ) or degree of saturation (S)] [31]. The soil-water characteristics (also known as the soil-water retention or desorption curve) can be described as a measure of the water holding capacity (i.e. storage capacity) of the soil as the water content changes when subjected to various values of suction. SWCC is an indication of the ability of the soil to release water for plants use. The soil-water characteristics are a conceptual and interpretative tool through which the behaviour of unsaturated soils can be understood. As the soil moves from the saturated state to drier states (unsaturated states), the distribution of the soil, water and air phases changes as the stress state changes. The relationships between these phases take on different forms and influence the engineering properties of unsaturated soils [32–34].

Generally, the curve is a function of soil texture and soil structure. The graph of SWCC can be used to obtain the field capacity (FC) and the permanent wilting point (PWP). The curve also explains how different soil structures will hold and release water. From the SWCC curve (**Figure 5**), it can be deduced that a fine-textured soil (like peats) will hold more water at FC and PWP than a coarse-textured soil (sand). The relationship between pore water suction and water content, as presented in a SWCC, is one fundamental relationship used to describe unsaturated behaviour of a soil. Suction is inversely proportional to the water content in a soil. Suction generally increases as the soil desaturates. Increasing suction generally results in high resistance to flow and increase in effective stress. Desiccation is a by-product of the increased effective stress [36]. Increasing suction in compacted clays due to decrease in water content modifies the flow behaviour of covers. During desiccation, the saturation of a liner is reduced and the remaining pore water is held at increasingly large suction.

The relationship between saturation and suction during desiccation is described using the SWCCs. Knowledge of suction and corresponding water content in the soil can be used to predict cracking potentials of liners. The onset and resulting amount of cracking can be correlated to a soil-specific critical suction level [31]. Hence, the SWCC provides critical input to the design of a compacted clay cover liner due to its potential impact on flow rates and the desiccation

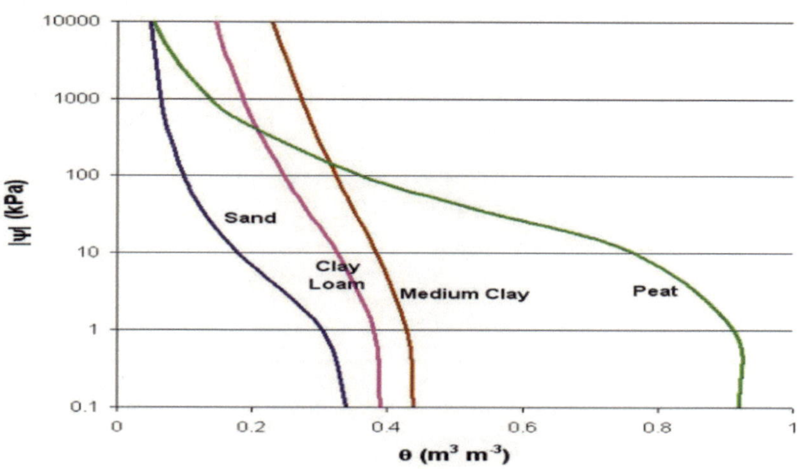

Figure 5. Soil moisture retention curve [35].

process. More so, Hillel [37] reported that the shape of the SWCC is a function of the soil type. Soils with smaller pores have higher air entry pressure (ψa). Soils with wider ranges of pore sizes exhibit greater changes in matric suction with water content. The SWCCs of compacted clay soils depend on the compaction water content, compactive effort and plasticity index [31].

5. Classification of soil water

Water occurs in the soil pores in varying proportions. Some of the definitions related to the water held in the soil pores are as follows:

1. *Gravitational water*: A soil sample saturated with water and left to drain the excess out by gravity holds on to a certain amount of water. The volume of water that could easily drain off is termed as the gravitational water (**Figure 6**). This water is not available for plant use as it drains off rapidly from the root zone.
2. *Capillary water*: The water content retained in the soil after the gravitational water has drained off from the soil is known as the capillary water. This water is held in the soil by

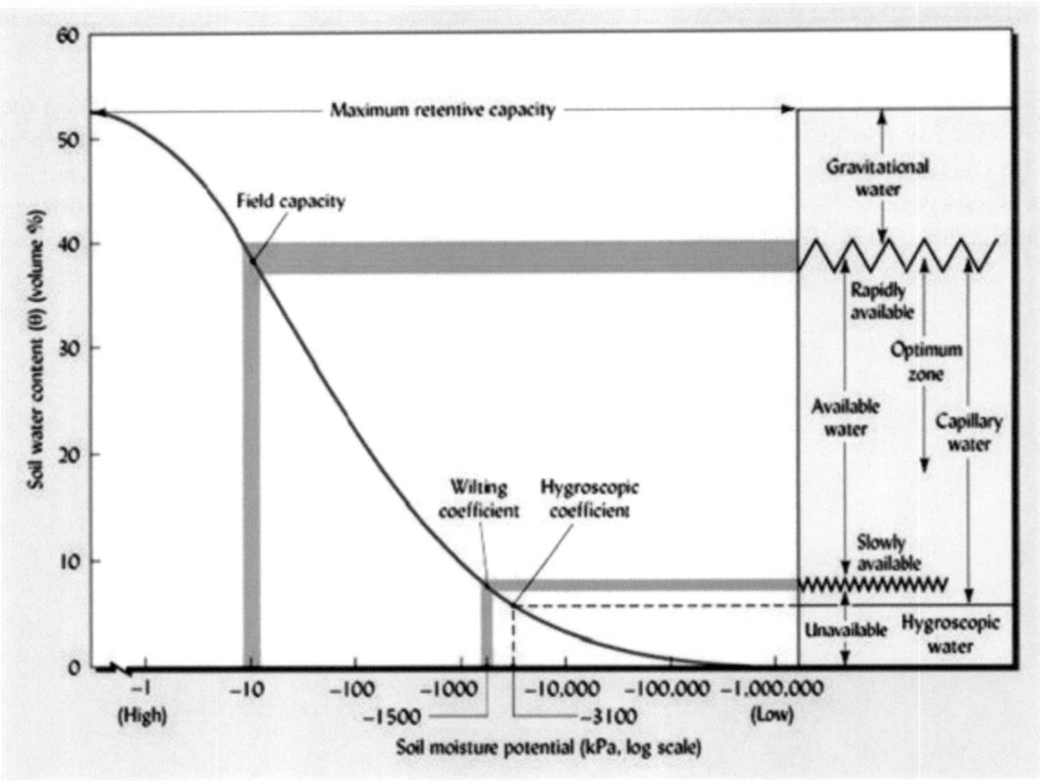

Figure 6. Soil-water movement [39].

surface tension. Plant roots gradually absorb the capillary water and thus constitute the principle source of water for plant growth.

3. *Hygroscopic water*: The water that an oven dry sample of soil absorbs when exposed to moist air is termed as hygroscopic water. It is held as a very thin film over the surface of the soil particles and is under tremendous negative (gauge) pressure. This water is not available to plants.

The above definitions of the soil water are based on physical factors. Some properties of soil water are not directly related to the above significance to plant growth. These are discussed next.

6. Soil-water constants

For a particular soil, certain soil-water proportions are defined which dictate whether the water is available or not for plant growth. These are called the soil-water constants. The soil-water constants refer to the different stages of moisture in the soil as one moves from a wet soil to a dry soil (i.e. from saturated to unsaturated soil condition). These constants are described below:

A. *Saturation capacity*: This is the total water content of the soil when all the pores of the soil are filled with water. It is also termed as the maximum water holding capacity of the soil. At saturation capacity, the soil moisture tension is almost equal to zero.

B. *Field capacity*: This is defined as the maximum amount of water that a soil will hold against the influence of gravity or gravitational force, when downward movement of water has ceased completely. It is the water retained by an initially saturated soil against the force of gravity. Hence, as the gravitational water gets drained off from the soil, it is said to reach the field capacity. At field capacity, the macropores of the soil are drained off, but water is retained in the micropores. Though the soil moisture tension at field capacity varies from soil to soil, it is normally between 0.1 (for clayey soils) and 0.3 (for sandy soils) atmospheres. Field capacity is the upper limit of soil available water.

C. *Temporary wilting point*: This denotes the soil water content at which the plant wilts at day time but recovers during night or when water is added to the soil. It is also referred to as *incipient wilting* or *partial wilting*. At this wilting point, the cassava plants fold their stomata through which water is lost to the atmosphere. This folding is pronounced in the day during periods of active solar radiation. Temporary wilting also denotes that the amount if water the plant is losing to the atmosphere (transpiration demand) is higher than the amount of water the plant roots are able to tap from the soil.

D. *Permanent wilting point*: This refers to the amount of water present in the soil when the soils can no longer supply water at fast sufficient rates to prevent the plants or crops from wilting permanently. Moisture under this condition is not available for the plant roots because the water is present in the latex and not the pores through which the roots elongate. The amount of water a soil at permanent wilting point is a function of soil texture (**Figure 5**). A fine-textured soil (like clay) will hold more water than sand (coarse textured) at this moisture condition. However, soil matric potential at permanent wilting point ranges from −10 bars to −20 bars with an average of −15 bars. A highly coarse-textured soil will have −10 bars, while a fine-textured soil will have −20 bars at the same moisture condition.

NB: For cassava (or any crop) production, it is imperative that for sustainable increase in production levels, the soil condition should never be allowed to reach the permanent wilting point, as this will be very detrimental to output returns. In addition, depending on the prevailing soil and climatic conditions, there may be need to incorporate organic matter and irrigation operations in order to raise the water table of the soil to the root depth of the crop and consequently meet the crop water requirement of the crop for optimum growth and development.

E. *Soil available water*: This is the zone at which water is made available to plant roots. It is the amount of water a soil holds between field capacity and permanent wilting point. The soil available water is the zone that is targeted during irrigation. As earlier maintained under field capacity and permanent wilting point, the soil texture influences greatly the amount of water available to plants at this moisture condition (**Figure 5**). Fine-textured soils will possess more soil available water than coarse-textured soils.

7. How soil holds water

Soil holds water in two ways: (1) as a thin film on individual soil particles and (2) as water stored in the pores of the soil. Water stored as a thin film on individual soil particles is held in place by adsorption forces. Adsorption involves complex chemical and physical reactions, but in simple terms, a thin film of water adheres to the outside layers of soil particle molecules. Water stored in the pores of the soil is stored by capillary forces. An example of the capillary force phenomenon would be to place one end of a glass capillary tube in a pan of water. Water in the tube will rise to a certain height, which depends on the diameter of the capillary tube. This phenomenon can act in any direction and is the key to water being stored in soil pores.

8. Soil-water tension

The soil-water tension simply refers to the amount of water retained [38]. The availability of soil moisture is now frequently described in terms of soil moisture tension, which is dependent upon surface forces, and in terms of total soil moisture stress, which includes surface and other forces arising from the presence of solutes in the soil solution. Soil-water tension (also known as soil-water potential) determines the ease by which water can be extracted from the soil. These are equivalent values, except for the sign (negative vs. positive), which might be thought of as either a push or a pull on the water. Water being held in pores by the capillary storage is held in the soil at a certain tension. The same is true for water held with the adsorption phenomenon.

As the soil dries, these tensions become larger. It is easier for a plant to extract water being held at lower tensions. Plants develop the tension, or potential, to move water from the soil into the roots and distribute the water through the plant by adjusting the water potential, or tension, within their plant cells. For water to move from soil, into roots, into stems, into leaves and finally into air, the water potential must always be decreasing.

Under waterlogged condition, the soil matric potential is zero ($\psi = 0$). This waterlogged condition is not favourable for cassava production. This is due to the inability of the crop to absorb nutrients from the soil as a result of poor respiration which tends to limit the crop energy potentials. Also the cassava plant cannot carry out other physiological processes like photosynthesis. At saturation, the soil-water tension is approximately 0.001 bars; it would therefore be easy for a cassava plant to extract water from a saturated soil [40]. Also, the dividing point between the available soil-water content and readily available soil-water content is named the *maximum allowable depletion*, or MAD, soil-water content. For most field crops like cassava, yam, etc., the MAD level is usually defined as about 50% available water. In some water-sensitive crops, such as vegetables and flowers, the MAD level may be less, such as 30% available water.

The water requirements of cassava at each developmental stage known as the *crop water requirement* greatly influence the need to adjust the water characteristics of the soil in order to ensure optimum growth and productivity of the crop. It has been already established that different soil types pose different soil-water properties, hence, ensuring that the adequate soil type which presents the best soil-water content at field capacity and wilting points is used for cassava production, if yield of the crop is to be improved upon, as water plays a major part in the first 4 months of the crop's development. Majumdar [41] reported that the optimum soil moisture requirement for tall wheat is from the field capacity to 50% of availability; the optimum soil moisture for barley ranges from the field capacity to 40% of availability; the optimum soil moisture for maize range is from 100 to 60% of availability in the maximum root zone depth which extends from 0.4 to 0.6 on different soil types, while for cassava, it ranges from field capacity to 50% of water availability in the maximum root zone, which extends to about 0.5–0.75 m in depth.

9. Soil, water and cassava production in tropical soils of Nigeria

Soil–plant-water relations refer to the physical properties of soil and plants that affect the movement, retention and use of water. These relations must be considered in designing an operation system. Soil is a store house of plant nutrients, a habitat for bacteria, an anchorage for plant and a reservoir that holds the water needed for plant growth. The amount of water a soil can hold in available form for plant use is determined by its physical properties. This amount determines the length of time a plant can survive without water being added. It determines both the frequency of irrigation and the capacity of irrigation system needed to ensure continuous crop growth. Soil is a three-phase system comprising the solid phase (made of mineral and organic matter and various chemical compounds), the liquid phase (called soil moisture) and the gaseous phase (called the soil air). They also contain variety of living an organisms.

Plant growth depends on two important natural resources—soil and water [38, 40]. The soil provides the mechanical support and nutrient reservoir necessary for plant growth. Water is essential for plant life processes. Soil acts like a reservoir that holds water and nutrients that plants need to grow. Some soils are large reservoirs with more holding capacity that releases water and nutrients easily to plants, while other soils have limited reservoirs (**Figure 5**). Effective management of these resources for cassava production in Nigeria necessitates a detailed understanding of the synergy between soils, water and plants by the crop producer. Knowledge of the soil physical and hydrological properties can influence the decision-making process, such as determining what crops to plant and when to irrigate. Cassava (like most other arable crops) has varying water needs at different stages of growth. While the plant is young, it requires less water than when it is in the reproductive stage. As a plant approaches maturity, its water needs drops.

Two important physical properties of soil, which differ the supply of water and air in soil, are texture and structure. Soil texture refers to the relative proportion of sand, silt and clay in a given volume of soil. The texture of the soil delineates the degree of fineness or coarseness of a soil as determined by the percentage of sand, silt and clay contained in the soil. The arrangement of these soil separates (sand, silt and clay) determines the size of pores in the soil. This lays credence to the reason why clay soil will hold more water than sandy soil in the same period. Soil structure on the other hand refers to the arrangement of soil separates into units called aggregates (which contain solids and pore spaces).

The arrangements of the soil separates into aggregates (known as soil structure) will determine the proportion of the macropores (large pores, non-capillary pores and/or air pores) and micropores (small pores, capillary pores and/or water pores) in the soil. Sandy soils with larger soil particle sizes will tend to have more preponderance of large or macropores due to the coarseness of the soil. This limits the amount of water it can hold. Conversely, clay soils that are characterised by high proportion of micropores which serve as pathways for water movements will tend to hold more water at the same period (**Figure 5**). Nevertheless, in tropical regions of sub-Saharan Africa, where the soils are highly heterogenous, the prevailing soil and environmental conditions will determine the amount of water available for cassava or any crop production. Cassava as a root crop requires a substantial amount of moisture in the first 3–4 months of its growth; hence, soils with high water holding capacity culminating in high water table for the root absorption are highly essential for optimum production. However, waterlogged conditions must be avoided.

Nigeria is a nation comprising various vegetation zones that differ in their soil and climatic characteristics. Cultivation of cassava in these diverse agro-ecologies has yielded varying quantities of cassava in different geopolitical zones in nation. The FAO statistics [8] noted that on a per capita basis, North-Central was the highest producing state at 0.72 tonnes per person, followed by South-East (0.56), South–South (0.47), South-West (0.34), North-West (0.10) and North-East (0.01).

Also, earlier reports from the International Institute of Tropical Agriculture in 2004 noted Cross River in the South–South geo-political zone as well as Benue and Kogi state in the North-Central zone were the largest producers of cassava in Nigeria (**Figure 7**). The study attributed the results to the management techniques and prevailing soil and climatic conditions. Cassava

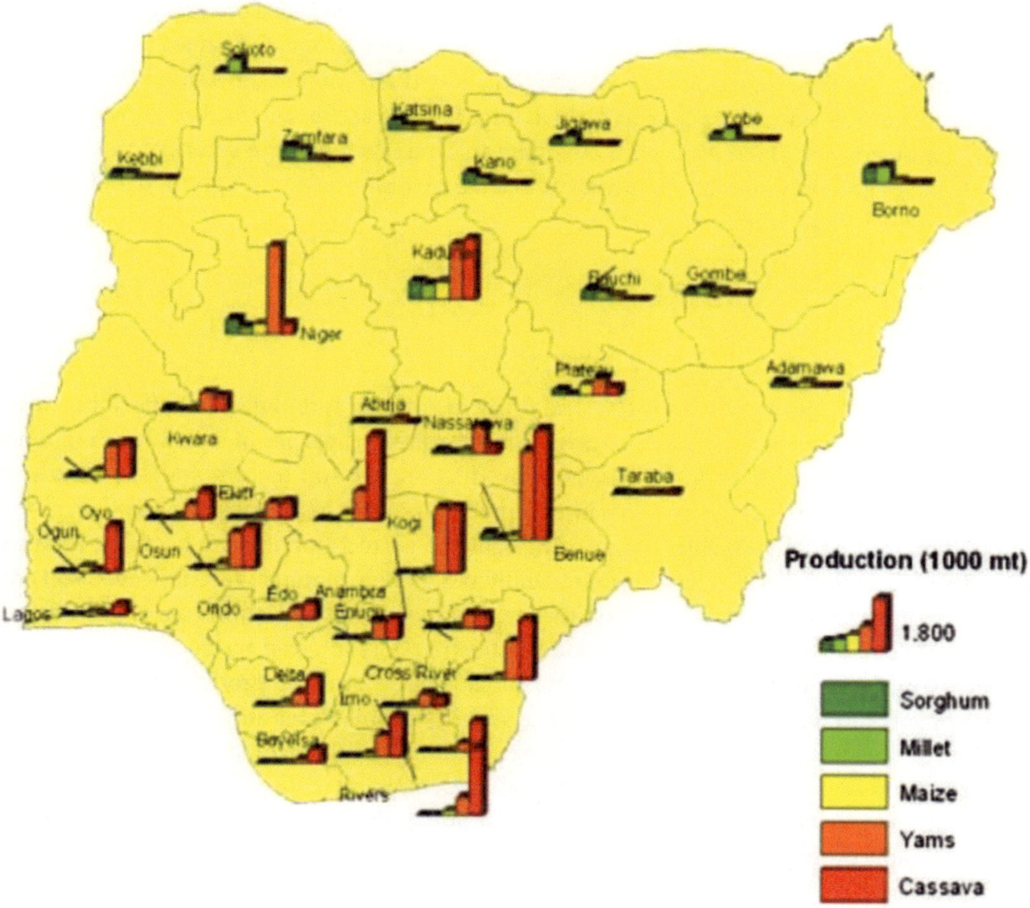

Figure 7. Crop production by Nigerian states [42].

Figure 8. Harvested cassava roots in sub-Sahara Africa [44].

yields favourably in acidic soils with high organic matter content. Hillel [43] maintained that soils with high organic matter content tend to hold more water with good aeration and contain more nutrients and beneficial microbial population, culminating in a good soil structure. This entails that such soils which are predominantly medium-textured soils are of high quality and presents the best inherent soil physical condition for cassava production in view meeting the food demands of the rising population in Nigeria. Hence, to combat the rising issues of food security in sub-Saharan Africa (particularly in Nigeria), the use of soils with good quality and water characteristics, juxtaposed with adequate soil management, are essential in obtaining sustainable cassava production (**Figure 8**) in this region of the world in light of the increasing scores of persons and animal population.

10. Summary

Soil-plant-water relation entails the physical properties of soil and plants that affect the movement, retention and use of water. Two important physical properties of soil, which influence the supply of water and air in soil for crop production, are soil texture and soil structure. Increasing cassava production outputs is extremely important in the face of a growing global population, but equally essential is ensuring that the correct quantity and quality are produced and distributed and in an environmentally sustainable manner with improved soil and water conditions. Cassava production in the wake of the brunt population increase in tropical regions of sub-Saharan Africa like Nigeria has been greatly affected by poor soil and water quality, which are intrinsic properties for sustainable optimum crop production. The varying effects of soil and water quality as well as the synergy of soil-water and crop play an important role in the lifecycle of cassava and ensure optimum production on a sustained basis. This chapter reviewed and discussed soil quality and its assessment methods, soil-water characteristics and the relationship between the soil, water and crop for soil quality management and the sustainable optimum production of cassava in tropical Nigeria.

Author details

Saurau O. Oshunsanya* and Nkem J. Nwosu

*Address all correspondence to: soshunsanya@yahoo.com

Department of Agronomy, University of Ibadan, Ibadan, Oyo State, Nigeria

References

[1] Gbadegesin AS. Variation in soil properties on cassava production in the coastal area of Southern Cross River State, Nigeria. Journal of Geography and Geology. 2010;**3**:94-103

[2] Oniya OO, Opadijo OO, Odedare KO, Oduyemi AT. Soil Suitability Analysis for Cultivation of Cassava in Ibarapa Central LGA: A GIS Approach. Japan: CIGR International Commission of Agricultural and BioSystems Engineering. 2009. p. 8

[3] Soil Science Society of America (SSSA). Glossary of Soil Science Terms. Madison, WI: Soil Science Society of America; 1996

[4] Lal R. Basic concept and global issues: Soil quality and agricultural sustainability. In: Lal R, editor. Soil Quality and Agricultural Sustainability. Chelsea, MI, USA: Ann Arbor Science; 1998. pp. 3-12

[5] El-Shal ZS, Soad ME. Response of cassava plant to water requirements and the levels of organic manure. Journal of Plant Production. 2012;**3**(7):2211-2222

[6] International Institute for Tropical Agriculture. Cassava in Tropical Africa: A Reference Manual. Wisbech, UK: Balding and Mansell International; 1990. 184 pp

[7] International Food Policy Research Institute. In: Sebastian K, editor. Atlas of African Agriculture Research and Development: Revealing Agriculture's Place in Africa. Washington, DC, USA: A Peer-Reviewed Publication of the International Food Policy Research Institute; 2014. 90 pp

[8] FAO. FAOSTATS—World Production of Cassava. Rome, Italy: Food and Agricultural Organisation of the United Nations; 2014

[9] Doran JW, Coleman DC, Bezdicek DF, Stewart BA. Defining soil quality for a sustainable environment. Soil Science Society of America Special (SSSA) Publication, No. 35. Madison, WI: Soil Science Society of America Incorporation and American Society of Agronomy Incorporated; 1994

[10] Larson WE, Pierce FJ. The dynamics of soil quality as a measure of sustainable management. In: Doran JW, Coleman DC, Bezdicek DF, Stewart BA, editors. Defining Soil Quality for a Sustainable Environment, SSSA Special Publication 35. Madison, WI, USA: Soil Science Society of America; 1994. pp. 37-51

[11] FAO. Strategic Environmental Assessment. Food and Agriculture Organisation Corporate Document Repository. Rome, Italy: FAO; 2017

[12] Ojiako IA, Tarawali G, Nze L, Ogundijo D, Edtet M, Audu B, Adenekan S. Root's supply response for smallholder farmers supplying cassava to commercial starch processors in Nigeria. Asian Journal of Rural Development. 2017;**7**:1-14

[13] Anyanwu CN, Ibeto CN, Ezeoha SL, Ogbuagu NJ. Sustainability of cassava (*Manihot esculenta Crantz*) as industrial feedstock, energy and food crop in Nigeria; 2015

[14] Gbadegesin AS, Abua MA, Atu JE. Variation in soil properties on cassava production in the coastal are of southern Cross River state. Nigerian Journal of Geographical Geology. 2011;**1**:94-103

[15] Anikwe MAN, Obi ME. Evaluation of an index model for quantifying the productivity of soils in southeastern Nigeria using maize as a test crop. Journal of Agriculture and Environment. 1999;**1**:80-89

[16] United State Department of Agriculture. Soil Quality Glossary. Washington, DC: USDA National Resource Conservation Service; 2006

[17] Papendick RI, Parr JF, van Schnilfgaarde J. Soil Quality: New Perspective for a Sustainable Agriculture. In Proceeding International Soil Conservation Organization. New Delhi, India; 1995

[18] Nwosu NJ, Okon PB. Impacts of climate change on soil quality of tropical acid-sands [Unpublished B.Sc project]. Nigeria: Department of Soil Science, University of Calabar; 2016. pp. 44-52

[19] United State Department of Agriculture. Indicators for Soil Quality Evaluation. Soil Quality Information Sheet. USDA National Resource Conservation Service. Washington, DC: USDA – NRC; 1996. 2 pp

[20] Sojka RE, Upchurch DR. Reservations regarding the soil quality concept. Soil Science Society of America Journal. 1999;**63**:1039-1054

[21] Anikwe MAN. Soil Quality Assessment and Monitoring: A Review of Current Research Efforts. Enugu, Nigeria: New Generation Books; 2006

[22] Adeoye GO, Agboola AA. Critical levels for soil pH, available P, K, Zn and Mn and maize ear-leaf content of P, Cu and Mn in sedimentary soils of South-West Nigeria. Fertilizer Research. 1985;**6**:65-75

[23] Wagenet RJ, Hutson JL. Soil quality and its dependence on dynamic physical processes. Journal of Environmental Quality. 1997;**26**:41-48

[24] Reganold JP, Palmer AS. Significance of gravimetric versus volumetric measurements of soil quality under biodynamic, conventional, and continuous grass management. Journal of Soil and Water Conservation. 1995;**50**:298-305

[25] Karlen DL, Eash NS, Unger PW. Soil and crop management effects on soil quality indicators. American Journal of Alternative Agriculture. 1992;**7**:48-55

[26] Abua MA. Suitability assessment of soils for cassava production in the coastal and hinterland areas of southern Cross River state—Nigeria. Journal of Science and Environmental Management. 2015;**6**(5):108-115

[27] Kennedy AC, Papendick RI. Microbial characteristics of soil quality. Journal of Soil Water Conservation. 1995;**50**:243-248

[28] Visser S, Parkinson D. Soil biological criteria as indicators of soil quality: Soil microorganisms. American Journal of Alternative Agriculture. 1992;**7**:33-37

[29] Kirkham MB. Principles of Soil and Plant Water Relations. USA: Elsevier Academic Press; 2005. pp. 1-39

[30] Singh R, Alderfer RB. Effects of soil—Moisture stress at different periods of growth of some vegetable crops. Soil Science. 1966;**1**(1):69

[31] Osinubi KJ, Bello AA. Soil-water characteristics curves for reddish Brown tropical soils. Electronic Journal of Geotechnical Engineering. 2011;**16**:1-25

[32] Fredlund DG, Rahardjo H. Soil Mechanics for Unsaturated Soils. New York: John Wiley and Sons; 1993. 560 pp

[33] Fredlund DG. The emergence of unsaturated soil mechanics. The Fourth Spencer Buchanan Lecture. College Station, Texas: Texas A and M University; 1996. 39 pp

[34] Vanapalli SK, Fredlund DG, Pufahl DE. The influence of soil structure and stress history on the soil-water characteristics of a compacted till. Geotechnique. 1999;**49**(2):143-159

[35] Hob D. Soil porosity and water retention. An e-publication of cactiguide.com. 2006. http://cactiguide.com/forum Yesiller /viewtopic.php?t=13438. [Accessed: 21 July 2017]

[36] Miller CJ, Yesiller N, Yaldo K, Merayyand S. Impact of soil type and compaction conditions on soil water characteristic. Journal of Geotechnical and Geoenvironmental Engineering, ASCE. 2002;**128**(9):733-742

[37] Hillel D. Fundamentals of Soil Physics. San Diego, CA: Academic Press Inc; 1980

[38] Oshunsanya SO. Soil Physics. Ibadan: Debank Publishers; 2011. 161 pp

[39] Lajos B. Soil Moisture Retention Curve. 2008. E-page: http://www.tankonyvtar.hu/en/tartalom/tamop425/0032_talajtan/ch07s05.html. [Accessed: 21 July 2017]

[40] Rogers DH, Aguilar J, Kisekka I, Barnes PI, Lamm FR. Soil, water and plant relationships. Kansas State University Research and Extension. Irrigation Management Series, L904; Kansas State, USA: Kansas State University Agricultural Experiment Station and Cooperative Extension Service; 2014. 8 pp

[41] Majumdar DK. Irrigation Water Management. India: Prentice Hall; 2000. 18 pp

[42] FAO. FAOSTATS. Rome, Italy: Food and Agricultural Organisation of the United Nations; 2002

[43] Hillel D. Introduction to Environmental Soil Physics. Amsterdam: Elsevier Academic Press; 2004. 494 pp

[44] FAO. FAOSTAT. Rome, Italy: Food and Agriculture Organisation of the United Nations; 2012

Section 4

Postharvest Processing of Cassava

Chapter 11

Baby Cassava: An Alternative Marketing Strategy for Freshly Cut Cassava

Kelem Silva Fonseca, Moab Torres de Andrade,
Daniel Gomes Coelho, Aline Ellen Duarte de Sousa,
Domingos Ferreira de Melo Neto,
Fred Augusto Lourêdo de Brito,
Rainério Meireles da Silva and
Adriano do Nascimento Simões

Additional information is available at the end of the chapter

http://dx.doi.org/10.5772/intechopen.70902

Abstract

There are many procedures for obtaining minimally processed fruits and vegetables, aiming at adding value and maintaining the quality for a longer period. Cassava is a root that adapts to minimum processing technology, because the tissues are more resistant, what helps in obtaining different cut shapes and formats. However, it is a root susceptible to browning and microbiological contamination. In this chapter, methodologies and procedures are described to obtain alternative formats for minimally processed cassava, which was generally denominated "*babycassava*", called "*babytolete*", "*cateto*", and "rubiene". Besides that, some preharvest and postharvest factors that influence the shape and quality of "*babycassava*" formats will be addressed. It was verified that preharvest factors could influence the quantitative and qualitative aspects, resulting in browning of the minimally processed root. Some of the factors studied seem to regulate key enzymes in which they mediate oxidative reactions that cause browning, such as polyphenol oxidase and peroxidase, and other enzymes that participate in the reactive oxygen species (ROS) elimination process. In this way, the turning stage of "*babycassava*" manufacturing removes the parenchyma, minimizing the effect of browning-related enzymes.

Keywords: *babycassava*, minimally processed, browning, quality, market

1. Introduction

In recent years, changes in eating standards have led to higher consumption of fruits and vegetables, and consumers are looking for quality foods that are healthy, safe, and practical.

Minimally processed products are an alternative to consumer demand, offering convenient, highly nutritious, and healthy products, while maintaining the freshness of fruits and vegetables *in natura*. Vegetables are considered minimally processed when they remain fresh, despite being physically altered.

Basically, minimum processing goes through the steps of selection and classification, peeling, cutting, sanitization, rinsing, spinning, and packaging. All these steps are aimed at providing the consumer with a fresh, healthy product, that is easy to prepare and that maintains high food quality, freshness, and safety. However, during and after minimal processing, plant tissues are more susceptible to biochemical and physiological stresses when compared to intact plants. Because of this, the proper handling of the vegetables from the harvest to the processing and the use of low temperatures are necessary to promote stability and maintenance of high quality in shelf life [1].

"*In natura*" roots are marketed in bark, usually dirty, or sometimes peeled, immersed in water or frozen. However, this type of product does not offer food safety to consumers. Therefore, maintaining the quality of cassava roots in markets or consumers' homes for days has been a great challenge for the agro-industrial development of this root.

The main technological challenge faced is to keep cut cassava roots without symptoms of browning. Deterioration is divided into two processes: primary or physiological deterioration and secondary or microbiological deterioration [2–5]. The postharvest physiological deterioration (PPD) profoundly affects processing, as well as root marketing. It is triggered by mechanical damage, an unavoidable result of the harvesting operations. The PPD then proceeds from the site of the damage, eventually causing general discoloration of the vascular parenchyma along the root [6, 7]. The physiological deterioration is usually the cause of reduced root acceptability. It may be observed by dark streaks in the root vascular tissue that subsequently spread and cause a more general brown discoloration, leading to unsatisfactory cooking quality and unpleasant odor and taste. The microbiological deterioration is caused under aeration conditions by the *Pseudomonas* sp. bacteria [8], and under low oxygen tension, Bacillus sp. are predominant, causing rot and increased acidity, fermentation and softening of the roots, and usually occurs when the roots have already become unacceptable because of physiological deterioration [9].

Alternative and innovative formats named "*babycassava*" are being studied and developed. Some investigations focusing factors pre and postharvest were made [10, 11]. The "*babycassava*" designation, proposed in this chapter, regards "*babytolete*", "*cateto*", and "*rubiene*" formats. In all cases, these shapes add value to cassava; they can make consumer's life easier for cooking faster and not requiring pressure cooker. Besides that, these are different and attractive shapes.

This chapter will show procedures for "*babycassava*" obtaining, and some factors before and after harvest that can influence "*babycassava*" quality.

2. Procedures for obtaining '*Babycassava*'

Cassava processing should be done in a cold ambience and with maximum agility, due to its susceptibility to browning. Based on this, a general flowchart is presented in **Figure 1**. The steps in red marked are stages considered critical in the process and that deserve attention in physiological aspect [12].

To obtain *"babycassava"*, the following steps are followed: once harvested, the cassava roots are transported to the minimum processing unit, where they are selected, weighed, and washed in running water with the help of a brush to remove dirt. The ends are removed and cooled in ice water for 5 minutes, and they are cut transversely into 6-cm pieces. The segments are peeled with the aid of a stainless steel knife (**Figure 2**). The 6-cm segments should be cut transversally to the middle, resulting in 3-cm long segments. A longitudinal cut must be made to obtain the *"babytolete"* format. To obtain the *"cateto"* shape, the edges of the 3 cm fragments should be removed, creating cube-shaped pieces. Then, the cassava cubes were turned for 120 seconds, finally getting into the shape called *"cateto"*. The *"rubiene"* format will be obtained from 3-cm long segments. A longitudinal cut is made, and soon after, the turning is performed for 120 seconds. The turning made in cassava is similar to that already known in carrots, *"babycarrots"* [13].

After obtaining the three types of *"babycassava"* (*"babytolete"*, *"cateto"*, and *"rubiene"*), the products are immersed in ice water for 10 seconds to perform the initial rinsing, then immersed in sanitizing solution with concentrations of 200 and 5 mg L^{-1} of active chlorine, for 10 minutes at each concentration to reduce microbial contamination.

After the sanitization, the centrifugation step in a domestic centrifuge is done. The centrifugation time for 200 g of *"babytolete"* is 30 seconds, and for 800 g of *"cateto"* or *"rubiene"* shapes, it is 60 seconds [12]. These times are sufficient to remove excess water acquired during the sanitization step. After that, they should be packed. Among the various types of packaging, it

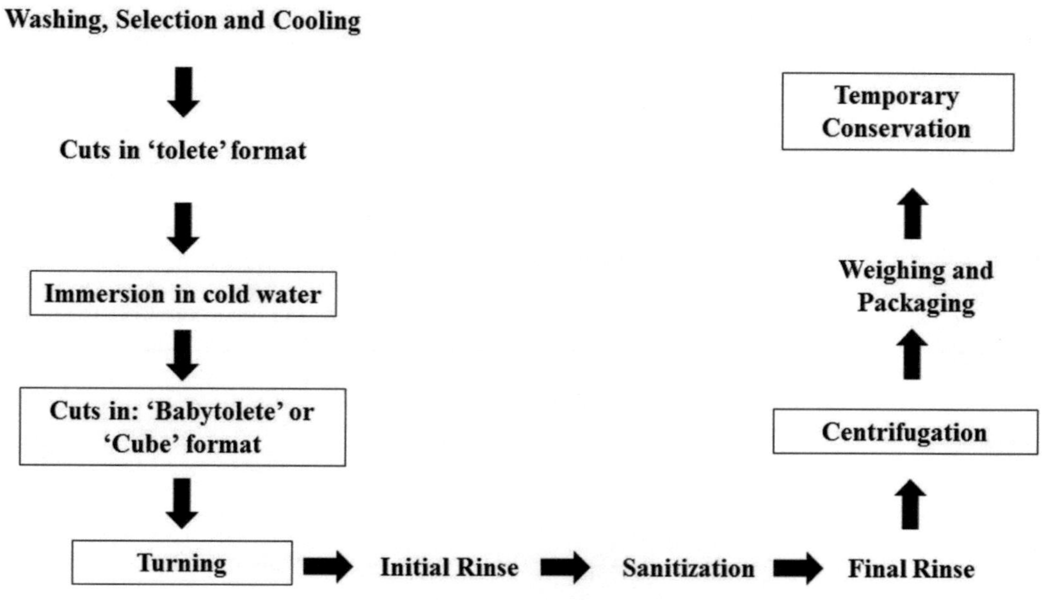

Figure 1. Minimum processing operational flowchart of cassava in *"babycassava"* format. Steps marked in red are considered critical in that they influence the browning.

is recommended to use polypropylene packaging. Once packaged, the products should be stored in refrigerated display units at 5 ± 2°C and 90 ± 5% RH.

In general, proposed flowcharts are alternatives to the formats studied and have been perfected for other roots, such as yam and sweet potato in the "baby" format. These formats make the product more attractive and different from what has already been found in the market, with the advantage of good yields (**Table 1**).

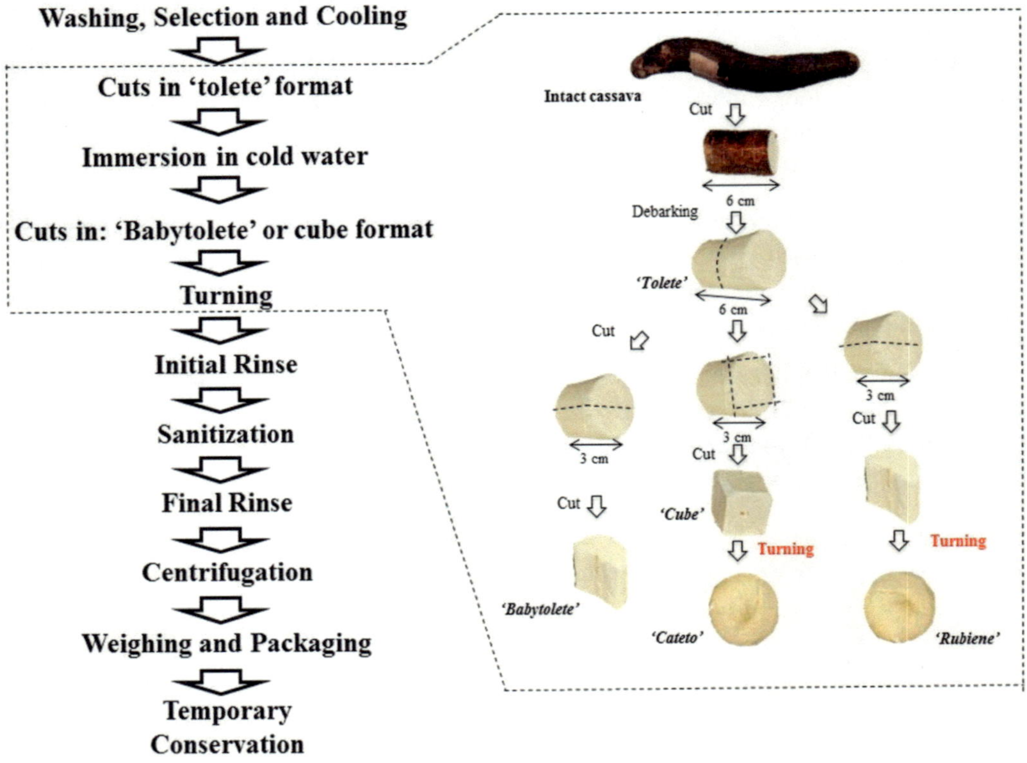

Figure 2. Cutting stages for "*babytolete*", "*cateto*", and "*rubiene*" formats obtaining.

Agroindustrial yield (%)		
Babytolete	**Cateto**	**Rubiene**
81 ± 5	40 ± 5	60 ± 5
Data of Freire et al. [10] and Brito et al. [11].		

Table 1. Yield agroindustrial of *babycassava: babytolete, cateto,* and *rubiene* formats.

3. Pre and postharvest factors that influence the quality of *'babycassava'*

3.1. Preharvest factors

It is believed that several factors influence cassava final quality. However, it is worth considering crop factors such as population density and harvest ages.

The population density seems to be more involved in root shape, with respect to morphology, number of roots per plant (**Figure 3A**), diameter of the roots (**Figure 3B**), and productivity (**Figure 3C**). Although it was reflected in thinner minimally processed roots, helping to obtain the *'babycassava'* (**Figure 4**), it did not influence the acceptance made by a sensorial panel composed of 50 people at Federal Rural University of Pernambuco (**Figure 4**). People preferred larger pieces to the detriment of smaller ones (**Figure 4**). This is due to the local custom of consuming large roots and not making the proposal of producing small roots for immediate consumption invalid.

On the other hand, age of harvest influences minimally processed root qualitative aspects in regard to browning. Based on the deterioration of the minimally processed root in the *"babytolete"* and *"cateto"* or *"rubiene"* formats, it was developed a subjective scale of grades ranging from 5 (best grade) to 1 (worst grade). Score 3 was set as an acceptance limit, and the overall score corresponds to the average of the scores for each *"babycassava"* [14] (**Figure 5**).

Based on the visual scale, it is observed that roots harvested later (360 and 420 days after planting), become more susceptible to browning, compared to those harvested earlier (300 days after planting) (**Figure 6**). One of the explanations for these results seems to be related to the ability of cassava tissues to enzymatically combat reactive oxygen species (ROS), through catalase and superoxide dismutase activity [15], as well as signaling control involving calcium and programmed cell death [16].

According to the data presented, it is verified that crop management, related to density and mainly age of harvest, are decisive factors for the quality of the root in which the objective is minimum processing. Voluntary impositions in the cuts cause browning of the tissue. Thus, roots harvested earlier are more suited to minimum processing.

3.2. Postharvest factors

Cassava roots have a short shelf life due to PPD. Fresh cassava roots are traditionally marketed without postharvest treatment and therefore should reach the consumer in a short time before deterioration becomes visible. PPD makes roots quickly unfit for consumption and therefore nontradable. The short lifespan of roots severely limits marketing options, increasing the likelihood of losses, and overall marketing costs. Extension of the useful life of cassava roots would result in large annual savings [17].

PPD begins when the root is separated from the parent plant and resembles the changes that occur in plant response to injury, thus triggering a series of biochemical reactions. These reactions may include the accumulation of several secondary metabolites, for example,

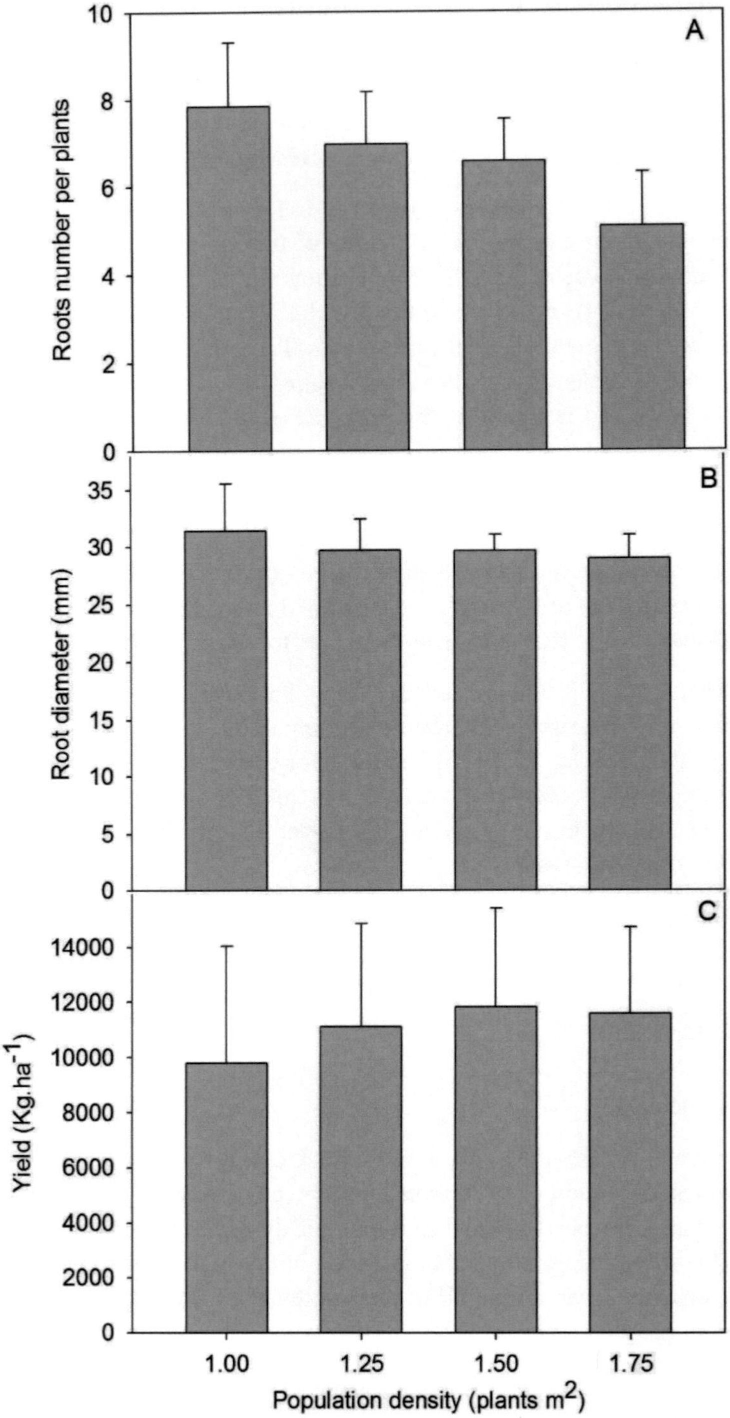

Figure 3. Number of roots per plant (A), root diameter (B), and yield (C) of cassava planted at densities: 1.0, 1.25, 1.5, and 1.75 plants m^{-2}.

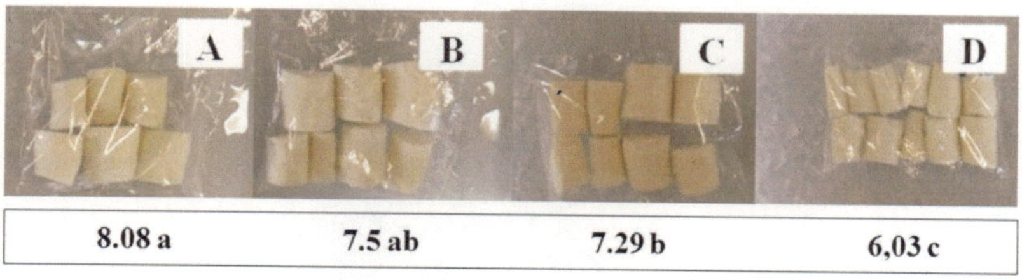

Figure 4. Cassava minimally processed planted at 1.0 (A), 1.25 (B), 1.50 (C), and 1.75 (D) plants m^{-2}. The numbers below of the figures are means of acceptance test conducted with 50 people at the Rural Federal University of Pernambuco, Brazil. Equal letters on the same line indicate that there was no significant difference between the samples at 5% probability level.

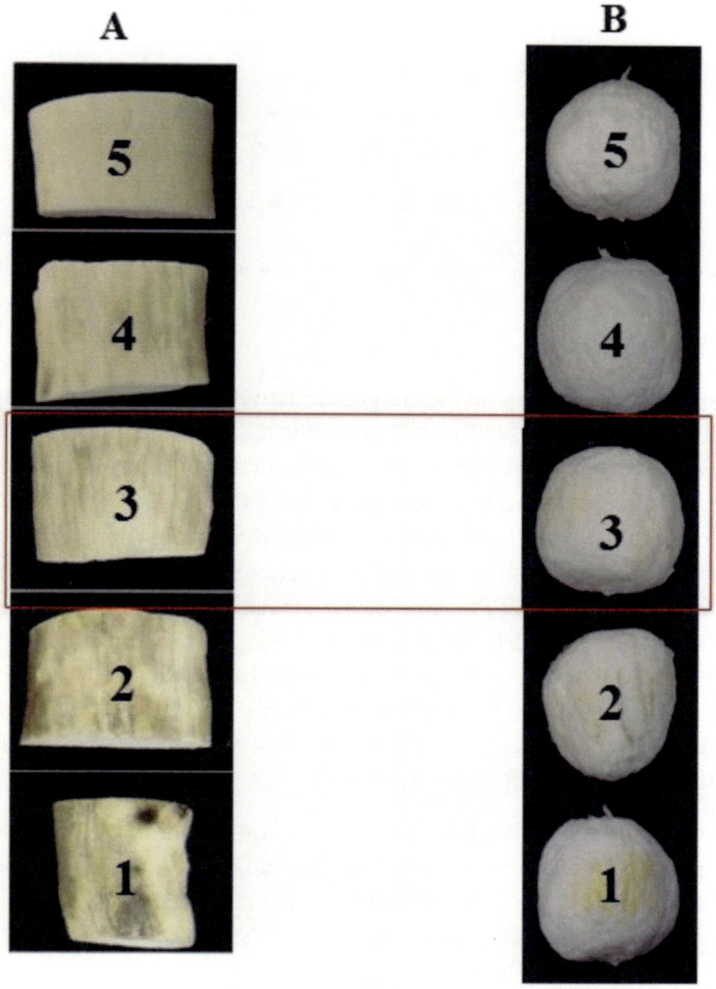

Figure 5. "*Babytolete*" and "*cateto*" formats in different grades in the subjective evaluation according to Freire et al. [10]. Note that the shape "*cateto*" (B) does not darken intensively compared to "*babytolete*" (B) format. The notes corresponding to three (3), marked in red, represent commercial acceptance limit.

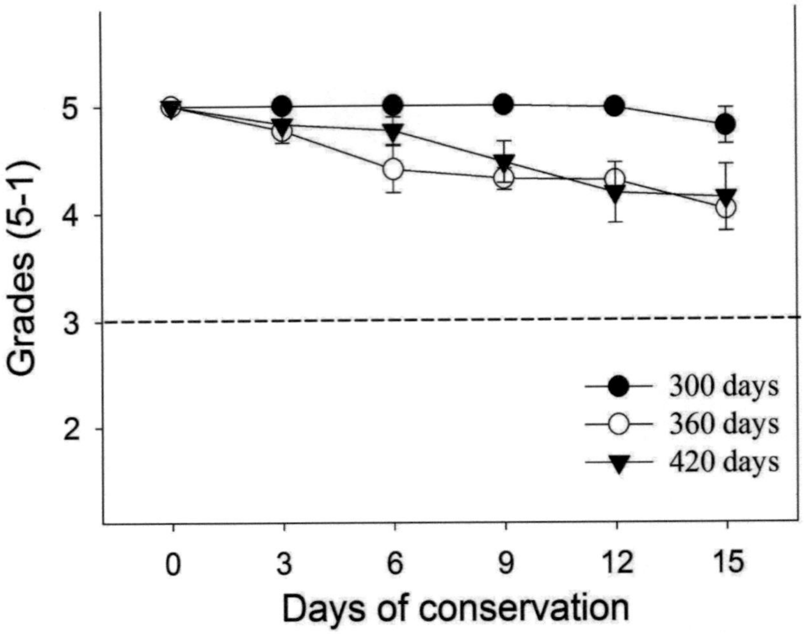

Figure 6. General appearance in minimally processed cassava harvested at different ages. All kept at 5°C for 15 days. The dotted line represents commercial acceptance score limit.

hydroxycoumarin, and an increase in the enzymatic activity of polyphenol oxidase. Secondary metabolites, enzymes, and polysaccharides are activated in response to physiological stress induced [3]. These compounds appear to play an important role in reducing or delaying the process of physiological deterioration. Ref. [3] studies the biochemical profiles of different cassava cultivars. These authors hypothesized that changes in the metabolism and enzymatic activities of tissues that were in wound-induced deterioration serve as indicators of tolerance or susceptibility of genotypes to PPD. One of the results obtained in this study was that the main hydroxycoumarin identified in cassava was scopoletin. Tolerant cultivars for PPD showed higher amounts of this compound. Scopoletin indices increased during PPD, suggesting that scopoletin should be involved in reducing the rate of deterioration in the early stage of PPD. Additionally, high levels of ascorbic acid, polyphenol oxidase, dry matter, and protein correlated with lower deterioration rates [8].

Among the postharvest factors that influence cassava conservation, it is important to note the handling during the minimum processing, emphasizing immersion in ice water; formats; turning and centrifuging (**Figure 1**). The handling stages after minimum processing are storage conditions, packaging, and storage temperature.

Immersion of the raw material in ice water (around 5°C) for 1 hour after harvesting and also in the sanitization and rinsing stages can be done to reduce the respiration transiently, as well as ethylene production and enzymatic activity in carrots [18]. In case of cassava, cooling also helps in the peeling and decreases the browning [12].

Therefore, the hydrocooling is a viable and effective alternative practice, proceeding before the minimum processing operations, to reduce the responses associated to stress, especially those caused by the field heat and those visible in the products. In the same way, roots must also be cooled, in order to remove the field heat and lower metabolic activity.

Turning is another step that is considered critical in the process. Turning removes the superficial tissue by abrasion. In carrots, the turning is done using two turners, one for the removal of the periderm and another to make the rounding of the edges, making them more attractive [14]. In the case of cassava, the periderm is removed with the aid of blades, only a turning machine is required whose purpose is to round the edges. However, it has been found that a short turning period causes surface browning in the conserved pieces (**Figure 7**). Possibly the most internal tissues are composed of differential cells without secondary phloem, in which they are the main site of browned striae. After 120 seconds of the turning process, approximately 5 mm is removed from the surface, making remaining surface cells less responsive to browning. This is evidenced by the high activity of enzymes that can cause browning, such as polyphenol oxidase (PPO) and peroxidase (POD) (**Figure 8**).

Thus, when the surface is removed, composed by the secondary phloem, the *"babycassava"* becomes more tolerant to the browning during the conservation (**Figure 9**). The results show that the turning technique can be an alternative to maintain quality of minimally processed cassava.

On the other hand, the turning must be carried out with maximum hygiene. One more stage is added to the flow chart, and the machine must be clean and properly sanitized, as it can become an inoculum source for contamination as seen in **Figure 8**, using a qualitative evaluation of *Pseudomonas* spp. (**Figure 10**).

Inadequate temperature, such as 10°C, can accelerate microbiological contamination, in relation to 5°C (**Figure 10**). Therefore, the conservation at 5°C is the most suitable for sale in retail, regardless of the format, as both reduce physiological changes as the browning, as well as possible symptoms of disorders of a microbiological nature.

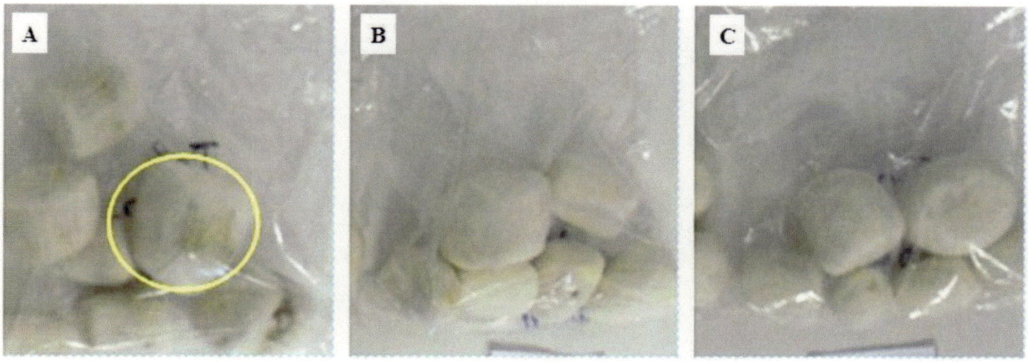

Figure 7. *"Rubiene"* from cv. Mossoró, harvested at 12 months and turned (1.5 kg) by 30 seconds (A), 60 seconds (B), and 120 seconds (C) and maintained for 11 days at 5 ± 2°C. The circles in yellow highlights the browning.

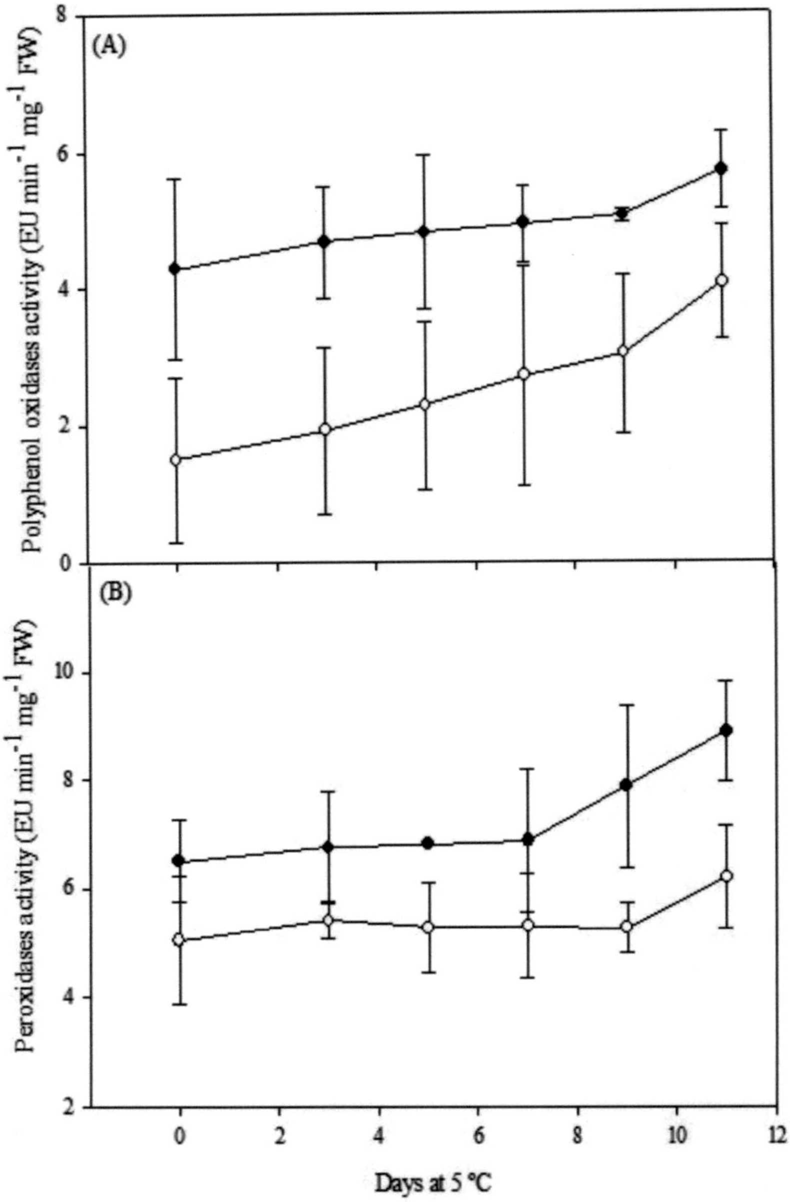

Figure 8. Polyphenoloxidases (A) and peroxidases activity (B) in roots of cassava cv. Mossoró harvested at 14 months and minimally processed in *"Babytolete"* (•) and *"Rubiene"* (◦) shapes stored at 5°C and 90 ± 5% RH for 0, 3, 5, 7, 9, and 11 days. The vertical bars represent the standard deviation from the mean and the minimal significant difference (MSD) at 5%. Data for three replications.

When *"babycassava"* is marketed at 10°C or without refrigeration, it should not exceed 12 hours until consumption [12]. This is important as in the case of the institutional market, that is, industrial kitchens, schools, companies, among others, in which consumption takes place in a few hours or even after a short period of transportation at room temperature.

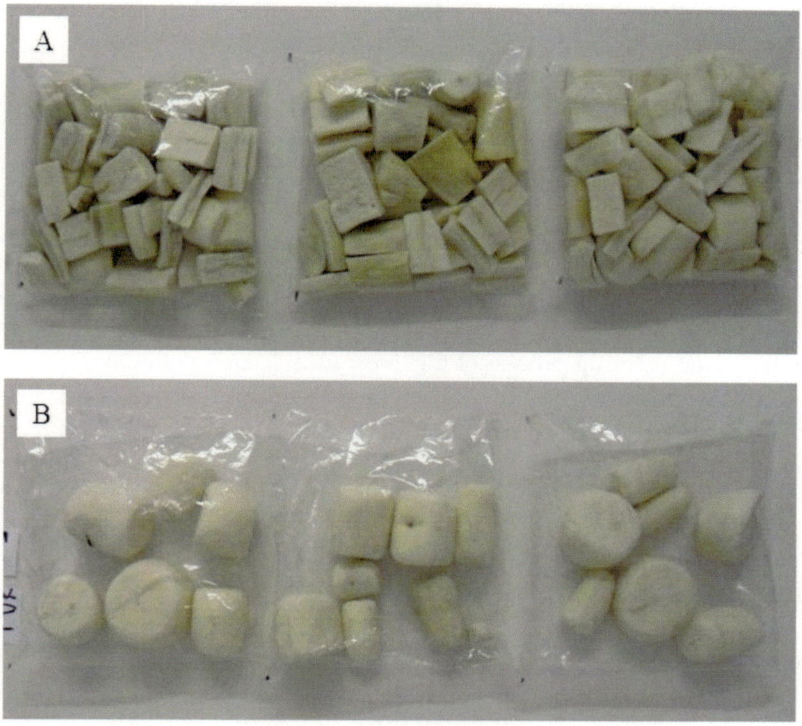

Figure 9. Pieces of *"babytolete"* (A, not turned) and *"rubiene"* (B, turned for 120 seconds), of sweet cassava cv. "Mossoró" at 7 days at 10°C.

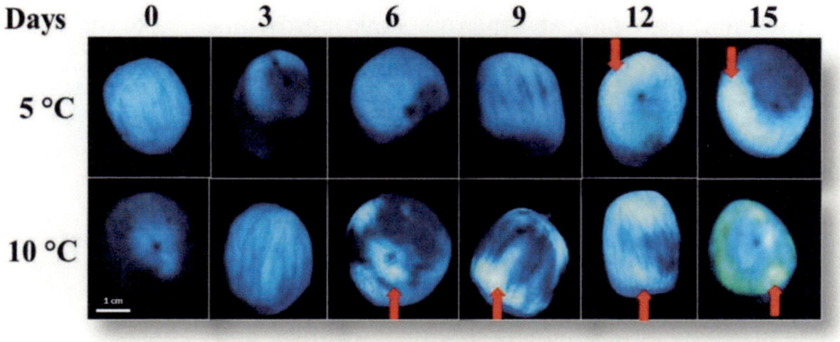

Figure 10. Fluorescence emission of *"rubiene"* pieces kept at 5 and 10°C for 15 days. The red arrows indicate the incidence of *Pseudomonas* sp. The photos were captured by a semi-professional digital camera (Nykon; D3100 14.2 megapixels) coupled to the darkroom (CN-6; Vilber Lourmart) with incidence of ultraviolet light, 365 nm, with filter 1 × 6 Watts and power 220 V 50/60 Hz (VL-6.L; Vilber Lourmart).

Conservation of minimally processed products for marketing is generally done in refrigerated shelves, whose temperatures are between 5 and 10°C. In addition, the minimally processed cassava can be transported and consumed for 12 hours at room temperature or 12 days at 5°C [11]. This can be an extremely strong inducer for precipitating darkening, by an increase of the activities of enzymes cited and increasing the susceptibility of microbial growth. The changes in the biochemical markers associated to phenolic compounds can be an important tool to coordinate browning and shelf life of minimally processed cassava.

This chapter showed that some factors of the medium, before harvest, during, and after minimal processing, in the temporary conservation, are modulators of enzymatic activity. The age of harvesting seems to be of extreme importance for an increase in the enzymatic activity. The later the harvest, the greater the PPO and POD activity of the roots in the conservation, suggesting that old roots are more sensitive to browning. This physiological response does not seem to occur in roots grown at different densities, because only diameter, shape, and other agronomic characteristics are modified with different densities of planting [19].

In addition, the manufacturing process of the '*babycassava*', such as the turning, in which with parenchyma removal by turning, reduced browning and the activity of associated enzymes to browning. This is also extended to temperature, packaging, among other postharvest factors not discussed in this review.

4. Summary and conclusions

Population density seems to have more influence on the productive aspect and root morphology. However, harvest time seems to be more related to quality, once that young roots are more tolerant at browning. This seems to be related to the oxidative metabolism involving the enzymes polyphenol oxidase (PPO) and peroxidase (POD), and other enzymes that take part in the process of elimination of reactive oxygen species (ROS).

In postharvest, the process of obtaining "*babycassava*" using turning, "*cateto*" and "*rubiene*" are alternatives to minimize browning and microbiological growth, allied to the storage temperature, help to keep "*babycassava*" quality for longer time.

Author details

Kelem Silva Fonseca[1], Moab Torres de Andrade[1], Daniel Gomes Coelho[1],
Aline Ellen Duarte de Sousa[2], Domingos Ferreira de Melo Neto[1],
Fred Augusto Lourêdo de Brito[3], Rainério Meireles da Silva[4] and
Adriano do Nascimento Simões[1]*

*Address all correspondence to: adriano.simoes@ufrpe.br

1 Academic Unit of Serra Talhada, Federal Rural University of Pernambuco, Serra Talhada, Pernambuco, Brazil

2 Department of Animal and Plant Production, Federal University of Amazonas, Manaus, Amazonas, Brazil

3 Department of Plant Biology, Federal University of Viçosa, Viçosa, Minas Gerais, Brazil

4 Faculty of Agronomic Engineering, Federal University of Pará, Altamira, Pará, Brazil

References

[1] Ghidelli C, Pérez-Gago MB. Recent advances in modified atmosphere packaging and edible coatings to maintain quality of fresh-cuts fruits and vegetables. Critical Reviews in Food Science and Nutrition. 2016;**28**:1-18. DOI: 10.1080/10408398.2016.1211087

[2] Ingram JS, Humphries JRO. Storage of cassava. A review. Tropical Science. 1972;**9**:277-350

[3] Uarrota VG, Maraschin M. Metabolomic, enzymatic, and histochemical analyzes of cassava roots during postharvest physiological deterioration. BMC Research Notes. 2015;**8**:648

[4] Acedo JZ, Acedo AL. Controlling postharvest physiological deterioration and surface browning in cassava (*Manihot esculenta* Crantz) roots with hot water treatment. Acta Horticulturae. 2013;**989**:357-362

[5] Njoku DN, Amadi CO, Mbe J, Amanze NJ. Strategies to overcome post-harvest physiological deterioration in cassava (*Manihot esculenta*) root: A review. Nigeria Agricultural Journal. 2014;**45**:51-62

[6] Beeching JR, Dodge AD, Moore KG, Philips HM, Whenhan JE. Physiological deterioration in cassava: Possibilities for control. Tropical Science. 1994;**34**:335-343

[7] Cereda MP, Vilpoux O. Conservação de raízes. In: Cereda MP, editor. Tecnologia, Usos e Potencialidades de Tuberosas Amiláceas Latino Americanas. São Paulo: Fundação Cargill; 2003. p. 13-29

[8] Uarrota VG, Nunes EdC, Peruch LAM et al. Toward better understanding of postharvest deterioration: Biochemical changes in stored cassava (*Manihot esculenta* Crantz) roots. *Food Science & Nutrition*. 2016;**4**:409-422. DOI: 10.1002/fsn3.303

[9] Lund DG, Petrini LA, Aleixo JAG, Rombaldi CV. Uso de sanitizantes na redução da carga microbiana de mandioca minimamente processada. Ciência Rural. 2005;**35**:1431-1435

[10] Freire CS, Simões AN, Vieira MRS, Barros-Júnior AP, Costa FB. Qualidade de raízes de mandioca de mesa minimamente processada nos formatos Minitolete e Rubiene. Revista Caatinga. 2014;**27**:95-102

[11] Brito FAL, Maia RKM, Coelho Júnior LF, Coelho DG, Vieira MRS, Silva SLF, Simões AN. Changes in phenol metabolism of minimally processed 'baby cassava' under different temperatures: An alternative to commercialization. African Journal of Biotechnology. 2017;**16**:928-936

[12] Brito FAL, Araújo MLP, Andrade DP, Puschman R, Simões AN. Influence of minimum processing procedures on the quality of sweet cassava. International Journal of Agriculture Innovations and Research. 2013;**2**:189-196

[13] Lana MM, Vieira JV, Silva JBC. Cenourete e Catetinho: Minicenouras brasileiras. Horticultura Brasileira, Brasília. 2001;**19**:376-379

[14] Freire CS, Simões AN, Barros-Júnior AP, Vieira MRS, Ferreira-Silva SL, Silva EF. Activity of oxidative enzymes involved in the browning of minimally processed sweet cassava (*Manihot esculenta* Crantz). Australian Journal of Crop Science. 2015;**9**:296-302

[15] Qin Y, Djabou ASM, An F, Li K, Li Z, Yang L, Wang X, Chen S. Proteomic analysis of injured storage roots in cassava (*Manihot esculenta* Crantz) under postharvest physiological deterioration. PLoS One. 2017;**12**:1-24. DOI: 10.1371/journal.pone.0174238 eCollection 2017

[16] Djabou ASM, Carvalho LJCB, Li QX, Niemenak N, Songbi Chen S. Cassava postharvest physiological deterioration: A complex phenomenon involving calcium signaling, reactive oxygen species and programmed cell death. Acta Physiologiae Plantarum. 2017;**39**:1-10. DOI: 10.1007/s11738-017-2382-0

[17] García JA, Sánchez T, Ceballos H, Alonso L. Non-destructive sampling procedure for biochemical or gene expression studies on post-harvest physiological deterioration of cassava roots. Postharvest Biology and Technology. 2013;**86**:529-535

[18] Simões AN, Costa FB, SILVA E, Puschmann R. Estratégias para reduzir o estresse do processamento mínimo. Visão Agrícola: Pós-colheita de Frutas e Hortaliças. 2007;**7**:92-97

[19] Andrade DP, Brito FAL, Carvalho-Sá MJB, Vieira MRS, Barros Júnior AP, Ferreira-Silva SL, Simões AN. Avaliação de cultivares de mandioca de mesa em diferentes idades de colheita. Interciencia. 2014;**39**:736-741

Chapter 12

Cassava in Central and Western Africa: Postharvest Constraints and Prospects for Research and Market Development

Robert Ndjouenkeu

Additional information is available at the end of the chapter

http://dx.doi.org/10.5772/intechopen.71507

Abstract

Cassava, one of the main components of the diets of the populations of Central and West Africa, conveys an image of the culture of the poor, due to structural and technological constraints that inhibit its industrial and commercial expansion. Technological constraints are reviewed in the context of food uses of the tuber. They mainly focus on the diversity of processing practices, the low technological level of the processing tools and/or their inadequacy, the lack of standardization of processes, and the quality of the products. Removing these constraints calls for technological research for which research and innovation tracks are raised. These mainly concern the characterization and control of existing or potential markets and the optimization of processing processes, in relation with the quality requirements of the products. This optimization approach must take into account the cultural diversity of the actors of the systems, which could be essential, if not crucial, in defining the forms and modes of perception and definition of the quality of cassava products.

Keywords: cassava, postharvest system, processes, quality, market, actors

1. Introduction

Cassava (*Manihot esculenta* Crantz), native in Brazil, is one of the main root and tuber crops grown in the world. The white and milky pulp of the tuberous root occupies an important place in the diet of several people of Africa, especially in Central Africa. In this regard, cassava is the third most important food production in the tropics after rice and maize [1]. Africa is the largest producer, accounting for more than 50% of the world production.

The strong expansion of cassava in the tropics, particularly in Africa, is due to the simplicity of its cultivation; its ability to grow on marginal lands that are difficult to use for other crops; its resistance to drought conditions, which, moreover, justifies its extension in the Sahelian zones; and the possibility of leaving it in the ground and harvesting it progressively, thus allowing for extensive management of its food consumption. As the main component of the food ratio of more than 25% of the African population for an average annual consumption of 100 kg of roots per inhabitant, cassava is, in Central and Western Africa, a crop destined totally and exclusively for human food. However, in emerging economies and cassava producers, as it happens in Brazil, Malaysia, Thailand, and South Africa, the development of cassava cultivation is supported by a process of upgrading the tuber in various food and nonfood industrial systems (animal feed, starches, sugar derivatives, glues, etc.), with highly significant capital gains [2]. In Central and Western Africa, only Nigeria has embarked on a policy of industrial valorization of cassava.

In addition, the nutritional importance of cassava is limited by nutritional and toxicological constraints. The tuber only supplies energy because of its high starch content but is deficient in lipids and proteins—which, in fact, are of poor quality, because they contain very few essential amino acids, minerals, and vitamins [3]. This deficiency also justifies the fact that maps of protein malnutrition (kwashiorkor) coincide with those where cassava predominates in the diet of children [1]. Another limitation of the tuber is the toxicity of certain varieties containing cyanogenic compounds.

Finally, in the countries of Central and Western Africa, cassava is closely linked to rural poverty, although it is not the cause of it; the tuber is sociologically perceived as the culture of the poor, because the marginal areas where cassava is grown are those where poor people generally live [4]. Moreover, the narrowness of individual crop areas, the isolation of production areas, and the low technological level of the processing system are all constraints which limit the productivity of the root and its access to the market and contribute to reducing the level of its industrial valorization.

The above observations have justified a series of reflections on the promotion of cassava development strategies, such as the forum on the Global Cassava Development Strategy (Rome, April 2000), which proposed an approach to make cassava more competitive on the market. This approach is based on identifying and developing the potential markets for cassava and its products and improving varieties and yields to supply these markets with quality tubers and, at competitive prices, technological valorization of the tuber in response to the needs of consumers. In fact, it is a matter of integrating cassava into the lucrative market of starch products, through the development of finished and semifinished products likely to contribute to the agricultural development and economic growth of the producing countries. The great variability of cassava peasant processing systems and products constitutes a foundation whose mastery of practices and associated constraints is the key element for the development of markets and the quality of products.

2. Cassava and its uses in Africa

The food use of cassava incorporates two main forms of consumption, the peeled and cooked tuber, which absorbs about 30% of the African cassava production, and then the remaining

70% is processed into various derived products (chips, flour, cooked pasta, gari, etc.) whose processes and denominations differ from one region to another, even within the same region [1] (**Table 1** and **Figure 1**). Fermented products are the main form of cassava use in Africa, accounting for almost 75% of the cassava-based foods [5]. The microorganisms of cassava fermentation are predominantly lactic bacteria (*Lactobacillus plantarum, Streptococcus faecium, Leuconostoc mesenteroides*) and bacilli [1, 6–8]. Their activity results in the reduction of cyanogenic compounds and in the production of pectinolytic enzymes which promote the softening of the pulp, thus facilitating subsequent manipulations of pressing, crumbling, and conditioning. At the same time, they develop characteristic flavors conferring an organoleptic typology to the cassava fermentation products.

Cassava leaves are also integrated into the diets of several African regions, particularly in Central and Eastern Africa. The young leaves are, depending on the case, fermented or not

Products	Processing
Chips/flour	Peeling roots, cutting into pieces, retting (optional), and drying
Gari	Pre-gelatinized cassava granules obtained by root peeling, grating, pressing/ fermentation, flaking, and pre-gelatinization by roasting in wood fires in large stoves with the addition of palm oil
Attiéké (Côte d'Ivoire)	Cassava semolina obtained as a result of operations analogous to those of the preparation of gari, except that pre-gelatinization is carried out by steam cooking
Fufu (Central Africa) Foo foo (Ghana) Ugali (East Africa)	Paste cooked from fresh roots (foo foo) or chips flour (ugali), or from fresh fermented (water fufu) or dried (fufu) pulp
Chikwangue or Kwanga (Congo), Miondo, Bobolo, Mintoumba (Cameroon) Mboung (Gabon) Mangbele (Central Africa Republic)	Fermented cassava paste cooked in vegetable leaves (banana, ginger). The different local names differ in terms of the fermentation conditions of the tubers during the retting and of the possible addition of other ingredients (palm oil in the case of Mintoumba)
Melongo or Medua-me-mbong (Cameroon) Cassadan (Gabon) Mpataka or jiboh or iwaukpu (Nigeria)	Household conservation form of cooked cassava. Cooked roots are diced or sliced and, then soaked in water with daily change of the soaking water. The product is consumed as snack.
Buvard or mapala or ipoti (Gabon)	Steaming of retted roots in Marantaceae or banana leaves. Conservation by drying or smoking. For consumption, the dried product is soaked in water, followed by a new steaming.
Mahiac (Gabon)	Slurry prepared from retted cassava flour into which grilled peanut paste is added. This preparation is comparable to corn gruel prepared with unroasted peanut paste in savanna regions. Interest: improvement of the nutritional value of the slurry by addition of fat
Nkonda (Cameroon)	Retted cassava paste mixed with groundnut peanut, and possibly crayfish, and then packaged and cooked under the same conditions as miondo, bobolo, and chikwangue.
Ntoba mbodi (Congo)	Sliced and fermented cassava leaves.

Table 1. Some forms of cassava food use in Africa.

Figure 1. Pictures of some traditional cassava products.

and then crushed and cooked in sauce with various condiments. Unlike roots, cassava leaves are characterized by a better nutritional value, with in particular nearly 30% of the proteins. In the Democratic Republic of Congo, cassava leaves have a better market value than roots [9] and account for almost 68% of the leafy vegetables produced in the country [10].

The diversity of forms of cassava use is integrated into the cultural diversity of the producing populations. The unit operations involved in their preparation constitute, regarding the conditions, tools, and means of their implementation, the framework of the constraints to be lifted in order to improve the use and market value of cassava products.

3. Constraints of the cassava production system in Africa

Production and processing constraints of cassava can be perceived both in relation to the modes and conditions of production and to the biological nature of the product and the limits of the processing system.

3.1. Postharvest constraints

Transport of roots from the field to markets and to processing workshops is the first bottleneck in the cassava sector. The bagged roots are transported on the head or on wheelbarrows and carts. Due to the remoteness of the markets and the isolation of production areas, only 15% of the root production arrives on the market or in processing structures [1], unless the latter are close to the fields or if the fields are in peri-urban areas.

The highly perishable character of cassava, due to its water load, limits its consumption period to a few days after harvest. Postharvest degradation of cassava can occur into physiological and microbial forms. Physiological degradation, characterized by internal discoloration of the tuber and known as vascular striatum, is manifested in bluish or brownish streaks along vascular vessels [11]. It is easily induced by mechanical lesions inherent to the conditions of harvesting and handling of the tubers [12]. These lesions serve as a gateway for microorganisms that initiate secondary degradation leading to fermentation phenomena. Various scientific studies dealing with the biochemical mechanisms of physiological and microbial deterioration of cassava highlight the enzymatic processes involved in reactions [13], their association with starch conversion and the accumulation of cyanogenic glycosides [14], as well as differentiation of degradation velocity and the intensity with cassava variety and/or cultivar [12, 15]. These studies provide indications to understand some secular practices of storage, such as the burial of tubers in trenches covered with earth [1], or the relatively recent fact of pruning the plant 2–3 weeks before the harvest [16]. The ancestral landfill practices of fresh cassava have been enhanced by technological research which has proposed stacking fresh tubers in bags, cartons, baskets, or crates filled with sand or sawdust [17, 18], in order to facilitate the transport and conservation of the product. The sand or the sawdust serves as an absorbent material, which helps to regulate the humidity of the medium, a critical parameter of conservation. This practice allows tubers to be kept for up to 2 months, which may be extended by the use of fungicides. Other more elaborate and efficient techniques, such as tuber refrigeration and freezing, have been successfully tested [18] but have the disadvantage of having a high cost of production for a low-cost farmers' system such as cassava, particularly in Africa.

The presence of cyanogenic compounds (linamarin and lotaustralin) in cassava may also be considered as postharvest stress, since these compounds, which may be hydrolyzed to cyanide, limit the food use of certain varieties of cassava, because of its toxicity. The manifestations of this toxicity, combined with a continuous consumption of cassava, concern the degradation of hemoglobin [19], goiter and cretinism [20], and konzo (paralytic disease) [21, 22]. Cassava varieties are arbitrarily classified according to their cyanogenic content [23] and in relation to their toxicity [24, 25]. However, there are still discrepancies in the toxicity threshold: Bolhuis [24] sets the toxicity threshold around 50 mg of HCN/kg fresh tuber, while Rosling [25] proposes 20 mg/kg. The bitter or sweet taste of cassava is also associated with its content of cyanogenic glycosides, bitter cassava being toxic, as it is rich in cyanogens, while sweet cassava is considered healthy. While this is generally true, it should be noted that no scientific cause-and-effect relationship has been established between the organoleptic flavor of cassava and its toxicity, especially since certain varieties of sweet cassava contain cyanogenic glycosides [26, 27], apparently resulting from their environmental conditions of cultivation. In addition, a bitter compound, different from cyanogenic glycosides, has been isolated from cassava [28].

The constraints inherent in the toxicity and perishability of cassava certainly justify the development of peasant farming practices which, for the most part, have higher health and conservation potential with limits at a different level.

3.2. Constraints of the processing system

The unit processing operations of cassava, as applied in peasant practices (**Figure 2**), reveal two main types of constraints, inherent in the induced work load and in the control of processes for the quality of products on the markets.

Most of the postharvest operations at the peasant scale are carried out manually, due to the lack or inadequacy of high-performance processing tools. In a case study of the cassava production and processing processes in Tanzania, Van Oirschot et al. [29] found that manual labor loads are mainly related to harvesting and transport (28%), peeling (35%), and grating (13%). In general, the postharvest workload of cassava is higher than that of other staple foods. The harvest and processing of 10 tons of cassava (average production of 1 hectare of plantation) require about 721 man-hours of work, of which 212 man-hours for harvesting, 156 man-hours for handling operations, and 353 man-hours for processing operations [1, 30]. Moreover, this workload is handled at almost 92% by women [1, 31], since men are involved only when the opportunities for mechanization and commercialization are established [32].

The manual workload induced by postharvest cassava operations results in as many losses during the various operations. The most important losses are induced by the various processing operations (23.2%), harvesting (13.6%), and handling operations (8.5%) [1]. This finding justifies the efforts made in recent years to develop equipment and protocols adapted to the postharvest system of cassava by various research and development institutes, with the aim of both reducing the hardship and improving process and product quality. In this respect,

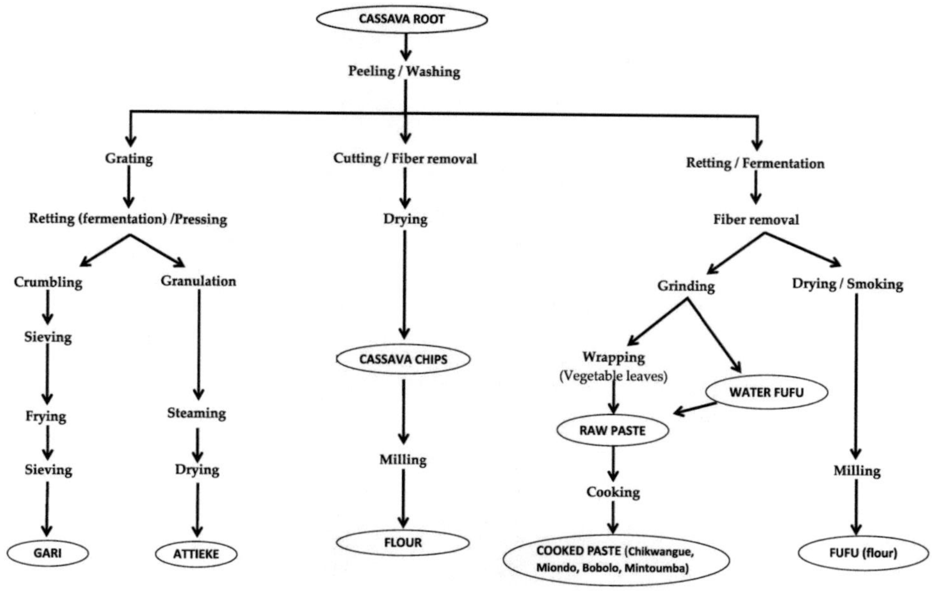

Figure 2. Main operations of traditional cassava processing in Africa.

the technological package proposed and tested by IITA, containing various improved equipments (peeler, rasping, press, mill, gari fryer, sieve), allows significant improvements both in terms of ease of operations and productivity (**Figure 3**). The development of processing equipment is therefore a priority issue for the improvement of the productivity of the cassava system. In this respect, various equipments more or less adapted are offered on the market, with various efficiencies [33], relating to cost, maintenance, and ease of use. Research needs therefore remain important and call for the development of operational approaches and tools to diagnose processing techniques and evaluate the effectiveness of the equipments proposed or to be developed [34–36]. In this logic, peeling, drying, and fermentation processes remain among the most common technological constraints of the cassava system and deserve special attention in terms of both improving and standardizing processes and the development or adaptation of equipment.

Processes for the processing of cassava incorporate, for a given product, unit operations which are substantially comparable from one region to another. However, many variants exist, on the one hand, in the conduct of these unit operations and on the other hand in the nature and the mode of use of the equipment involved. Although this variability in technical practices can be a wealth in terms of specification of product representative of the image of the terroirs, it nevertheless incorporates limits relating to a lack of standardization of techniques and/or products in a given space. Indeed, the same product, processed under comparable technical conditions between two space, or even on the same space with different actors, can present a variation of quality according to space or actors. This lack of standardization can be considered as a constraint whose lifting is likely to create conditions for the emergence of the postharvest cassava system for markets.

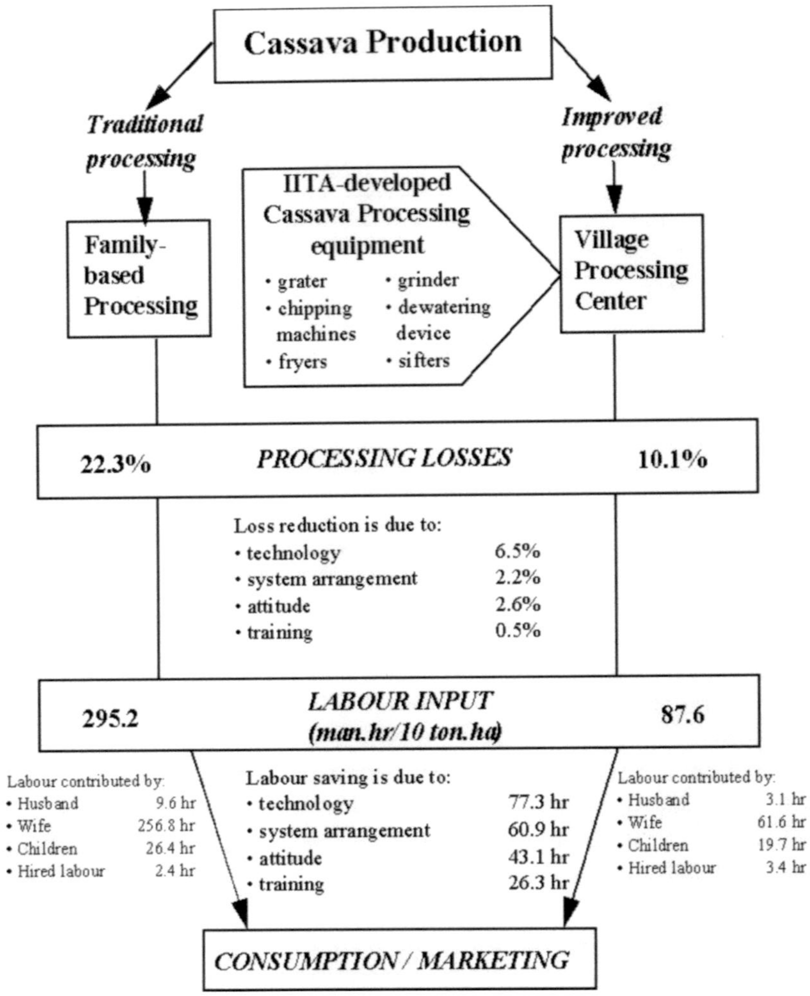

Figure 3. Impact of improved processes and equipments on workload and postharvest losses of the cassava processing system (source: [1, 30]).

The practice of fermentation constitutes one of the operations common to all practices of food processing of cassava. Three variants of this practice are common in Africa:

a. Fermentation of the grated tuber in bag, carried out by lactic bacteria [37–39] and used in the production of gari and attiéké.

b. Fermentation by soaking the tuber in water for 3–6 days depending on the practices and products concerned. This practice, used for the production of fufu and cooked cassava paste (chikwangue, bobolo, miondo, mintoumba), starts with a combined action of various bacteria (*Bacillus, Leuconostoc, Klebsiella, Corynebacterium, Lactobacillus, Aspergillus, Candida, Geotrichum*) and ends by a dominant action of lactic acid bacteria and yeasts [40].

c. The fungal fermentation obtained by piling up of fresh tubers. This practice, common in Tanzania, Uganda, and Mozambique [41, 42], involves microorganisms of the genus *Rhizopus*, *Mucor*, *Penicillium*, and *Fusarium* [43].

Fermentation practices are also applied on cassava leaves for food consumption. In Congo, the semisolid fermentation of cassava leaves results in a product called "*ntoba mbodi*." Though the fermentation of cassava roots is a lactic fermentation with *Lactobacillus* as dominant microflora, that of cassava leaves is an alkaline fermentation where *Bacillus* constitutes the main microflora [44, 45]. The advantage of fermentation is the detoxification of cassava by the degradation of cyanogenic compounds [46]; the improvement of the protein and vitamin value of the product; the development of characteristic texture, aroma, and flavor compounds [47, 48]; and digestibility of the products. The diversity of the microorganisms involved in the fermentation process is, to some extent, indicative of the diversity of the ecosystems in which the tuber is processed, which may justify the diversification of the quality of the products. Various scientific studies have been carried out to identify both the qualitative and quantitative differential functions of the microorganisms involved in the process [49–53] and to standardize the process for a given product [7, 54, 55]. The scientific opportunities offered by these various studies focus as much on the characterization and the control of the enzymatic processes involved in the fermentation operation as on the ecogeographical typology of the microorganisms. Such mastery is likely to allow objective territorial groupings of practices and a standardization approach based on coherent indicators that take into account the tools and technological factors of fermentation, the relationship with the physicochemical and organoleptic characteristics of products, as well as the conditions for their preservation.

If the fermentation practice is decisive for the physicochemical and organoleptic characteristics of cassava processing products, they are also influenced by other factors such as the cassava variety [56–58], the soil type, the planting season [59], the method of culinary preparation [60], and the conditions and means of preserving [61] and distributing the products. These factors must be taken into account in any process of standardization and improvement of the quality of cassava food products [62].

Various research efforts are also being carried out to valorize endogenous cassava processing techniques [63, 64], improve the nutritional value of products [65, 66], develop equipment [67, 68], understand the relationship between processes and quality in order to develop new products of food and industrial interest [69–72], and to attract private sector interest in investment in this sector. However, there are weaknesses in the proposal and the adoption of technological innovations that may favor the transition from self-consumption to industrial or market products. Moreover, beyond the significant amount of work on the physicochemical properties of cassava processing products, in relation to processes, control of the quality and functionality of these products remains relatively limited; this contributes to the weakness of the development of the industrial processing of cassava in tropical Africa.

It is true that local efforts to capitalize technological knowledge through the development of cassava processing SMEs have met with inadequate raw material in most countries, the latter being intended primarily for self-consumption. This fact has justified the implementation of the global cassava development strategy in various African countries, carried out by various international organizations (IFAD, FAO, CIAT, CIRAD, IITA, NRI). This involves not only developing cassava cultivation and improving yields but also exploring and capitalizing on all the technological and market opportunities likely to favor the industrial emergence of the postharvest system.

4. Prospects for the development of the cassava postharvest system in Africa

The recognition of cassava as a strategic culture in Africa, its deep set in local food traditions, and the diversity of its processing products offer so many research and development opportunities for the fight against poverty, improvement of nutritional situations, development of new products, and identification of market niches. The implication of technological research must be based primarily on steps to remove the constraints of the postharvest system and to promote the commercial emergence of products.

The overall cassava development strategy bases the emergence of the postharvest cassava system on the development of markets for agricultural production. In this regard, from the point of view of research, it is appropriate not only to characterize and control the structural and functional elements of these markets but also to correlate their requirements with the quality of cassava products. Therefore, controlling the postharvest processes, consistent with the use value of the products for the markets, constitutes a priority axis of research.

Central and Western Africa, regions with a high intensity of localized production and consumption of cassava, are characterized by relatively small national markets and processing practices of a relatively rudimentary technical level. The analysis of markets, processing processes, and the use value of products in these areas must be able to be carried out in an integrated way, taking into account, on a comparative basis, localized traditional knowledge. This include understanding the social and cultural determinants of local processing practices, the perception and management of product quality by the processing actors, and linking the indicators derived from those findings with the technological quality of the products. The correlation between social perception and management of product quality on the one hand and technological quality on the other hand may give rise to identify innovation opportunities to introduce in the system. An integrated methodology to undergo the above research and development initiative is proposed on **Figure 4**. This methodology starts with an interdisciplinary technical and socioeconomic diagnosis of the whole cassava postharvest sector, covering demand structure, market analysis, processing and consumption practices, actors' organization, processing workshops distribution, etc. Technologists, socioeconomists, geographers, sociologists, and even anthropologists are involved in this diagnosis which allows setting up the key elements for the understanding of the technical system and the boundaries of the activities to undergo further. From the results of the diagnosis, the specialists, later on, return in their specific specialties to study, on the one hand, the perception and the management practices of quality by actors and, on the other hand, the technological characteristics

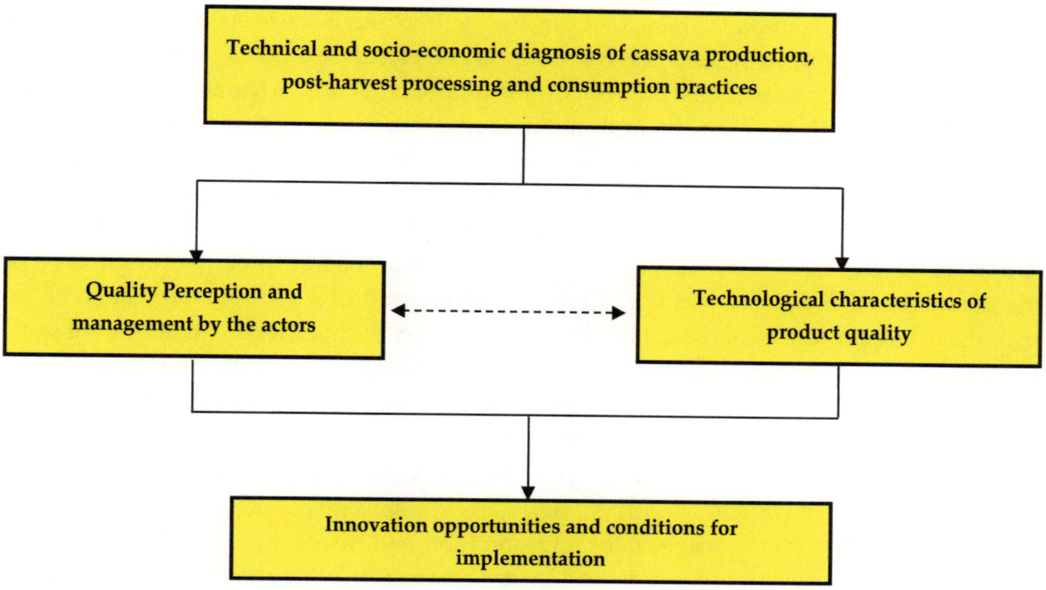

Figure 4. Operational approach for the analysis of production systems and the identification of innovations.

of that quality. Integration can, in this case, consist in a definition of the conditions for the transfer of technological knowledge on an intra- or interregional scale. Similarly, the definition of cost-effective operating conditions for postharvest systems, adapted to the small size of internal markets, may also be an interesting issue, given the geographical isolation of most of the production and processing systems.

From a purely technological point of view, the above market stake is accompanied by a perfect mastery of the unit operations involved in the various processing processes. This mastery goes hand in hand with that of the relationship between localized technological practices and the quality of products. In detail, this technological expertise integrates, as a priority, the optimization of fermentation processes, drying practices, types, and methods of packaging and distribution of products in relation to their physicochemical, nutritional, and functional characteristics. The search for optimization takes into account the analysis and the systematic identification of constraints and development opportunities for each segment of the sector and has as goals which are to choose or develop adapted tools (equipment, starters, etc.) for the standardization of processes. In this context, prioritization of farmers' initiatives, often driven by endogenous dynamics in response to production and processing constraints, must integrate the research approaches to be implemented. It is in fact, for research and development, to rely on the actors' perception of their production constraints, the mechanisms of endogenous reactions they implement in the face of constraints, and then to deduce from it, the opportunities for innovations which will have the advantage of being carried and better adopted by the internal dynamics of these actors. In this respect, and by way of example, a case study on the valorization of peasant retting practices in Central Africa has shown the interest of taking into account the processing initiatives of the actors [73].

The retting process, common to most root processing practices, is the basis of the organoleptic and sanitary quality of fermented cassava products and constitutes thus a critical operation of cassava processing. The primary purpose of this unit operation is to soften and detoxify the root, with softening being the main indicator of retting from the actors' point of view. For this reason, actors use different strategies to accelerate the softening of the root during retting, with regard to the high market demand of retted cassava pulp and products. Old fermented retting solution or preferably pre-retted cassava flour is used in this respect. The limit of this practice is that the result obtained varies from one actor to another and even on the productions of the same actor. Considering that this variability is the result of a non-standardization of the manufacturing process of starters, it has become necessary to characterize these peasant starters and to develop standardization functions. In a participating research with retting actors (effective activities with actors in their workshops, combined to laboratory development), the main standardization functions defined were pre-retting time, reactivation conditions of the starter microflora (reactivation solution, temperature, and time), and conditions of application of the starter (dilution conditions of the starter, starter/cassava ratio, retting temperature). This approach presented more prospects for appropriation and dissemination of the innovation by the actors of cassava processing system. In fact, actors from other cassava processing areas are asking for training on the preparation and use of retting starter, and local development projects have been set up in this respect [74]. The peasant will to optimize the retting process, though endogenous use of starter to accelerate the softening of the roots constitutes a priority axis of development both for scientific and technological mastery requirements and for economic purposes. The variability of the technical practices of retting, the diversity of the microbial strains involved in the process, the varietal diversity of the roots to be retted, or even the diversity of the finished products raises up the problem of standardization of the starter or of research for suitable formulations adapted to each case. These questions constitute converging research opportunities toward the optimization of the cassava root retting process.

It should be noted that agricultural research has developed a wide range of cassava varieties characterized mainly by yields, disease resistance, and adaptation to different ecosystems. In general, little information is available on their technological and food use values. This gap needs to be filled by technological research in view of the emergence of varieties developed for the industry and markets. Similarly, the diversity of cassava varieties, practices, and forms of processing and utilization corresponds, most of the time, to a cultural diversification of the actors, which can be associated with a diversification of perception of the quality of cassava products. In this respect, the following question can be addressed: on the basis of which criteria do the different actors (producers, processors, consumers) of the cassava production and processing system perceive the quality of the tuber and its products? It is possible, hypothetically, to envisage a territorial or cultural and even varietal differentiation of the quality of cassava and its processing products.

5. Conclusion

Since the development of the cassava system depends mainly on the conditions of its operational and effective integration into the market, the support initiatives related to it must take into account the real needs of the actors involved, the majority of whom being under the weight of poverty, so that the emergence of the root does not carry the risk of an extraversion toward new moneyed class actors, to the detriment of the actors of the peasant system, who would continue to convey the false image of cassava as a culture of the poor. Much of this development relates to the need to remove the logistical and technological constraints that hamper the development of the root market and its products, notably the relationship between technical practices of cultivation and processing and quality of products for the markets.

The finalization of such a goal implies the implementation of a dynamic that is based on:

- A thorough knowledge of the existing and potential markets for cassava and its products, taking into account the different mechanisms induced and consumption habits.
- A contextualized analysis of the local systems of processing and use of the roots, as well as their actual or potential relationship with markets, i.e., the quality required by these markets.
- A reflection on the dynamics to accompany or to implement for the operational integration of African cassava production, processing, and utilization systems in a wealth-generating market. It is important to integrate in the process, an approach of ownership of support initiatives by actors involved.

These are some of the issues whose answers are likely to allow the valorization of cassava and its integration into the market. The aim of technological research for the conduct of this dynamic is therefore the development and dissemination of innovations that are relevant and adapted to the requirements of product and market quality.

Acknowledgements

The present review has been written in the framework of an ongoing Central Africa Regional Research and Development project entitled, "*Production durable du manioc en Afrique Centrale et intégration au marché*," funded by the European Union (DCI-FOOD/2010/252–886) and coordinated by PRASAC (Pôle Régional de Recherche Appliquée au développement des Systèmes Agricoles d'Afrique Centrale).

Author details

Robert Ndjouenkeu

Address all correspondence to: rndjouenkeu@gmail.com

Department of Food Science and Nutrition, University of Ngaoundere, Ngaoundere, Cameroon

References

[1] Bokanga M. In: AGSI/FAO, editor. Cassava: Post-harvest Operations. 2001. http://www.fao.org/inpho

[2] FAO. FAO Database 2001. http://www.fao.org

[3] Bichon L. Le manioc. Développement et Santé. 1981;**35**:34-40

[4] Agueguia A, Apedo-Ama A, Fontem DA, Manga GA, Mboua JC, Mouen M., Ndzana X, Ngeve Mbua J, Poné Kamdem D, Tchuanyo M, Tetka J, Tiki Mpondo G, Wouamane M, Zok S. Le manioc un aliment pour tout le monde. In: Bell A, Mück O, Schule B, editors. Les plantes à racines et tubercules en Afrique : une contribution au développement des technologies de récolte et d'après-récolte. DES/ZEL; 2000. http://www.fao.org/inpho

[5] Westby A. Importance of fermentation in cassava processing. In: Proceedings of the Ninth Symposium of the International Society for Tropical Root Crops; Accra, Ghana. 20-26 October 1991

[6] Kastner S, Tschannen A, Assanvo J, N'zi Agbo G, Farah Z, Lacroix C, Meile L. Les microbes dans la fermentation du manioc-vers l'amélioration de la qualité et de la sécurité de l'attiéké. In: Actes de l'Atelier "Potentialités à la transformation du manioc en Afrique de l'Ouest"; Abidjan, Côte d'Ivoire. 4-7 Juin 2007

[7] Assanvo JB, Agbo GN, Behi YEN, Coulin P, Farah Z. Microflora of traditional starter made from cassava for "attiéké" production in Dabou (Côte d'Ivoire). Food Control. 2006;**17**:37-41

[8] Coulin P, Farah Z, Assanvo J, Spillmann H, Puhan Z. Characterisation of the microflora of attiéké, a fermented cassava product, during traditional small-scale preparation. International Journal of Food Microbiology. 2006;**106**:131-136

[9] Lutaladio NB, Ezumah HC. Cassava leaf harvesting in Zaire. In: Terry ER, Oduri KA, Caviness F, editors. Tropical Root Crops: Research Strategies for the 1980s. Ottawa, Canada: IDRC; 1981. pp. 134-138

[10] Tchibaka T, Lumpungu K. Trends and prospects for cassava in Zaire. Working paper n°4 on cassava. Washington: International Food Policy Research Institute; 1989

[11] Averre CW. Vascular streaking of stored cassava roots. In: Proc. 1st Int. Symp. Trop. Root Crops, Trinidad. 1967;**2**(4):31-35

[12] Booth R. Storage of fresh cassava (Manihot esculenta): I. Post-harvest deterioration and its control. Experimental Agriculture. 1976;**12**:103-111

[13] Wheatley CC, Schwabe WW. Scopoletin involvement in post-harvest physiological deterioration of cassava roots (Manihot esculenta Crantz). J. Exptl. Bot. 1985;**36**(166):783-791

[14] Kojima M, Iwatsuki N, Data ES, Villegas CDV, Uritani I. Changes of cyanide content and linamarase activity in wounded cassava roots. Plant Physiology. 1983;**72**:186-189

[15] Montaldo A. Vascular streaking of cassava root tubers. Tropical Science. 1973;**15**:39-46

[16] Marriott J, Been BD, Perkins C. The aetiology of vascular discoloration in cassava roots after harvest: Development of endogenous resistance in stored roots. Physiologia Plantarum. 1979;**54**:51-56

[17] Cock JHC. New potential for a neglected crop. Boulder, Colorado, USA: Westview Press; 1985

[18] Rickard JE, Coursey DG. Cassava storage. I. Storage of fresh cassava roots. Tropical Science. 1981;**23**(1):1-32

[19] Nartey F. Cassava cyanogenesis ultrastructure and seed germination. In: Denis R, Walter F, editors. Cassava. Copenhaggen: Munksgaard. New York. 1978. p. 234

[20] Ermans AM, Bourdoux P, Kinthaert J, Lagasse R, Luvivila K, Mafuta M, Thilly CH, Delange F. Role of cassava in the aetiology of endemic goitre and cretinism. In: Proceedings of a Workshop on Cassava Toxicity and Thyroid: Research and Public Health Issues. 1983; pp. 9-16. Ottawa, Canada, 31 May-2 June, 1982. Internat. Develop. Res. Centre Monogr. IDRC-207e

[21] Tylleskar T. The association between cassava and the paralytic disease konzo. Acta Horticulturea. 1994;**375**:321-331

[22] Howlett WP, Brubaker GR, Mlingi N, Rosling H. Konzo an upper motor neurone disease studied in Tanzania. Brain. 1990;**113**:223-235

[23] Hahn SK, Keyser J. Cassava: A basic food of Africa. Outlook on Agriculture. 1985;**14**(2)

[24] Bolhuis GG. The toxicity of cassava roots. Netherlands Journal of Agricultural Science. 1954;**2**:176-185

[25] Rosling AR. Linamarine. Phytochemistry. 1988;**24**:13-20

[26] Coursey DG. Cassava as food: Toxicity and technology. Chronic cassava toxicity. In: Proceedings of an interdisciplinary workshop, London, England. 1973; pp. 27-36. 29-30 January 1973. International Development Research Centre Monogr. IDRC-010e

[27] Pereira JF, Seigler DS, Splittstoesser WE. Cyanogenesis in sweet and bitter cultivars of cassava. Hortscience. 1981;**16**(6):776-777

[28] King NLR, Bradbury JH. Bitterness of cassava: Identification of a new apiosyl glucoside and other compounds that affect its bitter taste. Journal of the Science of Food and Agriculture. 1995;**68**(2):223-230

[29] Van Oirschot Q, White J, Westby A. Case study on the technical and economic aspects of small–scale cassava processing in a selected village in the Lake Zone of Tanzania. DFID Research for Development Document Record. 2002; 69 pages. http://www.dfid.gov.uk/r4d/SearchResearchDatabase.asp?OutputID=5531

[30] IITA. Improving Postharvest Systems: Archival Report. Ibadan, Nigeria: Crop Improvement Division. International Institute of Tropical Agriculture; 1996

[31] Nweke F. Processing potential for cassava production growth in sub-Saharan Africa. In: COSCA Working Paper No. 11. Collaborative Study of Cassava in Africa. Ibadan, Nigeria: International Institute of Tropical Agriculture (IITA); 1994

[32] Ugwu BO, Ay P. Seasonality of cassava processing in Africa and tests of hypotheses. In: COSCA Working Paper No. 6. Ibadan, Nigeria: International Institute of Tropical Agriculture (IITA); 1992

[33] Davies RM, Olatunji MO, Burubai WA. Survey of cassava processing machinery in Oyo State. World Journal of Agricultural Sciences. 2008;**4**(3):337-340

[34] Marouzé C, Da G, Dufour D. Inventaire des équipements de petite et moyenne capacités de production d'amidon de manioc par la voie humide. In : Amani G, Nindjin C, N'zue B, Tschannen A, Aka D, editors. Actes de l'Atelier "Potentialités à la transformation du manioc en Afrique de l'Ouest"; Abidjan, 4–7 Juin 2007; pp. 127–132

[35] Marouzé C, Da G. Outil de caractérisation d'une ligne de transformation alimentaire; cas de la production d'amidon de manioc. In : Amani G, Nindjin C, N'zue B, Tschannen A, Aka D. (eds), Actes de l'Atelier "Potentialités à la transformation du manioc en Afrique de l'Ouest"; Abidjan, 4–7 Juin 2007; pp. 133–139

[36] Smith DW, Sims BG, O'Neill DH, editors. Sélection et évaluation des équipements et machines agricoles. Principes et application. In: Bulletin des services agricoles de la FAO. Rome: FAO; 1998. p. 110

[37] Ngaba PR, Lee JS. Fermentation of cassava (*Manihot esculenta* Crantz). Journal of Food Science. 1979;**44**:1570-1571

[38] Abe MO, Lindsay R. Evidence for the lactic streptococci role in acidic cassava (*Manihot esculenta* Crantz). Journal of Food Science. 1978;**42**:781-784

[39] Okafor N. Microorganisms associated with cassava fermentation for gari production. Journal of Applied Microbiology. 1977;**42**:279-284

[40] Oyewole OB, Odunfa SA. Microbiological studies on cassava fermentation for lafun production. Food Microbiology. 1988;**5**:125-133

[41] Ndunguru GT, Thomson M, Waida TDR, Rwiza E, Jeremiah S, Westby A. Relationship between quality and economic value of dried cassava products in Mwanza, Tanzania. In: Akoroda MO, Terri J, editors. Food Security and Crop Diversification in SADC Countries: The Role of Cassava and Sweet Potato. Malawi: SARRNET; 1999. pp. 408-414

[42] Essers AJA. Removal of cyanogens from cassava roots: studies on domestic sun–drying and solid substrate fermentation in rural Africa (Ph.D. thesis). The Netherlands: Wageningen Agricultural University; 1995

[43] Essers AJA, Nout MJR. The safety of dark moulded cassava flour compared with white—A comparison of traditionally dried cassava pieces in North East Mozambique. Tropical Science. 1889;**29**:261-268

[44] Kobawila SC, Louembe D, Keleke S, Hounhouigan J, Gamba C. Reduction of the cyanide content during fermentation of cassava roots and leaves to produce bikedi and ntoba mbodi, two food products from Congo. African Journal of Biotechnology. 2005;**4**(7):689-696

[45] Nguie R, Moyen R, Mokemiabeka SN, Koumou CAGN, Kobawila SC, Louembe D. Physiological characterization of Staphylococci and micrococci isolated from fermented cassava leave (*Manihot esculenta* Crantz), Ntoba mbodi. International Journal of Current Microbiology and Applied Sciences. 2016;**5**(8):936-942

[46] Meuser F, Smolnik FD. Processing of cassava to gari and other foodstuffs. Starch. 1980;**32**(4):116-122

[47] Brauman A, Keleke S, Mavoungou O, Ampe F, Miambi E. Etude cinétique du Rouissage traditionnel du manioc en Afrique Centrale (Congo). In: Agbor ET, Brauman A, Grifon D, Treche S, editors. Transformation Alimentaire du manioc. France, Paris: Editions ORSTOM; 1995. pp. 287-305

[48] Phan TH, Mercier C. Amélioration de la composition et de la texture du plat traditionnel africain « le foufou » par addition de monoglycerides au manioc. Sciences des Aliments. 1984;**4**:109-120

[49] Daouda Nimaga, Akely PMT, Amani GN, Kakou C. Influence des caractéristiques microbiologiques du ferment sur la qualité organoleptique de l'Attiéké. In: Amani G, Nindjin C, N'zue B, Tschannen A, Aka D. (eds), Actes de l'Atelier "Potentialités à la transformation du manioc en Afrique de l'Ouest"; Abidjan, 4–7 Juin 2007; pp. 285–287

[50] Keleke S, Butel MJ, Rimbault A. Etude du phénomène de ramollissement des racines de manioc au cours du Rouissage. In: Amani G, Nindjin C, N'zue B, Tschannen A, Aka D, editors. Actes de l'Atelier "Potentialités à la transformation du manioc en Afrique de l'Ouest"; Abidjan, 4–7 Juin 2007; pp. 275–284

[51] Assanvo JB. Etude de la microflore du ferment de manioc pour la production d'attiéké. Enquêtes de consommation et de production d'attiéké réalisées à Abidjan, Dabou et Jacqueville. Revue internationale des sciences de la vie et de la terre, n° spécial. 2002. Acte du Colloque International, CSRS 27-29 Août 2001. Abidjan: Editions Universitaire de Côte d'Ivoire. pp. 1-9

[52] Ampe F, Brauman A. Origin of enzymes involved in detoxification and root softening during cassava retting. World Journal of Microbiology and Biotechnology. 1995;**11**:178-182

[53] Brauman A, Kéléké S, Malonga M, Miambi E, Ampe F. Microbiological and biochemical characterization of cassava retting: a traditional lactic acid fermentation for foo–foo (cassava flour) production. Applied and Environ. Microbiol. 1996;**62**(8):2854-2858

[54] Kastner S, Tschannen A, Assanvo J, N'zi Agbo G., Farah Z, Lacroix C, Meile L. Les microbes dans la fermentation du manioc – vers l'amélioration de la qualité et de la sécurité de l'attiéké. In: Amani G, Nindjin C, N'zue B, Tschannen A, Aka D. (eds), Actes de l'Atelier "Potentialités à la transformation du manioc en Afrique de l'Ouest"; Abidjan, 4-7 Juin 2007; pp. 252-255

[55] Djoulde Darman R. Mise au point d'un ferment destine à la bioconversion des tubercules de manioc cyanogène (Ph.D thesis). Cameroun: Université de Ngaoundéré; 2005

[56] Shittu TA, Sanni LO, Awonorin SO, Maziya-Dixon B, Dixon A. Effect of genotype on the flour making properties of some CMD resistant varieties of cassava. Food Chemistry. 2007;**101**:1634-1643

[57] Ehui FH, Amani NG, Djédji C, Sako A. Etude des propriétés fonctionnelles des amidons de variétés sélectionnées de manioc (*Manihot esculenta* Crantz). In: Amani G, Nindjin C, N'zue B, Tschannen A, Aka D. (eds), Actes de l'Atelier "Potentialités à la transformation du manioc en Afrique de l'Ouest"; Abidjan, 4–7 Juin 2007; pp. 195–197

[58] Ekissi AC, Nindjin C, Bakayoko S, Amani NG, Kamenan A. Caractérisation biochimique et sensorielle de variétés de manioc (*Manihot esculenta* Crantz) améliorées. In: Amani G, Nindjin C, N'zue B, Tschannen A, Aka D, editors. Actes de l'Atelier "Potentialités à la transformation du manioc en Afrique de l'Ouest"; Abidjan, 4–7 Juin 2007; pp. 198–200

[59] Hongbete F, Dedegbe D, Akissoe N, Mestres C, Nago M. Influence de la variété et de la saison de plantation sur la texture du manioc (*Manihot esculenta* Crantz). In: Amani G, Nindjin C, N'zue B, Tschannen A, Aka D, editors. Actes de l'Atelier "Potentialités à la transformation du manioc en Afrique de l'Ouest"; Abidjan, 4–7 Juin 2007; pp. 190–194

[60] Ayemou-Allou A, Nindjin C, Dao D, Girardin O, Assa A. Effet du type de sol sur les attributs sensoriels de trois produits de transformation à base de manioc (*Manihot esculenta* Crantz). In: Amani G, Nindjin C, N'zue B, Tschannen A, Aka D. (eds), Actes de l'Atelier "Potentialités à la transformation du manioc en Afrique de l'Ouest"; Abidjan, 4-7 Juin 2007; pp. 187-189

[61] Bindzi JM, Guiama VD, Ndjouenkeu R. Combined smoking and drying kinetics of retted cassava pulp and effects on the physicochemical and pasting properties of the extracted starch. Food and Bioprocess Technology. 2014;**7**:2749-2758

[62] Johnson Paa-Nii T, Tomlins Keith I, Oduro-Yeboah C, Tortoe C, Quayson ET. Une étude de cas visant à élaborer un système d'assurance de qualité adaptée à deux aliments prêt–à–servir à base de manioc. In: Amani G, Nindjin C, N'zue B, Tschannen A, Aka D, editors. Actes de l'Atelier "Potentialités à la transformation du manioc en Afrique de l'Ouest"; Abidjan, 4–7 Juin 2007; pp. 293–297

[63] Rasoanantoandro-Gothard MC. L'amélioration des techniques de transformation dumanioc en cossettes en RCA : Rouissage en vase clos et séchage en en cage thermique. In : Jamin JY, Seiny Boukar L, Floret C, éditors. Savanes africaines: des espaces en mutation, des acteurs face à de nouveaux défis. Actes du colloque, mai 2002, Garoua, Cameroun. Prasac, N'Djamena, Tchad-Cirad, Montpellier, France; 2003

[64] Aboua F. Optimum conditions for cooking attiéké. Centre Ivoirien de Recherche Technologique. Tropical Sciences. 1998;**38**:220-223

[65] Egounlety M, Adjakidje AS, Segbedji CM, Yao AA, Dortu C, Kostinek M, Franz CMAP, Thonart P, Holzapfel WH, Mengu M. Vers l'industrialisation des aliments fermentés traditionnels africains : une étude de cas du gari fortifié au Bénin. In: Amani G, Nindjin C, N'zue B, Tschannen A, Aka D. (eds), Actes de l'Atelier "Potentialités à la transformation du manioc en Afrique de l'Ouest"; Abidjan, 4–7 Juin 2007; pp. 242–247

[66] Amani NG, Kamenan A. Nutritional potentialities and traditional process of starchy crops in the Côte d'Ivoire. 2nd International workshop. Food-based approaches for a healthy nutrition in west Africa: the role of food technologists and nutritionists. Ouagadougou, Burkina Faso. 2003; 152 pp

[67] Sanni L, Alenkhe B, Edosio R, Patino M, Dixon A. Technology transfer in developing countries: Capitalizing on Equipment Development. Journal of Food, Agriculture & Environment. 2007;**5**(2):88-91

[68] Tetevi K, Sedzro K. Vulgarisation d'une râpe à manioc motorisée. In: Amani G, Nindjin C, N'zue B, Tschannen A., Aka D. (eds), Actes de l'Atelier "Potentialités à la transformation du manioc en Afrique de l'Ouest"; Abidjan, 4–7 Juin 2007; pp. 140-144

[69] Touko, AB, Egue K, Goto CE, Sedzro K, Tougnon K, Amouzou A. Promotion de la production du manioc par la diversification de ses formes d'utilisation. In: Amani G, Nindjin C, N'zue B, Tschannen A, Aka D. (eds), Actes de l'Atelier "Potentialités à la transformation du manioc en Afrique de l'Ouest"; Abidjan, 4–7 Juin 2007; pp. 120-122

[70] Onabolu A, Bokanga M. Promotion of high quality cassava flour (HQCF) for the food industry in Nigeria. Proc. 6th Sym. ISTRC-AB. 1995; pp 263–269

[71] Udensi EA, Ukozor AUC, Ekwu FC. Effect of fermentation, blanching, and drying temperature on the functional and chemical properties of cassava flour. International Journal of Food Properties. 2005;**8**(1):1171-1177

[72] Niba LL, Bokanga MM, Jackson FL, Schlimme DS, Li BW. Physicochemical properties and starch granular characteristics of flour from various *Manihot esculenta* (cassava) genotypes. Journal of Food Science. 2002;**67**(5):1701-1705

[73] Ze NN, Kamdem SS, Agnia HB, Mengo F E'o, Ngang JJE. Assessment of cassava (*Manihot esculenta* Crantz) fermentation and detoxification using previously cassava-fermented chips. British Journal of Applied Science & Technology. 2016;**13**(3):1-10

[74] CTA Valorisation et diffusion de starters paysans de rouissage du manioc au Cameroun. Projet ENSAI N° 1–1–13–011–6–Contrat 2017–130. 2017

Cassava Residues Could Provide Sustainable Bioenergy for Cassava Producing Nations

Sammy N. Aso, Arthur A. Teixeira and Simeon C. Achinewhu

Additional information is available at the end of the chapter

http://dx.doi.org/10.5772/intechopen.72166

Abstract

Many cassava producing nations lack suitable energy availability and sufficiency. Just 10% of the population in Haiti receive power from the national electric grid. The proportion is 7% for Mozambique. In both countries, deforestation is extensive due to dependence on wood and charcoal for 70 and 85% of energy requirement respectively. In the case of Ghana, although biomass accounts for 64% of national energy supply, the dependence on primary biomass energy sources like wood and charcoal has led to increased loss of biodiversity, soil erosion and health problems. Prospects for the use of cassava peeling residues as a source of biomethane to mitigate these constraints have received little attention. In this chapter, the advantages and benefits of biomass energy, along with the potential for cassava as a feedstock and rationale for anaerobic digestion are highlighted. Depending on the quantity of cassava root processed by individual countries, the energy recovered from cassava peeling residues could satisfy up to 100% of national energy requirements.

Keywords: biomass, cassava residues, anaerobic digestion, biomethane, renewable energy

1. Introduction

In July 2015, world population was estimated at over 7.3×10^9 persons and will exceed 9.7×10^9 persons in July 2050 [1]. At the same time, planet earth's capacity to sustain life is diminishing. Issues such as land use conflicts, rural poverty, food insecurity, energy insecurity and environmental pollution are posing serious threats to humanity. Global energy supply is dependent on fossil fuels, which account for over 78% of final energy consumption [2]. Fossil fuels are depleting non-renewables, and their use exacerbates anthropogenic forcing of environmental perturbations including carbon dioxide and other greenhouse gas emissions, acid rain, biodiversity and ozone layer depletions. Advancing global energy supply system toward renewable bioenergy could constrain these adverse impacts.

The poor economic development and progress in developing countries have been attributed in part to inadequate suitable energy supply. Mainly developing countries produce cassava. However, most of the production occurs in rural areas where fuel/electricity availability is limited. In Mozambique for instance, only 7% of households (1% in rural areas) has access to electricity [3], and 85% of total energy consumed comes from firewood and charcoal [4]. In Haiti, 10% of the population receive power from the national electric grid while wood and charcoal account for 70% of the nation's energy use [5, 6]. These circumstances have led to extensive deforestation and soil erosion in both countries; with just 1.5% of land forested in Haiti [6], and 219,000 hectares of land deforested per year in Mozambique [4]. In the case of Ghana, although biomass accounts for 64% of national energy supply, the dependence on primary biomass energy sources like wood and charcoal has led to increased deforestation, land degradation, loss of biodiversity, soil erosion, and health problems [7]. Creative use of cassava as an energy crop would help to mitigate environmental degradation and energy paucity issues, as well as minimize the health problems associated with the combustion of firewood and charcoal in cassava producing nations.

Biomass energy should be of interest to developing and developed countries. This is because biomass alleviates reliance on limited fossil fuel sources, creates employment, and contributes to economic development and revitalization of rural communities. Biomass is a clean energy source that dramatically improves the environment by generating far less air emissions than fossil fuels, reduces the amount of waste sent to landfills, and decreases reliance on chemical fertilizer. Moreover, biomass energy is renewable and therefore sustainable. Renewable energy supplied 19% of global energy consumption in 2012 and in 2013, accounted for more than 56% of net additions to global power capacity with about 6.5 million people employed [2]. In 2015, renewable energy sales in Europe was 150 billion euros (≈ US$ 178 billion) [8].

These trends demonstrate the growing utility, benefits and advantages of renewable energy of which biomass energy is a major component. However, the use of edible biomass (food crops such as sugarcane, corn (maize), soybean, palm oil, etc.) for biofuel (bioethanol, biodiesel, etc.) production has raised ethical concerns about competition and diversion of land and food to fuel production. Perhaps a reasonable alternative is biofuel production from biomass originating from nonfood sources such as, agricultural residues, food processing residues, lignocelluloses, and microalgae. Nonfood cassava peeling residues (CPR) could come to the rescue.

2. Potential of cassava as feedstock for bioenergy production

Cassava (*Manihot esculenta Crantz*) is a mostly vegetatively propagated perennial root crop that grows well in tropical climates. Nevertheless, the roots (main reason for growing cassava) are very perishable once taken from the soil and go to waste unless processed in some way soon after harvest. Most processing requires removal of peels (cortex and periderm), head, and tail ends. These components usually discarded as waste, engender environmental pollution. In this chapter, the components are referred to collectively as cassava peeling residues (CPR), and instead of being discarded as waste, would be put to bioenergy production function. The CPR is generated during production of numerous cassava root based food products like akpakpuru, attieke, casabe, chickwangue, farina (farinha de mandioca), fufu, fuku, gaplek, gari, ijapu, konkonte, lafun, landang, peujeum, and thundam [9–22]. Because more than 65% of global annual cassava output is processed for human consumption, enormous quantity of CPR is generated. This nonfood organic matter is potential good feedstock for anaerobic digestion (AD) processes that generate bioenergy.

There are numerous other reasons for the attraction of cassava crop as source of food and bioenergy.

- Cassava provides economic and subsistence value for 800–1000 million people in more than 90 countries including Angola, Barbados, Brazil, Cambodia, China, Cook Islands, Democratic Republic of Congo, Dominica, Ghana, Haiti, India, Indonesia, Lao Peoples Democratic Republic, Mozambique, Nigeria, Suriname, Thailand, Uganda, United Republic of Tanzania, and Vietnam [23–25].

- It is the fourth most important food crop in developing nations. Cassava is also world's third largest source of food carbohydrates and the top food energy supplier for tropical and subtropical regions. About 30% of all calories consumed in Mozambique come from cassava [26]. In Zaire, cassava roots provide 60% of the daily caloric intake, while 20% of protein come from cassava leaves [27]. In addition, Cassava can be biofortified with vitamin A, iron and zinc to eliminate hidden hunger and improve the nutritional status of vulnerable groups.

- Cassava presents numerous agro-climatic advantages and benefits as well. First, it has high biological efficiency as the edible root portion lies underground and does not require support from stems and branches. It is easily cultivated by stem cuttings for multiplication and planting purposes, and requires minimum agricultural inputs (fertilizers, pesticides, etc.). With the possible exception of sugarcane, cassava's productivity in terms of calories per unit land area per unit of time is significantly higher than that of other staple food crops; and its production requires energy input that constitutes just 5–6% of the energy output of the entire cassava biomass [28].

- Cassava can be planted most time of the year and is available all year long with more than 2 years harvest window. Cassava is adaptable to various farming systems. It can be intercropped with beans, yams, and other annual crops. It is tolerant of various climatic conditions (e.g., high drought; temperature: 8–33°C; rainfall: 500–6000 mm per annum; relative humidity: 15–90%; and elevation: sea level–2500 m). Cassava is also productive on soils with pH of 3–9.5. It can thus be cultivated on marginal lands where other crops such as corn, wheat, rice and sugarcane cannot be grown well [29, 30]. Cassava has high efficiency of photosynthetic CO_2 assimilation. The photosynthetic rate of cassava is 40–50 μmol CO_2 m^{-2} s^{-1} under high solar radiation. That of rice is around 20 μmol CO_2 m^{-2} s^{-1} [31].

- Cassava root is endowed with high starch content of excellent functional and structural qualities. The cassava starch can be transformed into products with huge industrial applications and is of major economic importance in Brazil, India, Indonesia, Philippines, China, Thailand, South East Asia, and in the tropical regions of the world.

- Cassava is a major ingredient for livestock feeds.
- Cassava is important in the provision of bioenergy such as bioethanol and biogas. For instance, the yield of bioethanol from cassava (6000 kg/ha) is higher than that of sugarcane (4900 kg/ha), carrot (4500 kg/ha), sweet sorghum (2800 kg/ha), Rice (2250 kg/ha), Maize (2050 kg/ha), and wheat (1560 kg/ha) [32]. A feasibility case study in Kenya using biogas engine for backup power generation showed ample savings over the use of diesel engine. Biogas engine saved 17 tons of carbon dioxide emissions, 18% reduction in net present cost, 20% reduction in levelized cost of electricity, and 30% reduction in capital cost [33].

Energy recycling from biomass residues and wastes is increasingly attractive because the sustainability of analyzed feedstock favors biomass waste flows over dedicatedly cultivated

S/N	Feed stock	BFP yield	BFP units	References
1.	Cassava peeling residue	377	L CH_4/(kg VS)	[47]
2.	Cassava peeling residue	180–310	L CH_4/(kg VS)	[10]
3.	Cassava peeling residue	280	L CH_4/(kg VS)	[48]
4.	Cassava peeling residue	87.1	L biogas/(Total mass of slurry)	[49]
5.	Cassava peeling residue	68.7	L biogas/(Total mass of slurry)	[50]
6.	Cassava starch extraction wastewater	360	L biogas/(kg COD removed)	[51]
7.	Cassava starch extraction wastewater	130–325	L biogas/(kg dry matter)	[52]
8.	Cassava starch extraction wastewater	134–316	L CH_4/(kg VSS Day)	[53]
9.	Cassava starch extraction wastewater	140	Nm^3 per Mg dry mass of COD	[54]
10.	Cassava starch extraction wastewater	11.3	L CH_4/(kg VSS Day)	[55]
11.	Cassava starch extraction wastewater	0.52–3.70	L biogas/(L wastewater Day)	[51]
12.	Cassava starch extraction wastewater	0.40–0.55	L CH_4/(L effluent Day)	[53]
13.	Cassava stillage	215–380	L CH_4/(kg VS)	[56]
14.	Cassava stillage	132–259	L CH_4/(kg VS)	[57]
15.	Cassava stillage	249	L CH_4/(kg VS)	[58]
16.	Cassava stillage	158–248	L CH_4/(kg VS)	[59]
17.	Cassava stillage	220–230	L CH_4/(kg COD added)	[60]
18.	Cassava tubers	660	L biogas/(kg VS)	[48]
19.	Cassava tubers	475–510	L biogas/(kg VS)	[61]
20.	Cassava stem residues after starch extraction	153	Nm^3 per Mg dry mass of stem residues	[54]
21.	Cassava flour and meal industry effluent	14.5	L biogas/Day	[62]

Table 1. Biofuel potentials (BFP) of cassava feedstocks.

energy crops [34]. Therefore, utilization of nonfood cassava processing residues such as CPR in biomethane production via the anaerobic digestion technology is prudent and beneficial. Nevertheless, in order to properly assess and quantify the value and contribution of CPR to the energy mix of cassava producing nations, establishment of Biofuel Potential (BFP) of CPR is necessary. Relatively very few studies have been published on biomethane production from cassava feedstocks. Most of the studies utilized cassava starch extraction wastewater. Other cassava feedstocks used were stillage (wastewater) from cassava ethanol production; cassava stem residue; whole cassava root; effluent from cassava flour and meal industry; and cassava peeling residue (CPR). However, CPR constitutes about 19% fresh weight of the root and is perhaps the most abundant residue from cassava root processing. It is easy to generate and does not require water usage. Therefore, analyses of energy impact of cassava crop in this chapter will use CPR as the feedstock of choice in renewable biomethane production. **Table 1** summarizes the biofuel potentials of CPR and other cassava feedstocks.

3. Rationale for anaerobic digestion technology

Anaerobic digestion (AD) is a biochemical process that converts organic matter to biogas (a mixture of methane and carbon dioxide). This is achieved through the action of a mixed culture of naturally occurring microorganisms under near oxygen free ambient environmental conditions. The following attributes are among the numerous advantages and benefits of AD technology:

a. Flexible technology; energy efficient; prevents emission of volatile hazardous compounds (air pollution control); biotransformation and biodegradation of xenobiotics; treatment of seasonal effluents (e.g., wastewaters from sugar and fish processing industries); system stability and minimal operational difficulties such as bulking and biomass washout; higher loading rates and concentrations operations, from 20 to 40 kg BOD removed/m^2 per day; reduced mass and volume of waste sludge; high waste stabilization; and Low construction, treatment and maintenance costs are typical examples.

b. AD can accommodate tighter restrictions on sludge disposal site location, air pollution, hazardous waste disposal, odor control, and other environmental regulations.

c. AD is attractive as a means of generating alternative energy such as biogas used for electricity and heat production, and to feed gas networks.

d. Among biofuel systems, AD is a highly energy positive process. AD generates energy as methane; with about 3.53 kWh/(kg COD) produced as biogas while aerobic treatment operations consume 0.5–2.0 kWh/(kg O_2) [35].

e. AD is used to produce hygienic digestate; a good source of soil organic amendment, compost and biofertilizer that can be sold for income generation.

f. AD is a low cost, low technology energy source for developing countries. It can be used to achieve more sustainability and energy justice in society [36, 37].

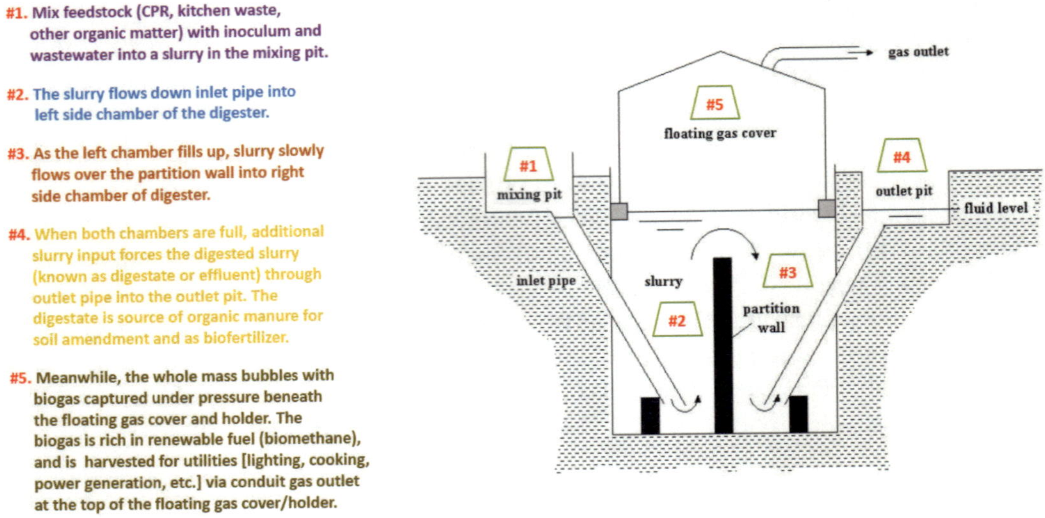

Figure 1. Basic architecture and operating principles of the floating cover anaerobic digester. The sketch to the right was adapted with permission from Ref. [46].

g. In addition, AD is versatile, with commercial equipment in varied types, shapes, sizes and operating modes. These include BIOCEL, Bioferm, GICON and SEBAC (sequential batch anaerobic composting); as well as the ABR (anaerobic baffled reactor), AF (anaerobic filter), CSTR (completely stirred tank reactor), EGSB (expanded granular sludge bed), UASB (upflow anaerobic sludge bed), fixed dome, floating cover, and balloon/tube digesters. **Figure 1** highlights the operating principles of a simple floating cover anaerobic digester.

4. Technical feasibility of anaerobic digestion of CPR

Information obtained from available literature were analyzed to determine critical values relevant to CPR biogasification characteristics. **Tables 2** and **3** summarize estimates and assumptions concluded from the analyses. These data and assumptions were also used to perform the mass balance computations presented in **Figure 2**.

To compute the energy obtained from methane generated by AD of CPR, the following equations were used.

$$T_{me} = M \times HHV \tag{1}$$

$$E_{me} = T_{me} \times \epsilon = M \times HHV \times \epsilon \tag{2}$$

S/N	CPR mass fraction (%)	References
1.	18	[10]
2.	18	[63]
3.	30	[47]
4.	16	[64]
5.	17	[65]
6.	17	[66]
Mean: ≈ 19		

Table 2. CPR mass fractions of fresh cassava root.

where T_{me} = Thermal energy content of methane (MJ), M = Mass of methane (kg), HHV = Heat of combustion of methane (MJ/kg), E_{me} = Electrical energy equivalent of T_{me} (MWh), ϵ = Conversion efficiency; thermal energy to electrical energy (%), Note: This work used HHV = 55.53 MJ/kg, ϵ = 25% and 3600 MJ = 1 MWh.

Table 4 presents the 2014 cassava output and energy consumption patterns of cassava producing countries. Many of these countries are net energy importers; lacking in local energy capacity and sufficiency. However, in 2014, the global cassava output was over 268 million tonnes. Based on the equations, mass balances and analyses already posited in this chapter,

S/N	Variable of interest	Unit	Value assumed	Explanation/justification	Source/ references
1.	CPR mass fraction of root	(%)	19	Derived from literature data	**Table 2**
2.	CPR moisture content, wet basis	(%)	67	Mean of four replications	[10]
3.	CPR methane capacity	(L CH_4/ (kg VS))	303	Derived from **Table 1**	[10, 47, 48]
4.	Methane obtained from CPR generated by processing 1 tonne (1000 kg) of roots	(kg)	12.55	Derived from 252 L CH_4/(kg VS CPR) = 10.44 kg CH_4	[10]
5.	Proportion of cassava root output processed	(%)	66 & 100	More than two–thirds of total production processed for human food	[67]
6.	Quantity of cassava root output processed	(kg)	Varies per country	Based on 66 and 100% of individual country's 2014 cassava output (see **Table 4**)	[24, 67]

Table 3. Values assumed for variable parameters used in analytical modeling.

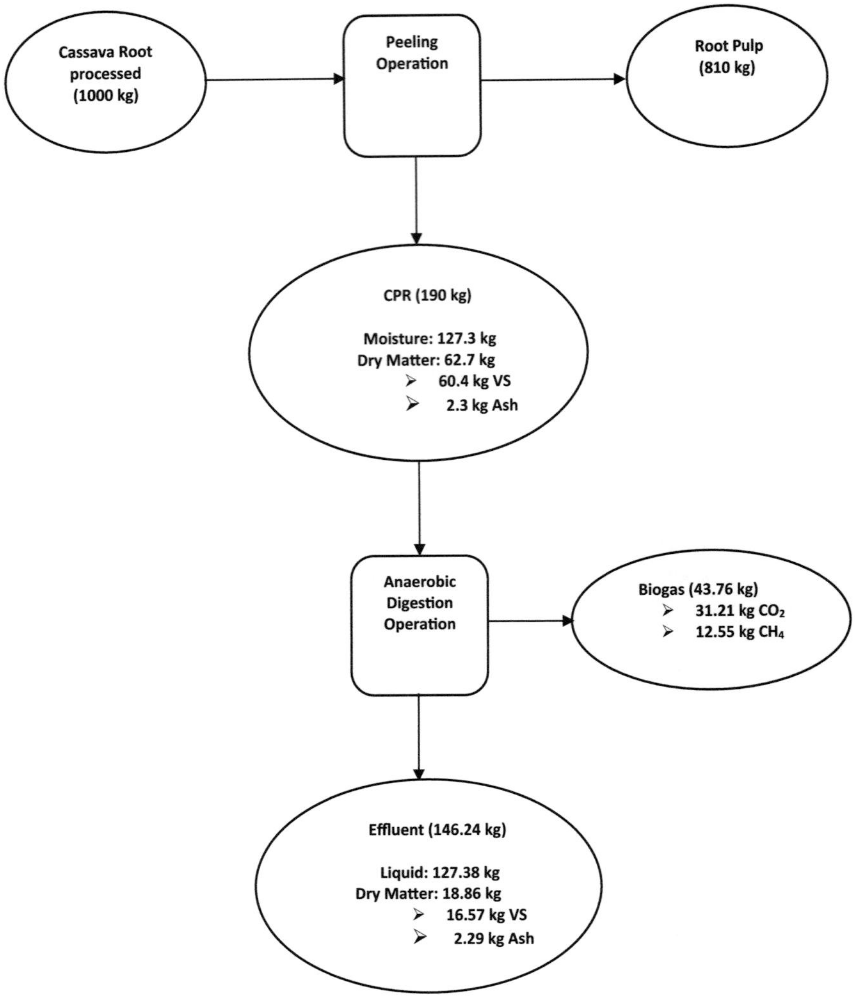

Figure 2. Mass balance for the anaerobic digestion of CPR from 1 tonne (1000 kg) of cassava root for biomethane production.

1 tonne of cassava root yielded 190 kg CPR. This CPR is transformed to 12.55 kg of methane; producing 697 MJ thermal energy or about 174 MJ electrical energy (≈ 0.0484 MWh). Therefore, the 268 million tonnes global cassava root output in 2014 would produce 51 million tonnes of CPR. This quantity of CPR would generate 3363.4 million kg of bio-methane which translates to 186.8×10^9 MJ of thermal energy; equivalent to 46.7×10^9 MJ of electrical energy ($\approx 13 \times 10^6$ MWh). This is an enormous quantity of energy that could satisfy all the yearly energy needed by Slovenia or Turkmenistan. This energy should be recovered for the benefit and rescue of cassava producing nations. The foregoing analyses were applied to individual cassava producing nations to estimate the energy recoverable from their CPR. The results obtained are also presented in **Table 4**. It could be seen that the ability of recoverable energy from CPR to provide national energy requirement depends on the quantity or proportion of national cassava root output processed.

http://dx.doi.org/10.5772/intechopen.72166

S/N	Country	2014 Cassava output (×10^9 kg) [b]	2014 National energy consumption (MW.h/Y) [c]	Energy from CPR if 100% of national cassava output was processed (MW.h/Y)	Potential of CPR to provide national energy requirement if 100% national cassava output was processed (%)	Energy from CPR if 66% of national cassava output was processed (MW.h/Y)	Potential of CPR to provide national energy requirement if 66% national cassava output was processed (%)
1.	Nigeria	54.8316	24,000,000	2653301.124	11.05542135	1751178.742	7.296578091
2.	Thailand	30.022052	164,000,000	1452767.096	0.885833595	958826.2835	0.584650173
3.	Indonesia	23.436384	195,000,000	1134086.622	0.581582883	748497.1704	0.383844703
4.	Brazil	23.253514	518,000,000	1125237.542	0.217227325	742656.778	0.143370034
5.	Ghana	16.524	9,200,000	799596.36	8.691264783	527733.5976	5.736234757
6.	Dem. Rep. of Congo	14.683266	9,300,000	710523.2417	7.640034857	468945.3395	5.042423006
7.	Viet Nam	10.209882	125,000,000	494056.19	0.395244952	326077.0854	0.260861668
8.	Cambodia	8.325098	4,100,000	402851.4922	9.825646152	265881.9849	6.48492646
9.	India	8.13943	1,001,191,000	393867.0177	0.039339848	259952.2317	0.0259643
10.	Angola	7.63888	8,100,000	369645.4032	4.563523496	243965.9661	3.011925508
11.	Mozambique	5.304188	12,000,000	256669.6573	2.138913811	169401.9738	1.411683115
12.	Malawi	5.012763	2,100,000	242567.6016	11.55083817	160094.617	7.623553192
13.	United Rep. of Tanzania	4.992759	5,000,000	241599.608	4.83199216	159455.7413	3.189114826
14.	Cameroon	4.917544	6,100,000	237959.9542	3.900982855	157053.5697	2.574648684
15.	China	4.659481	5,919,800,000	225472.2856	0.003808782	148811.7085	0.002513796
16.	Côte d'Ivoire	4.239303	5,800,000	205139.8722	3.536894348	135392.3156	2.33435027
17.	Sierra Leone	4.135064	200,000	200095.747	100.0478735	132063.193	66.0315965
18.	Benin	4.066711	1,000,000	196788.1453	19.67881453	129880.1759	12.98801759
19.	Rwanda	3.159551	500,000	152890.6729	30.57813458	100907.8441	20.18156882

S/N	Country	2014 Cassava output ($\times 10^9$ kg) [b]	2014 National energy consumption (MW.h/Y) [c]	Energy from CPR if 100% of national cassava output was processed (MW.h/Y)	Potential of CPR to provide national energy requirement if 100% national cassava output was processed (%)	Energy from CPR if 66% of national cassava output was processed (MW.h/Y)	Potential of CPR to provide national energy requirement if 66% national cassava output was processed (%)
20.	Paraguay	3.06	9,700,000	148073.4	1.526529897	97728.444	1.007509732
21.	Madagascar	2.929743	1,300,000	141770.2638	10.90540491	93568.37409	7.197567238
22.	Uganda	2.812	2,700,000	136072.68	5.039728889	89807.9688	3.326221067
23.	Philippines	2.540254	66,000,000	122922.8911	0.186246805	81129.1081	0.122922891
24.	Burundi	2.242352	400,000	108507.4133	27.12685332	71614.89276	17.90372319
25.	Colombia	2.186207	60,000,000	105790.5567	0.176317595	69821.76744	0.116369612
26.	Lao People's Dem. Rep.	1.629805	3,900,000	78866.26395	2.022211896	52051.73421	1.334659851
27.	Congo	1.334881	900,000	64594.89159	7.177210177	42632.62845	4.736958717
28.	Guinea	1.264078	900,000	61168.73442	6.796526047	40371.36472	4.485707191
29.	Peru	1.195926	39,000,000	57870.85914	0.148386818	38194.76703	0.0979353
30.	Togo	1.153109	1,100,000	55798.94451	5.072631319	36827.30338	3.347936671
31.	Zambia	0.919497	11,000,000	44494.45983	0.404495089	29366.34349	0.266966759
32.	Kenya	0.858461	7,600,000	41540.92779	0.546591155	27417.01234	0.360750162
33.	Centr. Afric. Rep.	0.699764	200,000	33861.57996	16.93078998	22348.64277	11.17432139
34.	Haiti	0.615	400,000	29759.85	7.4399625	19641.501	4.91037525
35.	Liberia	0.534810	300,000	25879.4559	8.6264853	17080.44089	5.693480298
36.	Myanmar	0.485	11,000,000	23469.15	0.213355909	15489.639	0.1408149
37.	Cuba	0.435772	15,000,000	21087.00708	0.140580047	13917.42467	0.092782831
38.	Venezuela (Boli. Rep.)	0.357876	78,000,000	17317.61964	0.022202076	11429.62896	0.01465337

http://dx.doi.org/10.5772/intechopen.72166

S/N	Country	2014 Cassava output ($\times 10^9$ kg) [b]	2014 National energy consumption (MW.h/Y) [c]	Energy from CPR if 100% of national cassava output was processed (MW.h/Y)	Potential of CPR to provide national energy requirement if 100% national cassava output was processed (%)	Energy from CPR if 66% of national cassava output was processed (MW.h/Y)	Potential of CPR to provide national energy requirement if 66% national cassava output was processed (%)
39.	Sri Lanka	0.301548	11,000,000	14591.90772	0.132653707	9630.659095	0.087551446
40.	Senegal	0.257259	3,000,000	12448.76301	0.414958767	8216.183587	0.273872786
41.	Gabon	0.247889	2,100,000	11995.34871	0.571207081	7916.930149	0.376996674
42.	Bolivia (Plu. State of)	0.245808	7,500,000	11894.64912	0.158595322	7850.468419	0.104672912
43.	Zimbabwe	0.235052	8,000,000	11374.16628	0.142177079	7506.949745	0.093836872
44.	Nicaragua	0.231658	4,412,000	11209.93062	0.25407821	7398.554209	0.167691619
45.	Argentina	0.186944	116,000,000	9046.22016	0.007798466	5970.505306	0.005146987
46.	Dominican Republic	0.178327	15,140,000	8629.24353	0.056996325	5695.30073	0.037617574
47.	Costa Rica	0.1755	9,200,000	8492.445	0.092309185	5605.0137	0.060924062
48.	Chad	0.166888	200,000	8075.71032	4.03785516	5329.968811	2.664984406
49.	Papua New Guinea	0.148213	3,000,000	7172.02707	0.239067569	4733.537866	0.157784596
50.	Niger	0.133099	1,200,000	6440.66061	0.536721718	4250.836003	0.354236334
51.	South Sudan	0.126244	694,100	6108.94716	0.880124933	4031.905126	0.580882456
52.	Ecuador	0.111743	21,000,000	5407.24377	0.02574878	3568.780888	0.016994195
53.	Somalia	0.090233	300,000	4366.37487	1.45545829	2881.807414	0.960602471
54.	Fiji	0.075277	800,000	3642.65403	0.455331754	2404.15166	0.300518957
55.	Equatorial Guinea	0.071673	91,140	3468.25647	3.805416359	2289.04927	2.511574797
56.	Comoros	0.068733	40,920	3325.98987	8.128029985	2195.153314	5.36449979

S/N	Country	2014 Cassava output ($\times 10^9$ kg) [b]	2014 National energy consumption (MW.h/Y) [c]	Energy from CPR if 100% of national cassava output was processed (MW.h/Y)	Potential of CPR to provide national energy requirement if 100% national cassava output was processed (%)	Energy from CPR if 66% of national cassava output was processed (MW.h/Y)	Potential of CPR to provide national energy requirement if 66% national cassava output was processed (%)
57.	Mali	0.052152	1,400,000	2523.63528	0.180259663	1665.599285	0.118971377
58.	Malaysia	0.051911	131,000,000	2511.97329	0.001917537	1657.902371	0.001265574
59.	Guinea-Bissau	0.045392	31,620	2196.51888	6.946612524	1449.702461	4.584764266
60.	El Salvador	0.036026	5,700,000	1743.29814	0.030584178	1150.576772	0.020185557
61.	French Guiana	0.029906	–	1447.15134	–	955.1198844	–
62.	Timor-Leste	0.029485	125,300	1426.77915	1.138690463	941.674239	0.751535706
63.	Honduras	0.025526	5,300,000	1235.20314	0.02330572	815.2340724	0.015381775
64.	Panama	0.018802	7,800,000	909.82878	0.011664472	600.4869948	0.007698551
65.	Mexico	0.018135	238,000,000	877.55265	0.00036872	579.184749	0.000243355
66.	Guatemala	0.017498	8,915,000	846.72822	0.009497793	558.8406252	0.006268543
67.	Jamaica	0.016549	2,800,000	800.80611	0.028600218	528.5320326	0.018876144
68.	Taiwan, China Rep	0.013017	249,500,000	629.89263	0.000252462	415.7291358	0.000166625
69.	Gambia	0.011555	300,000	559.14645	0.18638215	369.036657	0.123012219
70.	Micronesia (Fed. States)	0.008891	178,000	430.23549	0.241705331	283.9554234	0.159525519
71.	Tonga	0.007862	46,500	380.44218	0.818155226	251.0918388	0.539982449
72.	Suriname	0.007127	1,900,000	344.87553	0.018151344	227.6178498	0.011979887
73.	Guyana	0.006781	800,000	328.13259	0.041016574	216.5675094	0.027070939
74.	Burkina Faso	0.004105	1,200,000	198.64095	0.016553413	131.103027	0.010925252
75.	Cabo Verde	0.003847	300,000	186.15633	0.06205211	122.8631778	0.040954393

http://dx.doi.org/10.5772/intechopen.72166

S/N	Country	2014 Cassava output ($\times 10^9$ kg) [b]	2014 National energy consumption (MW.h/Y) [c]	Energy from CPR if 100% of national cassava output was processed (MW.h/Y)	Potential of CPR to provide national energy requirement if 100% national cassava output was processed (%)	Energy from CPR if 66% of national cassava output was processed (MW.h/Y)	Potential of CPR to provide national energy requirement if 66% national cassava output was processed (%)
76.	French Polynesia	0.003805	700,000	184.12395	0.026303421	121.521807	0.017360258
77.	Trinidad and Tobago	0.003194	9,100,000	154.55766	0.001698436	102.0080556	0.001120968
78.	Brunei Darussalam	0.00306	3,766,000	148.0734	0.003931848	97.728444	0.00259502
79.	Solomon Islands	0.003025	79,050	146.37975	0.185173624	96.610635	0.122214592
80.	Wallis and Futuna Islands	0.001874	—	90.68286	—	59.8506876	—
81.	New Caledonia	0.001777	2,000,000	85.98903	0.004299452	56.7527598	0.002837638
82.	Sao Tome and Principe	0.001349	65,100	65.27811	0.100273594	43.0835526	0.066180572
83.	Guadeloupe	0.001235	—	59.76165	—	39.442689	—
84.	Saint Lucia	0.001233	300,000	59.66487	0.01988829	39.3788142	0.013126271
85.	Dominica	0.001217	90,210	58.89063	0.065281709	38.8678158	0.043085928
86.	Bahamas	0.000938	1,600,000	45.38982	0.002836864	29.9572812	0.00187233
87.	Belize	0.000927	400,000	44.85753	0.011214383	29.6059698	0.007401492
88.	Cook Islands	0.000869	31,620	42.05091	0.13298833	27.7536006	0.087772298
89.	St. Vin./Gren.	0.000721	100,000	34.88919	0.03488919	23.0268654	0.023026865
90.	Barbados	0.000553	900,000	26.75967	0.002973297	17.6613822	0.001962376
91.	Mauritius	0.000466	2,600,000	22.54974	0.000867298	14.8828284	0.000572416

S/N	Country	2014 Cassava output ($\times 10^9$ kg) [b]	2014 National energy consumption (MW.h/Y) [c]	Energy from CPR if 100% of national cassava output was processed (MW.h/Y)	Potential of CPR to provide national energy requirement if 100% national cassava output was processed (%)	Energy from CPR if 66% of national cassava output was processed (MW.h/Y)	Potential of CPR to provide national energy requirement if 66% national cassava output was processed (%)
92.	Samoa	0.000424	100,000	20.51736	0.02051736	13.5414576	0.013541458
93.	Puerto Rico	0.000377	19,000,000	18.24303	9.60159E-05	12.0403998	6.33705E-05
94.	Mauritania	0.00025	800,000	12.0975	0.001512188	7.98435	0.000998044
95.	Seychelles	0.000232	300,000	11.22648	0.00374216	7.4094768	0.002469826
96.	Grenada	0.000217	200,000	10.50063	0.005250315	6.9304158	0.003465208
97.	Antig./ Barbuda	0.000127	300,000	6.14553	0.00204851	4.0560498	0.001352017
98.	Ame. Samoa	0.000087	100,000	4.20993	0.00420993	2.7785538	0.002778554
99.	Niue	0.000044	3720	2.12916	0.057235484	1.4052456	0.037775419
100.	Réunion	0.000036	—	1.74204	—	1.1497464	—
101.	Cay. Islands	0.000007	600,000	0.33873	0.000056455	0.2235618	3.72603E-05
102.	Maldives	0.000006	300,000	0.29034	0.00009678	0.1916244	6.38748E-05
103.	Singapore	0.000001	47,180,000	0.04839	1.02565E-07	0.0319374	6.76927E-08

[a]Using HHV (heat of combustion) = 55.53 MJ/(kg CH_4) and ϵ (conversion efficiency from thermal energy to electrical energy) = 25%.

[b]Source: [24].

[c]Source: [68, 69].

Table 4. Year 2014 cassava production output of nations, their energy consumption capacity and potential of CPR generated from cassava processing to provide national energy requirements [a]

5. Applications, utilizations and dividends of biomethane from CPR

The anaerobic digestion of CPR would generate biogas, which could be used as is or upgraded to obtain more efficient biomethane. The energy content of either fuel could be put to various applications and utilities. These include:

- Fuel for stoves in cooking; boiling, frying, roasting, etc.
- Fuel for lamps in lighting; illumination, reading, playing, etc.
- Fuel for transportation; cars, trucks, sea vessels, etc.
- Electrical power in processing operations; drying, grinding, heating, pumping, refrigeration, washing, etc.
- The digester effluent (digestate) could be utilized for soil amendment and/or serve as biofertilizer for enhanced crop production. This was demonstrated to increase potato yield [38].
- Perhaps the critical humane benefits are the reduction of drudgery and burden on the one hand and the improvement of health conditions on the other hand. This is due to reduced time spent on fetching firewood and charcoal for domestic fuel, and the reduced exposure to their combustion products.
  - Women and children may carry on their head 10 kg of firewood for distances up to 8 km, spending 5–6 hours per trip [39]; 2–6 hours per day [40].; or 5 hours per week [38]

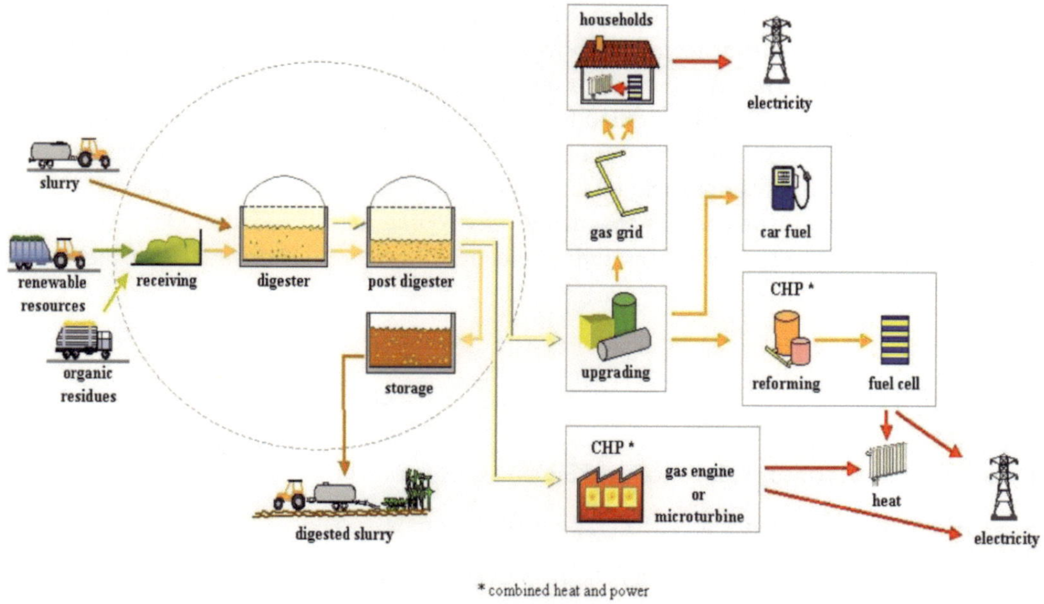

Figure 3. Schematics of biomethane production by anaerobic digestion of renewable feedstocks and pathways of the biomethane utilization. Adopted with permission from Ref. [46].

- Domestic combustion of the firewood releases health-impairing pollutants like carbon monoxide, hydrocarbons, smoke and other particulate matter. These combustion products may cause nausea, sneezing, eye and respiratory irritations [41]; pneumonia, lung cancer, and respiratory infections [42]; and reduced birth weight [43].
- Biomethane utilization reduced firewood consumption by 74% in China [44] and 84% in Sri Lanka [45], thereby minimizing the drudgery, burden, and health hazards associated with use of firewood for domestic energy.

Figure 3 presents pathways of the production and utilization of biomethane from CPR and other renewable feedstocks.

6. Conclusions

Global energy security, sustainability and renewability could be enhanced by harnessing non-food biomass. The work presented in this chapter demonstrated that anaerobic digestion of cassava peeling residue (CPR) generated biomethane that could come to the energy rescue of cassava producing nations. Depending on the specific country and proportion of national cassava output processed, recovered biomethane from CPR could provide up to 7% national energy requirement in Haiti; 8% in Comoros; 10% in Cambodia; 11% in Nigeria; 31% in Rwanda; and 100% in Sierra Leone. The biomethane could be put to various applications and utilities that minimize the drudgery and burden of gathering wood and charcoal for domestic fuel. As additional dividends, use of the biomethane would prevent implications of the combustion products of these solid fuels that degrade air quality and impair human health. The time saved from fetching firewood may be put to economic, educational and social activities. Furthermore, the digester effluent (digestate) could be sold for soil amendment and as organic fertilizer, or applied to agricultural land for increased crop yield. Either way more revenue is generated for economic empowerment. Therefore, anaerobic digestion of CPR would help cassava producing nations to not only mitigate their energy insufficiency, but also address issues such as climate change, environmental degradation, poverty alleviation, rural development, and the sanitation and health hazards associated with the use of wood and charcoal as fuel.

Author details

Sammy N. Aso[1]*, Arthur A. Teixeira[2] and Simeon C. Achinewhu[3]

*Address all correspondence to: sammyasso@yahoo.com

1 Food Engineering Laboratory, Rivers State University, Port Harcourt, Rivers, Nigeria

2 University of Florida, Gainesville, Florida, USA

3 Rivers State University, Port Harcourt, Rivers, Nigeria

References

[1] UNDESA: United Nations, Department of Economic and Social Affairs, Population Division. World Population Prospects: The 2015 Revision, DVD Edition. 2015. http://esa.un.org/unpd/wpp/Download/Standard/Population/

[2] REN21: Renewable Energy Policy Network for the 21st Century. Renewables 2014 Global Status Report. 2014. http://www.ren21.net/Portals/0/documents/Resources/GSR/2014/GSR2014_full%20report_low%20res.pdf

[3] World Bank. Mozambique Agricultural Development Strategy: Stimulating Smallholder Agricultural Growth. Report No. 32416-MZ, World Bank Agriculture, Environment, and Social Development Unit. 2006. http://siteresources.worldbank.org/MOZAMBIQUEEXTN/Resources/Moz_AG_Strategy.pdf

[4] Zvinavashe E, Elbersen HW, Slingerland M, Kolijn S, Sanders JP. Cassava for food and energy: Exploring potential benefits of processing of cassava into cassava flour and bio-energy at farmstead and community levels in rural Mozambique. Biofuels, Bioproducts and Biorefining. 2011;**5**(2):151-164

[5] Lansing S, Bowen H, Gregoire K, Klavon K, Moss A, Eaton A, et al. Methane production for sanitation improvement in Haiti. Biomass and Bioenergy. 2016;**91**:288-295

[6] World Bank. Haiti: Strategy to Alleviate the Pressure of Fuel Demand on National Wood-fuel Resources. Energy Sector Management Assistance Program (ESMAP) Technical Paper 112/07. Washington, D.C.: The International Bank for Reconstruction and Development/THE WORLD BANK; 2007 https://www.esmap.org/sites/esmap.org/files/TR_11207_Haiti%20Strategy%20to%20Alleviate%20the%20Pressure%20of%20Fuel%20Demand%20on%20National%20Woodfuel%20Resources_112-07.pdf

[7] Duku MH, Gu S, Hagan EB. A comprehensive review of biomass resources and biofuels potential in Ghana. Renewable and Sustainable Energy Reviews. 2011;**15**(1):404-415

[8] EurObserv'ER. The state of renewable energies in Europe. 16th EurObserv'ER Report. 2016 Edition. 2017. https://www.eurobserv-er.org/

[9] Aso SN. Food engineering stratagem to protect the environment and improve the income opportunities of gari processors. Journal of Nigerian Environmental Society (JNES). 2004;**2**(1):31-36

[10] Aso SN. Synergistic enzymatic hydrolysis of cassava starch and anaerobic digestion of cassava waste. PhD Dissertation. Gainesville, Florida: University of Florida, Department of Agricultural and Biological Engineering; 2013

[11] Ayankunbi MA, Keshinro OO, Egele P. Effects of methods of preparation on the nutrient composition of some cassava products – Garri (Eba), 'Lafun' and 'Fufu'. Food Chemistry. 1991;**41**(3):349-354

[12] Balagopalan C, Padmaja G, Nanda SK, Moorthy SN. Cassava in Food, Feed, and Industry. Boca Raton, Florida: CRC Press, Inc; 1988

[13] Bourdoux P, Seghers P, Mafuta M, Vanderpas J, Vanderpas-Rivera M, Delange F, Ermans AM. Cassava products: HCN content and detoxification processes. In: Delange F, Iteke FB, Ermans AM, editors. Nutritional Factors Involved in the Goitrogenic Action of Cassava. Ottawa, Canada: International Development Research Centre (IDRC) Report No. IDRC–184e, IDRC; 1982. pp. 51-58

[14] Collard P, Levi S. A two-stage fermentation of cassava. *Nature*. 1959;**183**(4661):620-621

[15] Congocookbook.com. Baton de Manioc & Chikwangue. 2013. http://congocookbook.com/staple_dish_recipes/baton_de_manioc_and_chikwangue.html

[16] Dufour DL. "Bitter" cassava: Toxicity and detoxification. In: Ortiz R, Nassar N, editors. Proceedings of First International Meeting on Cassava Breeding, Biotechnology and Ecology. Brasilia: University of Brasilia; 2007. pp. 171-184. http://www.geneconserve.pro.br/site/pags/meeting2/proceedings.pdf#page=171

[17] Hahn SK. An overview of traditional processing and utilization of cassava in Africa. Cassava as livestock feed in Africa: Proceedings of the IITA/ILCA/University of Ibadan Workshop on the Potential Utilization of Cassava as Livestock Feed in Africa, 14-18 November 1988, Ibadan, Nigeria. 1992. pp. 16-27. http://www.fao.org/Wairdocs/ILRI/x5458E/x5458e05.htm

[18] Lancaster PA, Ingram JS, Lim MY, Coursey DG. Traditional cassava–based foods: Survey of processing techniques. Economic Botany. 1982;**36**(1):12-45

[19] Mahungu NM, Yamaguchi Y, Almazan AM, Kahn SK. Reduction of cyanide during processing of cassava into some traditional African foods. Journal of Food and Agriculture. 1987;**1**(1):11-15

[20] Ross J. Attieke, a "new" staple. Minnesota Monthly, Monday 17th September 2012. 2012. http://www.minnesotamonthly.com/media/Blogs/Twin-Cities-Taste/September-2012/Attieke-a-New-Staple/

[21] Seigler DS, Pereira JF. Modernized preparation of casave in the *llanos orientales* of Venezuela. Economic Botany. 1981;**35**(3):356-362

[22] Sokari TG. Improving the nutritional quality of Ogi and Gari. In: Applications of Biotechnology in Traditional Fermented Foods. Washington, DC: National Academy Press; 1992. pp. 93-99

[23] FAO: Food and Agriculture Organization of the United Nations. Cassava for Food and Energy Security - Investing in Cassava Research and Development could Boost Yields and Industrial Uses. FAO, Rome: Posted by African Press Organization; 2008. http://appablog.wordpress.com/2008/07/25/cassava-for-food-and-energy-security-investing-in-cassava-research-and-development-could-boost-yields-and-industrial-uses/

[24] FAO: Food and Agriculture Organization of the United Nations. FAOSTAT. Online Statistical Database for Food and Agricultural Commodities Production. 2017. http://www.fao.org/faostat/en/#data/QC/visualize

[25] Nassar N, Ortiz R. Breeding cassava to feed the poor. Scientific American. 2010;**302**(5):78-84. DOI: 10.1038/scientificamerican0510-78 http://stopogm.net/sites/stopogm.net/files/webfm/plataforma/breedingcassava.pdf

[26] Donovan C, Haggblade S, Salegua VA, Cuambe C, Mudema J, Tomo A. Cassava commercialization in Mozambique (Working Paper No. 120). Michigan State University, Department of Agricultural, Food, and Resource Economics. 2011; http://ageconsearch.umn.edu/bitstream/120744/2/idwp120.pdf

[27] Koch BM, Sibbesen O, Swain E, Kahn RA, Liangcheng D, Bak S, Halkier BA, Moller BL. Possible use of a biotechnological approach to optimize and regulate the content and distribution of cyanogenic glucosides in cassava to increase food safety. Acta Horticulturae. 1994;(375):45-60. DOI: 10.17660/ActaHortic.1994.375.2

[28] Jansson C, Westerbergh A, Zhang J, Hu X, Sun C. Cassava, a potential biofuel crop in (the) People's republic of China. Applied Energy. 2009;**86**(S1):S95-S99

[29] CGIAR: Consultative Group on International Agricultural Research. Report on the Inter-Centre Review of Root and Tuber Crops Research in the CGIAR. Rome: FAO; 1997 http://www.fao.org/Wairdocs/TAC/X5791E/x5791e00.htm

[30] Rogers DJ. Some botanical and ethnological considerations of *Manihot esculenta*. Economic Botany. 1965;**19**(4):369-377

[31] El-Sharkawy MA, De Tafur SM, Cadavid LF. Potential photosynthesis of cassava as affected by growth conditions. Crop Science. 1992;**32**(6):1336-1342

[32] Wang, W. (2007). Cassava production for industrial utilization in China – Present and future perspectives. In: Howeler, RH, editor. Cassava Research and Development in Asia: Exploring New Opportunities for an Ancient Crop. Proceedings of the Seventh Regional Workshop, Bangkok, Thailand. October 28–November 1, 2002. pp. 33-38

[33] Sigarchian SG, Paleta R, Malmquist A, Pina A. Feasibility study of using a biogas engine as backup in a decentralized hybrid (PV/wind/battery) power generation system–case study Kenya. Energy. 2015;**90**(Part 2):1830-1841

[34] Pierie F, van Someren CEJ, Benders RMJ, Bekkering J, van Gemert WT, Moll HC. Environmental and energy system analysis of bio-methane production pathways: A comparison between feedstocks and process optimizations. Applied Energy. 2015; **160**:456-466

[35] Lema JM, Omil F. Anaerobic treatment: A key technology for a sustainable management of wastes in Europe. Water Science and Technology. 2001;**44**(8):133-140

[36] Lettinga G. Digestion and degradation, air for life. Water Science and Technology. 2001;**44**(8):157-176

[37] Sovacool BK, Dworkin MH. Energy justice: Conceptual insights and practical applications. Applied Energy. 2015;**142**:435-444

[38] Garfi M, Ferrer-Marti L, Velo E, Ferrer I. Evaluating benefits of low-cost household digesters for rural Andean communities. Renewable and Sustainable Energy Reviews. 2012;**16**(1):575-581

[39] Masekoameng KE, Simalenga TE, Saidi T. Household energy needs and utilization patterns in the Giyani rural communities of Limpopo Province, South Africa. Journal of Energy in Southern Africa. 2005;**16**(3):4-9

[40] DFID: Department for International Development. Energy for the Poor: Underpinning the Millennium Development Goals. 2002. https://www.ecn.nl/fileadmin/ecn/units/bs/JEPP/energyforthepoor.pdf.

[41] Oguntoke O, Opeolu BO, Babatunde N. Indoor air pollution and health risks among rural dwellers in Odeda area, south-western Nigeria. Ethiopian Journal of Environmental Studies and Management. 2010;**3**(2):39-46

[42] Ezzati M, Lopez AD, Rodgers A, Vander Hoorn S, Murray CJ, Comparative Risk Assessment Collaborating Group. Selected major risk factors and global and regional burden of disease. The Lancet. 2002;**360**(9343):1347-1360. DOI: 10.1016/S0140-6736(02)11403-6

[43] Boy E, Bruce N, Delgado H. Birth weight and exposure to kitchen wood smoke during pregnancy in rural Guatemala. Environmental Health Perspectives. 2002;**110**(1):109-114

[44] Remais J, Chen L, Seto E. Leveraging rural energy investment for parasitic disease control: Schistosome ova inactivation and energy co-benefits of anaerobic digesters in rural China. PLoS One. 2009;**4**(3):e4856

[45] De Alwis A. Biogas–a review of Sri Lanka's performance with a renewable energy technology. Energy for Sustainable Development. 2002;**6**(1):30-37

[46] Plochl M, Heiermann M. Biogas farming in central and northern Europe: A strategy for developing countries? Invited overview. Agricultural Engineering International. 2006;**8**(8):1-15

[47] Cuzin N, Farinet JL, Segretain C, Labat M. Methanogenic fermentation of cassava peel using a pilot plug flow digester. Bioresource Technology. 1992;**41**(3):259-264

[48] Jekayinfa SO, Scholz V. Laboratory scale preparation of biogas from cassava tubers, cassava peels, and palm kernel oil residues. Energy Sources, Part A: Recovery, Utilization, and Environmental Effects. 2013;**35**(21):2022-2032. DOI: 10.1080/15567036.2010.532190

[49] Ukpai PA, Nnabuchi MN. Comparative study of biogas production from cow dung, cowpea and cassava peeling using 45 litres biogas digester. Advances in Applied Science Research. 2012;**3**(3):1864-1869

[50] Ofoefule AU, Uzodinma EO. Biogas production from blends of cassava (*Manihot utilissima*) peels with some animal wastes. International Journal of Physical Sciences. 2009;**4**(7):398-402

[51] Colin X, Farinet JL, Rojas O, Alazard D. Anaerobic treatment of cassava starch extraction wastewater using a horizontal flow filter with bamboo as support. Bioresource Technology. 2007;**98**(8):1602-1607

[52] Manilal VB, Narayanan CS, Balagopalan C. Anaerobic digestion of cassava starch factory effluent. World Journal of Microbiology and Biotechnology. 1990;**6**(2):149-154

[53] Sun L, Wan S, Yu Z, Wang Y, Wang S. Anaerobic biological treatment of high strength cassava starch wastewater in a new type up-flow multistage anaerobic reactor. Bioresource Technology. 2012;**104**:280-288

[54] Zhu W, Lestander TA, Örberg H, Wei M, Hedman B, Ren J, et al. Cassava stems: A new resource to increase food and fuel production. Global Change Biology (GCB) Bioenergy. 2015;**7**(1):72-83

[55] Rajbhandari BK, Annachhatre AP. Anaerobic ponds treatment of starch wastewater: Case study in Thailand. Bioresource Technology. 2004;**95**:135-143

[56] Wang W, Xie L, Chen J, Luo G, Zhou Q. Biohydrogen and methane production by co-digestion of cassava stillage and excess sludge under thermophilic condition. Bioresource Technology. 2011;**102**(4):3833-3839

[57] Zhang Q, He J, Tian M, Mao Z, Tang L, Zhang J, Zhang H. Enhancement of methane production from cassava residues by biological pretreatment using a constructed microbial consortium. Bioresource Technology. 2011;**102**(19):8899-8906

[58] Luo G, Xie L, Zou Z, Wang W, Zhou Q, Shim H. Anaerobic treatment of cassava stillage for hydrogen and methane production in continuously stirred tank reactor (CSTR) under high organic loading rate (OLR). International Journal of Hydrogen Energy. 2010;**35**:11733-11737

[59] Zhang Q, Tang L, Zhang J, Mao Z, Jiang L. Optimization of thermal-dilute sulfuric acid pretreatment for enhancement of methane production from cassava residues. Bioresource Technology. 2011;**102**(4):3958-3965

[60] Luo G, Xie L, Zhou Q. Enhanced treatment efficiency of an anaerobic sequencing batch reactor (ASBR) for cassava stillage with high solids content. Journal of Bioscience and Bioengineering. 2009;**107**(6):641-645

[61] Anunputtikul W, Rodtong S. Laboratory scale experiments for biogas production from cassava tubers. In: Proceedings of the Joint International Conference on "Sustainable Energy and Environment (SEE), Hua Hin: Thailand; 2004. pp. 1-3

[62] Paixão MA, Tavares CR, Bergamasco R, Bonifácio AL, Costa RT. Anaerobic digestion from residue of industrial cassava industrialization with acidogenic and methanogenic physical separation phases. Applied Biochemistry and Biotechnology. 2000;**84-86**(1-9):809-819

[63] Cooke RD, de la Cruz EM. The changes in cyanide content of cassava (*Manihot esculenta Crantz*) tissues during plant development. Journal of the Science of Food and Agriculture. 1982;**33**(3):269-275

[64] Dufour DL. Cyanide content of cassava (*Manihot esculenta,* Euphorbiaceae) cultivars used by Tukanoan Indians in Northwest Amazonia. Economic Botany. 1988;**42**(2):255-266

[65] Gomez G, Valdivieso M. The effect of variety and plant age on cyanide content, chemical composition and quality of cassava roots. Nutrition Reports International. 1983; **27**(4):857-865

[66] Gomez G, Valdivieso M, Noma AT. The influence of cultivar and plant age on the chemical composition of field-grown cassava leaves and roots. Qualitas Plantarum - Plant Foods for Human Nutrition. 1985;**35**(2):109-119

[67] Tonukari NJ. Cassava and the future of starch. Electronic Journal of Biotechnology. 2004;**7**(1):5-8

[68] CIA: Central Intelligence Agency. The World Factbook. Country Comparison: Electricity-Consumption; 2017. https://www.cia.gov/library/publications/the-world-factbook/rankorder/2233rank.html

[69] Wikipedia. List of Countries by Electricity Consumption. 2017. https://en.wikipedia.org/wiki/List_of_countries_by_electricity_consumption

Improving Cassava Quality for Poultry Feeding Through Application of Biotechnology

Apeh Akwu Omede, Emmanuel Uchenna Ahiwe,
Ze Yuan Zhu, Fidelis Fru-Nji and Paul Ade Iji

Additional information is available at the end of the chapter

http://dx.doi.org/10.5772/intechopen.72236

Abstract

The continuous increase in cost of conventional energy sources caused by inadequate supply and stiff competition between human, animals and various industries for many decades has resulted to the need to source for suitable, readily available and cheap energy sources for poultry production globally. One such alternative is cassava. A native to South America, cassava is now found in abundance in most tropical countries. Due to lack of excellent post-harvest technologies, large quantities of cassava are wasted. An increased use of cassava in poultry feeding will go a long way to reduce this wastage and also reduce the high cost of poultry feed. However, the utilisation of cassava in poultry nutrition has been hindered by its lower nutritional value, especially protein and amino acids, presence of some ANF and dustiness when poultry feed is produced with cassava meal. Traditional processing methods have only succeeded in taking the inclusion level of cassava to 40% in some poultry diets. Researchers and poultry nutritionists have become interested in developing multi-pronged technologies and processing methods to increase cassava utilisation in poultry nutrition to reduce wastage, improve its nutritional value and maximise production. This chapter highlights the application of different technologies and the importance of biotechnology in improving the quality of cassava and increasing its utilisation for poultry feeding.

Keywords: biotechnology, cassava, feeding, nutrition, poultry

1. Introduction

For many decades, global poultry production has been negatively affected by high and increasing cost of feed. This persistent increase in the cost of poultry feed is caused majorly by the escalating cost of conventional energy and protein feed ingredients used to compound poultry diets. The continuous increase in the cost of these ingredients could be blamed on the food-feed competition between man, animal and various industries for conventional and available energy and protein sources. This incessant upsurge in the cost of conventional energy and protein source has been and will continue to be a source of dilemma to most poultry feed industries and farmers, if not checked. Therefore, the escalating price and seasonal fluctuation in the supply of conventional feed ingredients require alternative energy sources to be explored to ensure optimum performance of poultry at least cost [1].

Alternatives to conventional energy and protein feed ingredients for poultry must be cheap or cost-effective, readily available, have suitable nutrient composition and should have minimal negative effects on birds. Bearing this in mind, researchers, animal feed industries and poultry producers have over the years directed their attention towards finding alternative energy sources that meet the above-mentioned requirements [2–5]. One such alternative is cassava (*Manihot esculenta*). The nutrient composition of cassava differs according to the variety, the age of the harvested crop and soil and climatic conditions during cultivation. Generally, it is a staple/root crop tuber that is rich in carbohydrates, calcium, vitamins B and C and essential minerals and is considered to be a suitable alternative to corn as an energy source in poultry diets [6]. However, the use of cassava in poultry production is limited by its low protein content, unbalanced amino acid profile, dustiness and presence of anti-nutritional factors, e.g., cyanogenic glucoside. However, these shortcomings can be moderately remedied through adequate processing and use of feed additives. In 2016, the global production of cassava was estimated to be about 288.4 million tons [7]. The crop is grown in over 90 countries and is the third most important source of calories in the tropics, after rice and maize. It is a staple for half a billion people in Africa, Asia and Latin America.

Cassava is also a source of commercial animal feed. Over half of the cassava crop is grown in Africa, with a third in Asia and 14% in Latin America. Nigeria is the largest producer, growing 38 million metric tons in 2010. Other major producers are Brazil, Indonesia, Thailand, and the Democratic Republic of Congo [6]. The availability and the nutritional advantage of cassava make it a potential alternative to conventional maize in poultry feed. It is worthy to note that the use of cassava as poultry feed is not new. However, cassava root meal has not been fully adopted by the animal feed industry due to inconsistencies in animal response when it is included in diets. Over the years, various techniques, such as soaking, sun drying, boiling, ensilage and fermentation, have been employed to improve the nutritional value of cassava root meal with varying degrees of success achieved. However, none of these methods has successfully eliminated all the nutritional deficiencies inherent in this staple crop to make it possible for the root to be added at a 100% replacement of maize in poultry diet. Currently, cassava can only be added at levels of 30–40% in nutritionally balanced, pelleted diets. There is a great need to identify more effective processing techniques to improve the material for animal feeding [1]. In recent time, newer innovative processing methods such as the combination of various processing methods, application of genetic plant breeding techniques and the use of biotechnical methods have been tested in cassava processing.

Biotechnology is the use of living systems and organisms to develop or make products, or it involves any technological application that uses biological systems, living organisms or derivatives thereof to make or modify products or processes for specific use [8]. The use of biotechnical means to improve the feeding value of diets used for poultry is gradually gaining ground, and the results have been very encouraging so far. Research geared towards the improvement of cassava through biotechnological means has been reported by various authors. This chapter seeks to review the short falls and giant strides that have been achieved when cassava is processed with biotechnological methods as well as the effects of diets compounded with biotechnologically processed cassava meal when fed to poultry.

2. Nutritional composition of cassava

Information on the nutrient composition of cassava tuber is abundant. The variability of each major nutrient component is large. This variability is caused by several factors such as varieties (sweet or bitter varieties), geographical location, moisture content, processing technology and the age affects the chemical and nutritional composition of cassava root [9]. Various values for macro and micronutrients have been reported by various researchers as shown in **Table 1**. Cassava is high in carbohydrate; however, it has low protein and fat contents. The contents of some minerals and vitamins in cassava tuber are also low compared to the values reported in cereals and legumes [10].

2.1. Macronutrients found in cassava root

The main macronutrients found in cassava roots include carbohydrates, proteins, fat and fibre [11]. Raw cassava contains 60% water [12]. The carbohydrate content on dry weight basis has been reported by various authors to be within the range of 80–90% [13, 14], typically around 83.42–87.35%. On fresh weight basis, the carbohydrate content range reported by several authors is 32–35% [14] and as low as 29% [11]. Cassava root is rich in starch and digestible energy. The carbohydrate content of cassava is mostly made of starch of about 70–80% [15]. The starch in cassava root consists of two water-insoluble homoglucans, amylose (15–17%) and amylopectin (83%) [16, 17]. Compared to maize and barley, however, the starch content of cassava is lower. The percentage sucrose found in the bitter varieties of cassava has been reported to be lower than 17% sucrose found in the sweet varieties while small quantities of fructose and dextrose have been reported.

Cassava roots contain low levels of protein, with the protein content of both sweet and bitter varieties of cassava ranging from 1.17 to 5.13% [11, 17, 18]. These values are lower than those of cereal grains such as corn (8.8%) and wheat (11.3%) [19]. The crude protein content of cassava root consists of 50% whole protein, and the remaining portion consists of the free amino acids (predominantly glutamic and aspartic acids) and non-protein components such as nitrite, nitrate and cyanogenic compounds. The restricted use of cassava root meal in poultry feed formulation and feeding is due to its low protein content as well as a relative deficiency in essential amino acids [20]. Although high in arginine, glutamic acid and aspartic acid, other amino acids such as methionine, cysteine, phenylalanine, threonine, proline and isoleucine in cassava are low compared to maize [17]. The fibre content in cassava roots depends on the

Nutrient	Nutrient content per 100g	References
Macronutrients		
Water (g)	59.68	[11]
Energy (kcal)	160	[11]
Carbohydrates (g)	36.06	[11, 13, 14]
Protein (g)	1.36	[11, 17]
Total fat (g)	0.28	[11]
Crude fibre (g)	1.38–3.75	[11, 18]
Micro nutrients		
Vitamins		
Folates (μg)	27	[11]
Niacin (mg)	0.854	[11]
Riboflavin (mg)	0.048	[11]
Thiamin (mg)	0.087	[11]
Vitamin A (IU)	13	[11]
Vitamin C (mg)	20.6	[11]
Vitamin K (μg)	1.9	[11]
Minerals		
Sodium (mg)	14	[11]
Potassium (mg)	271	[11]
Calcium (mg)	16	[11, 18, 23]
Iron (mg)	0.27	[11]
Magnesium (mg)	21	[11]
Manganese (mg)	3	[13]
Phosphorus (mg)	27	[11]
Zinc (mg)	0.34	[11]
Cyanogenic glucoside		
HCN (mg)	1.34–2.78	[18, 22, 24]

Table 1. Concentrations of nutrients in raw cassava root (*Manihot esculenta*).

variety and the age of the root. The fibre content of cassava root is 4.4% for the sweet variety and 4.60% for the bitter cassava variety compared to 7.3 and 2.7% reported for maize and wheat, respectively [10, 21]. The fibre content in cassava root depends on the age and variety of the cassava. As the cassava matures, the fibre content increases while the sweet cassava root tends to contain less fibre. The fat or lipid content of cassava is very low; on fresh weight basis, the lipid content of cassava roots ranges from 0.1 to 1.5% compared to maize and wheat that have lipid contents of 4.7 and 2.47%, respectively [11, 21].

2.2. Micronutrients in cassava root

Cassava root contains several micronutrients, including water-soluble and fat-soluble vitamins. Cassava contains low amounts of the valuable B-complex group of vitamins and most of these are lost during processing. Compared to maize, the vitamin content in cassava root is low. Cassava root also contains substantial amounts of some essential minerals such as calcium, iron, potassium, magnesium, copper, zinc and manganese. Cassava root has been reported to contain (values in 100 g) 27 μg folate, 0.854 mg niacin, 0.048 mg riboflavin, 0.087 mg thiamin, 13 IU vitamin A, 20.6 mg vitamin C, 0.088 mg vitamin B6 and 1.9 μg vitamin K. The mineral content of cassava root compared to maize is usually low, except potassium (271 mg). The mineral content in cassava root includes calcium (16 mg), magnesium (21 mg), phosphorus (27 mg), sodium (14 mg), iron (0.27 mg) and zinc (0.34 mg) [11, 17].

3. Anti-nutritional factors in cassava root

Depending on the amount ingested, cassava contains anti-nutrients that can have adverse effects on health, especially when raw. The range of anti-nutritional factors found in cassava root includes cyanide (46–65 mg/kg), phytate (2160–3040 mg/kg), oxalate (2200–4400 mg/kg), tannins (40–60 mg/kg), saponin, trypsin inhibitors (100–400 mg/kg) and total alkaloids (10–52 mg/kg) [25]. Except for cyanide, the levels of these anti-nutrients in some varieties are below the normal toxic level [25]. Cassava variety and the amount of moisture it contains are major factors that determine the concentration of HCN contained in the root. The cyanogenic potential of known cassava cultivars ranges from 10 mg HCN/kg to more than 500 mg HCN/kg [26]. The bitter varieties have been reported to contain higher concentrations of HCN, whereas the sweet varieties contain lower levels of HCN [23]. The use of cassava as food and feed for human and animals is greatly compromised by its level of HCN. Cassava toxicity may be acute and/or chronic. Acute toxicity results from ingestion of a lethal dose and death is caused by the inhibition of cytochrome oxidase of the respiratory chain by cyanide [27].

The performance of broilers, and layer production and egg quality are negatively affected by cyanide levels of 100 mg HCN/kg and 25 mg HCN/kg, respectively [28]. The World Health Organisation set the safe level of cyanogen in cassava flour at 10 ppm [29]. When the cassava root is bruised or during the root peeling and grinding, two cyanogenic glucosides, linamarin (93%) and lotaustralin (7%), present in the cassava and synthesised from amino acids, isoleucine and valine, are hydrolyzed by an enzyme, linamerase, thereby releasing HCN [30]. When such cassava is ingested by human or animals, an enzyme, β-glucosidase, produced by intestinal microorganisms, converts the HCN to hydrocyanic acid [31], which is toxic to both humans and animals. Acute cyanide intoxication, which results from excess cyanide residue caused by ingesting unprocessed cassava, has been reported to be associated with ataxia (a neurological disorder that affects the ability to walk). In an attempt to reduce or eliminate the toxic effect of hydrocyanic acid produced in the system, the liver and the red blood cells produce two enzymes, thiosulphate cyanide sulphur transferase (rhodanase) and mercaptopyruvate cyanide sulphur transferase, respectively. These enzymes are derived mainly from sulphur-containing amino acids (cysteine, cystine and methionine). With the influence of vitamin B_{12} (hydroxycobalamin), thiosulphate cyanide sulphur transferase (rhodanase) and mercaptopyruvate cyanide sulphur transferase help to convert toxic hydrocyanic acid to a more harmless thiocyanate, which is then passed out of the body in urine [27, 32].

4. Physical limitation to the use of cassava meal

Apart from anti-nutritional factors and nutrient deficiencies inherent in raw and unprocessed cassava root, the physical characteristics of cassava root meal such as dustiness, poor pelleting quality and poor pigmentation tend to also limit the use of cassava as feed ingredient in animal diets. These physical limitations have been reported to reduce feed intake and affect body weight gain and feed conversion ratio, especially in poultry. Crop impaction and irritation in respiratory tract have also been observed in animals fed cassava-based diet that does not contain oil or fed as mash [33]. Dustiness of a cassava-based diet is usually related to the form at which the feed is presented to the animal. Mash feed containing high level of cassava meal tends to result in dusty feed. Issues of dustiness in cassava-based mash diet could be ameliorated through adequate pelleting, resulting to improved feed consumption and poultry performance. According to Hahn et al. [34], pelleting cassava-based diet results into a diet that is denser, uniform and less dusty. Pelleting decreases the bulkiness of cassava-based diets by about a third, thereby overcoming issues related to dustiness. However, in farms where pelleting equipment is lacking, strategies such as the addition of oil or molasses can be employed to reduce issues relating to dustiness in unpelleted poultry feed. Wet mashed feed can also be fed to the birds to prevent dustiness; however, wet mashed feed should not be stored for long to avoid contamination and spoilage. Another physical attribute that limits the use of cassava as feed ingredient for animals especially poultry is the issue of lack of pigmentation. Cassava root meal is white in colour (it does not have any pigmentation). Feeding high levels of cassava root meal to layers and broilers has been reported to result in light-coloured egg yolks and pale meat, respectively. These eggs and chicken meat have been reported to attract low price because of low consumer appeal. When a high level of cassava meal is to be used, leaf meal or other pigmentation agents should be added to the diet in order to improve the quality of the product. At least 30–50 g leaf meal per kg of poultry diet can be used to prevent issues relating to lack of pigmentation in cassava root meal. Leaf meals such as cassava leaves, sweet potato leaves, ipil-ipil (leucaena) leaves and young grass have been reported to be effective [35].

5. Effect of feeding raw/unprocessed cassava to poultry

There are several reasons why it is important to process cassava before using it in feed for poultry. Apart from the challenge that cassava root begins to spoil from 2 days after harvest due to physiological changes and microbial activity, unless kept under special storage condition [36], fresh cassava peel (*M. esculenta* Crantz) contains phytates and large quantities of toxic cyanogenic glycosides [37]. Feeding of fresh cassava roots may lead to cyanide toxicity, depending on the cyanide content in the roots, and thus should be processed in order to reduce cyanogenic and phytate contents [38] before being used as feed for poultry. Cyanide content in excess of 100 mg/kg diet impairs broiler performance, while laying hens may be affected by levels as low as 25 mg/kg diet [39]. In a study with local chickens [40], it was reported that birds lost 2 g/day when feed fresh feed materials, including cassava roots and duckweed, and attributed the poor performance mostly to the presence of ANF in the fresh feed materials. When unprocessed (peeled and unpeeled cassava root meal) was fed to broilers, Akapo et al. [41] reported depressed feed intake, weight gain and feed-to-gain ratio, mostly with birds fed the unpeeled cassava root meal. The study concluded that up to 100 g/kg dietary inclusion of the meals poses a threat to growth and health status of broiler chicks [41].

6. Physical processing of cassava for poultry feeding

Various physical processing methods can be applied to cassava and affect the nutritional quality of ingredient and its components for use in poultry feeding. Drying is the most common physical processing method for cassava. However, the type of drying method used seems to depend on the final cassava product being made. For example, sun drying has been reported to be more effective in reducing or eradicating cyanide in cassava than oven drying because with the former method the cyanide is in contact with linamarase for a longer period [42], with about 90% of cyanide content eliminated by sun drying alone [43]. Soaking of cassava before cooking or fermenting [44] or boiling for up to 15 min [45] is also important process applied in order to reduce the cyanide content of cassava before use for poultry feeding.

Wet fermentation (soaking over 2–3 days) causes a reduction in starch content while increasing total soluble and reducing sugar levels within the first 36 h and 24, respectively [46]. To be able to replace conventional energy source like corn, a high starch yield by cassava is important. High starch-yielding cassava can be achieved by using suitable drying conditions of raw materials. Olomo and Ajibola [47] found that oven-drying chips and flour resulted in a higher starch yield compared to sun drying. Wet milling of dry cassava chips has been reported to cause greater crumbling of cell content, resulting in a large fraction of fine fibre, whereas dry milling of the chips to 200 mesh size resulted in flour with highly varying particle size comprising very fine and coarse materials. However, wet or dry milling of dried roots had little effect on the composition [48]. While composition is not affected, particle size is a key factor that affects feeding and performance in broiler chickens [49, 50]. However, it is worthy to note that a high yield of starch does not necessarily mean the starch would have the desired quality for any specific application [51]. Researchers have also shown that blending different native starch sources can give rise to an extensive array of properties, possibly obviating the need for chemical processing and minimising undesired properties of gels of individual starches (e.g., excessive cohesiveness in cassava and exudate in yam and corn) [52, 53].

According to different authors [54, 55], because cassava root has high moisture content (62–68%) and easily fermentable carbohydrate, it may be suitable for silage. As silage, cassava roots have high dry matter content, crude protein, crude fibre and ash [56]. Similarly, cassava leaf has been improved for use in poultry feeding through ensiling, which has been found to be efficient in reducing cyanide content in cassava leaf meal by 62% [57], while improving its digestibility [58]. To partially mitigate the reduction in feed intake by poultry due to the bulkiness and dustiness of cassava products, Morgan and Choct [59] suggested that poultry diets based on cassava products should be further processed through pelleting or addition of molasses or fat to improve texture, while simultaneously supplying essential fatty acids. However, it is important to note that there is no one physical processing method for cassava that gives the best quality cassava product, rather a combination of different methods is recommended based on nutritional objectives and variety of cassava used.

7. Chemical processing of cassava for poultry feeding

There is scarcity of literature on chemical processing of cassava for poultry feeding. Previous studies on cassava starch focused on starch-hydrolyzing enzymes such as α-amylase, amyloglucosidase and pectinase to achieve maximum hydrolytic efficiency of about 98%, resulting in 160 g/L of total reducing sugar [60], although this seems not have a direct application in processing cassava or its products for poultry feeding. However, this processing method has potential for use in the poultry feed industry. Recently, Olanbiwoninu and Odunfa [61] hydrolysed cassava peel into fermentable sugars using organic acid pre-treatment before enzyme hydrolysis. Because cassava peel is high in cellulose and hemicellulose (34.4%), this form of pre-treatment may be important in aiding the initial breakdown process before enzymatic hydrolysis is applied for it to be used for poultry feeding.

8. Microbial processing of cassava for poultry feeding

In recent times, the use of biotechnology in the form of microbial process in the improvement of cassava or its component for use in poultry feeding has become common. A major microbial process is solid-state fermentation, which involves application of fungal cultures to further enrich some local cassava products. Studies have examined different species of *Bacillus* and *Aspergillus*, although a comparative study showed that cassava starch was less susceptible to alpha-amylase of *Bacillus subtilis* and amyloglucosidase of *Aspergillus niger* hydrolysis compared with corn starch [62]. However, *A. niger* possesses the ability to produce fibre-degrading enzymes (hemicellulases, hydrolases, pectinases), proteases, amylases, lipases and hydrolysed tannins as well as agents that can hydrolyse tannins [63]. A study was conducted where *A. niger* was used to improve the nutritional quality of cassava peel meal in order to use it as a carbohydrate source for broiler chickens [64]. The findings showed that hydrolyzing cassava peal meal with *A. niger* reduced the influence of cyanide in the peel. Enzymatic hydrolysis of cell wall polysaccharides improves the detoxification of cassava roots [65], hence making it a non-toxic and possible candidate as an energy source in the production of feeds for broiler chickens. The fatty acid functional group of lauric and palmitic acids can be imparted to cassava starch by bacterial and fungal lipases [66, 67]. Fermentation with *A. niger* was shown to increase the hemicellulose and amylopectin contents and metabolizable energy value of unpeeled cassava root meal for birds [68], suggesting that fermenting fungal organisms are able to release digestive enzymes which pre-digest the substrate, hence increasing the availability of nutrients [69]. Higher crude protein content and increased starch digestibility were recorded by [70] after fermenting cassava pulp with *Aspergillus oryzae*.

Cassava chip is an energy source with low crude protein, which when fermented with yeast could increase from 1–3% to 30.4% crude protein [71]. To obtain cassava products with high protein content and a relatively balanced amino acid profile for use in poultry feeding, cassava

Amino acid	Quantity (mg/100 g)
Alanine	70.5
Arginine	5.0
Aspartic acid	69.9
Cystine	5.0
Glutamic acid	189.4
Glycine	52.7
Histidine	55.4
Hydroxylysine	5.0
Hydroxyproline	5.0
Isoleucine	130.8
Leucine	201.5
Lysine	481.1
Methionine	16.3
Phenylalanine	167.8
Proline	47.9
Serine	29.8
Threonine	21.0
Tryptophan	15.1
Tyrosine	87.2
Valine	92.6

Source: [73].

Table 2. Chemical composition of yeast-fermented cassava chip products.

chip or other forms of cassava roots could be successfully fermented with yeast (*Saccharomyces cerevisiae*) [72–74]. Furthermore, a recent research showed that protein in cassava chips can be enriched by yeast fermentation up to 47.5% crude protein, which can be used as a replacement for soybean meal [75]. Cassava chips fermented with yeast have been shown to have high levels of lysine, glutamic acid, leucine and phenylalanine (**Table 2**). In contrast to fermentation with yeast, fermentation of cassava products with fungal inoculum has been shown to result in about 42% reduction in total amino acids [68, 76].

Sourcing cassava from different genotypes and harvesting at different times do not affect their susceptibility to glucoamylase microbial enzyme digestion, suggesting that cassava product quality can remain unchanged irrespective of the selected cultivars and growth conditions [77]. However, further supplementation of fermented unpeeled cassava root meal with exogenous enzyme has been reported to increase metabolizable energy values, indicating possible breakdown of fibrous content by the enzyme to release more energy [68]. Hence, microbial processes could be used to improve the quality of cassava and its products for use in poultry feeding.

9. Improving the feeding value of cassava through processing

For many decades, various processing methods have been used to enhance the feeding value of cassava for human and poultry use. These methods are all geared towards eliminating various ANF such as hydrogen cyanide, phytate, saponin and alkaloids that are inherent in raw cassava [78]. These processing methods have also been utilised to improve the nutrient deficiencies as well as various physical limitations such as dustiness and lack of pigmentation that tend to reduce the performance and product quality and increase mortality when unprocessed cassava is used as food or feed ingredient. Cassava processing methods can be classified into traditional [79] and modern [80]. Traditional cassava processing methods used in improving the nutritional composition and reducing the anti-nutrient content include drying, boiling, parboiling/cooking, steaming, frying, roasting, addition of oil, molasses, leaf meal and application of natural fermentation processes. These processes result in HCN losses ranging from 25 to 98% [30, 81]. Modern methods of cassava processing include addition of feed additives such as nutrient supplementation with amino acids, vitamins and minerals; addition of pigmentation agents; pelleting; synthetic enzyme supplementation; microbial fermentation of cassava root and genetic modification of cassava plant [82].

9.1. Traditional processing methods

Traditional methods of cassava processing, such as drying, boiling, soaking and natural fermentation, have been used for many years to reduce or eliminate ANF inherent in cassava root, thereby improving the feeding value of cassava for humans and animals. Drying has been reported to improve the shelf life and also reduce several ANF present in cassava roots [83]. There are two types of cassava drying processes—sun drying and oven drying [84, 85]. Sun drying has been reported to effectively eliminate more HCN than oven-drying process during cassava processing due to the prolonged contact time between linamarase and glucosides during the sun-drying process [73]. It has been reported that the combination of soaking followed by boiling tends to eliminate more HCN than single/individual action of soaking or boiling alone [79]. According to Nambisan [30], up to 80% of the glucosides are removed by boiling and sun drying, while only about 20% are eliminated by frying, baking and steaming. Reduction in total cyanogens is effected by enzymatic decomposition of cyanogenic glucosides and/or leaching of cyanogens in the water in which the roots are boiled, with liberation of volatile HCN. Nambisan [30] concluded that crushing and pounding fresh cassava roots, followed by sun drying eliminates as much as 95% of the cyanogens and is the most efficient traditional method used in reducing the ANF inherent in cassava root. Soaking of cassava chips in water for about 24 h prior to sun drying reduced the HCN in three cassava varieties from 108.37 to 10.83 ppm (reduced by 90%), from 66.45 to 13.33 ppm (reduced by 79.94%) and from 58.63 to 15.0 (reduced by 74.42%) [86]. The authors highlighted the importance of soaking of cassava chips for at least 24 h prior to sun drying for a safe level of HCN in the flour. Natural fermentation of cassava pulp for 96 h reduced the HCN by 22 ppm (52.4%). Soaking of the sliced cassava tissue for 24 h prior to sun drying resulted in 16 ppm (38.1%) and 15 ppm (38.4%) HCN reduction for two cassava varieties studied. The HCN loss during sun drying was 6 ppm (14.3%) and 5 ppm (12.8%) for the two cassava cultivars used in the trial [78].

9.2. Modern processing methods

With the established fact that cassava root is a rich energy source but has low protein, amino acid, mineral and vitamin contents and also contains ANF (most notably high cyanogen glucoside level), the need for adequate processing cannot be overemphasised. Traditional methods of processing, such as drying, have been used for many decades; however, results have been inconsistent and are not able to improve the nutrient composition and reduce the ANF content to below the United Nations' recommendation. This presents the need for more modern, efficacious and multi-targeted technologically driven approaches. In order to achieve safe levels of 10 μg CN/g in cassava products, new methods of processing, especially for cassava containing more than 250 μg CN eq./g, are needed [30]. In recent times, the use of biotechnological method of processing cassava aimed at improving the nutrient composition and reducing ANF inherent in the crop has received a lot of attention from several researchers. Biotechnology involves the use of living biological systems and organisms or derivatives to develop, modify or make products or processes for specific use [87]. The use of biotechnology (depending on the focus of its application) often cuts across other closely related fields such as bio-engineering, biomedical engineering, molecular engineering, bio-manufacturing and bioprocessing [87]. The use of bio-engineering, bioprocessing and bio-manufacturing techniques in crop improvement has been gaining a lot of research interest and is the main focus of this write-up. With reference to cassava plant, the use of biotechnology involves the exploitation of biological processes, especially the genetic manipulation of microorganisms for the production of hormones and enzymes, and manipulation of genes with the aim of producing disease-resistant varieties, improving the nutrient composition, as well as eliminating any ANF inherent in cassava crop. Several studies have shown the efficacy of biotechnologically targeted processing methods in improving the feed value of cassava for both humans and animals, for instance, in a study to ascertain the feasibility of enhancing root cyanide assimilation into protein, optimally overexpressed Arabidopsis CAS and NIT4 genes in cassava roots resulted in up to a 50% increase in root total amino acids and a 9% increase in root protein accumulation [88]. It has been hypothesised that cyanogen toxicity in cassava foods can be accelerated by cyanogenesis and cyanide volatilization during food processing [89]. To achieve this objective, the authors expressed the leaf-specific enzyme, hydroxynitrile lyase (HNL), in roots. This enzyme catalyses the breakdown of acetone cyanohydrin to cyanide. Expression of HNL in roots accelerated cyanogenesis by more than threefold, substantially reducing the accumulation of acetone cyanohydrin during processing thereby reducing the HCN content.

According to Siritunga and Sayre [89], cyanogen-free cultivars were generated by selective inhibition of CYP79D1/D2 gene expression. The CYP79D1/D2 enzymes catalyse the first dedicated step in cyanogen synthesis. Tissue-specific inhibition of CYP79D1/D2 expression in leaves leads to a 99% reduction in root cyanogen levels, indicating that the cyanogenic glycoside, linamarin, is synthesised in leaves and transported to roots. Zvauya and Muzondo [81] concluded that there was a marked improvement in protein level following microbial fermentation of cassava with *A. oryzae*. There was a 13.5% increase in protein and a marked reduction in HCN after cassava root pulp was fermented by *S. cerevisiae* in solid-liquid media fermentation conditions during 132 h and dried at 30°C [71].

A combination of microbial fermentation and other processing methods such as sun drying and milling led to a decrease in the total cyanogen level by 40% (158 mg/kg dry weight to 54.2 mg/kg dry weight). Oboh and Akindahunsi [90] reported that protein content of cassava flour improved when *S. cerevisiae* was added to raw and unprocessed cassava. The authors further observed that the cyanogen content also decreased. In an experiment aimed at enhancing starch production in cassava through genetic modification, Ihemere [91] reported an increase in carbohydrate/starch content in transgenic cassava due to enhanced root ADP-glucose pyrophosphorylase (AGPase). Leyva-Guerrero [92] reported that a biotechnological program known as BioCassava Plus employed modern biotechnologies to produce novel cassava germplasm with increased nutrient levels. The program demonstrated that cyanogens play a central role in cassava nitrogen metabolism and that strategies employed to increase root protein levels result in reduced cyanogen levels in roots. Furthermore, the program also demonstrated that enhancing root iron uptake has an impact on the expression of genes that regulate iron homeostasis in multiple tissues. These observations demonstrate the complex metabolic interactions involved in enhancing targeted nutrient levels in cassava plant. Available research results have demonstrated that micronutrient-enrichment traits are available within the genomes of these major staple food crops, including cassava that could allow for substantial increases in the levels of Fe, Zn and provitamin A carotenoids (as well as other nutrients and health-promoting factors) through biotechnological means without negatively impacting crop yield [93]. According to Refs. [46, 94], it was recommended that the target of biotechnological application in cassava processing should be the development of appropriate starter culture for cassava processing that will effectively produce linamarase enzymes for detoxifying cassava, break down starch to the simple sugars needed for acid production, improve the protein content of the products, reduce processing time and yield products with stable desired qualities. In addition, the authors concluded that *B. subtilis* produces amylase and other enzymes that are necessary for the breakdown of starch to sugars, which are needed for the growth of other fermenting microorganisms. The process of increased proliferation of lactic acid bacteria leads to a corresponding increase in nutrients and a reduction in ANF in cassava root that has undergone microbial fermentation. Lactic acid bacteria convert cassava sugars to lactic and other acids that contribute to the flavour in addition to having preservative effects. Strains of *Lactobacillus plantarum* that is capable of producing amylase and linamarase have been used to improve starch hydrolysis and achieve a 98% cyanide detoxification of cassava. Linamerase obtained from *Lactobacillus delbrueckii* NRRL B-763 when added to raw cassava tuber resulted in a 95% reduction of HCN [95].

9.3. Amino acid supplements

As cassava products are low in protein and/or deficient in amino acids [96–98], supplementation of amino acids, especially methionine and lysine, has been reported to be a viable method for improving the quality of diets containing cassava and its products in poultry feeding [99]. It has been reported that cyanide can be detoxified to thiocyanate by the enzyme, rhodanase, with methionine as the sulphur donor [100], which makes this amino acid a limiting factor in cassava-based diets. In addition to boosting amino acid content of such diets, an extra dose of methionine may contribute to cyanide detoxification [43]. A study on the effect of levels of methionine supplementation (0.2 and 0.4%) in cassava peel-based diets for broilers concluded that up to 15% cassava peel meal can be substituted in broiler diets using 0.2% methionine, while up to 20% cassava peel meal inclusion requires 0.4% methionine supplementation for birds to achieve the desired productive performance in terms of weight gain and feed conversion [101].

9.4. Miscellaneous supplements

The scope for improving the utilisation of cassava products is wide. Researchers have attempted to use various supplementations other than enzymes and amino acids to enhance the use of cassava products for poultry feeding. For example, feeding cassava chips supplemented with *Moringa oleifera* leaf meal at 5 and 10% levels enabled the chips to replace maize at 55.56 and 83.33% in the diets of broilers, with no negative effect on productivity and blood function when 5% *M. oleifera* leaf meal was added [102]. Furthermore, Tesfaye et al. [103] tested cassava root chips supplemented with *M. oleifera* in layers. They reported that body weight gain, egg weight and hatchability were higher when birds were fed diets containing up to 50% cassava root chips as a full substitute for corn, supplemented with 5% *M. oleifera. Moringa oleifera* has high levels of vitamins and minerals and a rich amino acid profile [104], hence supplementing it in diet containing low-protein cassava product may enhance the nutritional value of the diet. Again, because cassava root meal is deficient in carotene and other carotenoids, one of the ways to overcome this deficiency is to supplement cassava root meal with cassava leaf meal with its high carotene and protein, hence combining a high-energy/low protein and high protein/low energy ingredients to meet the nutritional requirement of birds. In a study where maize was replaced with diets containing varying levels of cassava root meal and cassava leaf meal mixture showed that feed intake was improved, mortality was low and feed cost per weight gain was least when broilers were fed up to 75% of such mixture in the diet [23]. Due to dustiness, researchers have supplemented cassava meal-based diets with oil to reduce this limitation; while in some cases, palm oil supplementation is used to balance energy in the diet. However, Ukachukwu [105], in a study involving composite cassava meal supplemented with or without palm oil and/or methionine, reported that such supplementations increased body weight and feed intake, while improving feed conversion. In another study by Kana et al. [106], they concluded that cassava meal with 3% palm oil and 1% cocoa husk supplementation can replace maize up to 75% in the diets for broilers without any adverse effect on bird performance.

10. Conclusion

There is an increasing demand on food and feed resources by man due to rising global population. To avoid future food crisis and loss of animal protein, animal nutritionists have continued to explore alternative feed resources to meet the needs of both man and farm animals. Several crops stand out as possible alternatives for feeding poultry. One such crop is cassava, a very abundant crop in tropical regions of Africa, Asia and South America. However, the use of cassava and its products is limited due to several reasons—short shelf life, low protein and amino acid contents, presence of cyanide and other toxic substances. To maximise the use of cassava and its products for poultry feeding, the application of different biotechnological techniques is needed, in order to enhance its preservation, improve its nutritional value and improve its utilisation in the poultry feed industry. In recent years, several researchers have used many of these techniques, ranging from different processing methods to supplementations with different feed additives. While there are no consistent agreements in some of the techniques, most of them hold promise for increased utilisation of cassava and its products with careful application of these techniques. It is necessary to continue to explore other methods and their applications so as to sustain the progress achieved in the biotechnological improvement of cassava usage in poultry feeding.

Author details

Apeh Akwu Omede[1,2], Emmanuel Uchenna Ahiwe[1,3], Ze Yuan Zhu[4], Fidelis Fru-Nji[4] and Paul Ade Iji[1]*

*Address all correspondence to: pauladeiji@gmail.com

1 School of Environmental and Rural Science, University of New England, Armidale, NSW, Australia

2 Department of Animal Production, Kogi State University, Anyigba, Kogi State, Nigeria

3 Department of Animal Science and Technology, Federal University of Technology, Owerri, Imo State, Nigeria

4 DSM Nutritional Products, Animal Nutrition and Health, Mapletree Business City, Singapore, Singapore

References

[1] Ravindran V. Poultry feed availability and nutrition in developing countries alternative feedstuffs for use in poultry feed formulations. In: Poultry development review. Rome: FAO; 2013. pp. 60-63

[2] Okeye FC, Ugwuene MC, Ubaeduonu LC. Effects of the replacement of maize with graded levels of sorghum malt (*Sorghum bicolor*) on the performance of broiler chicks. Agricultural Journal. 2006;**1**:77-80

[3] Kellems RO, Church DC. Livestock Feeds and Feeding. Boston, Massachusetts, USA: Prentice Hall; 2010

[4] Uchegbu MC, Omede AA, Chiedozie IC, Nwaodu CH, Ezeokeke CT. Performance of finisher broilers fed varying level of raw (sun-dried) cocoyam (*Xanthosoma sagittifolium*) corm meals. Report and Opinion. 2010;**2**:22-25

[5] Okereke CO. Utilization of cassava, sweet potato, and cocoyam meals as dietary sources for poultry. World Journal of Engineering, Pure and Applied Science. 2012;**2**:67

[6] CGIAR. Cassava. Available from http://www.cgiar.org/our-strategy/crop-factsheets/cassava/ [Assessed 2017-17-07]

[7] Food and Agricultural Organization. Food outlook. Biannual report on global food markets. Rome, Italy: FAO; 2016. Page 4. ISSN: 0251-1959 (print) ISSN 1560-8182 (Online). http://www.fao.org/3/a-i6198e.pdf

[8] EuropaBio. What is Biotechnology? 2014. Available from http://www.europabio.org/what-biotechnology [Accessed 2017-08-17]

[9] Tewe OO, Lutaladio N. Cassava for Livestock Feed in Sub-Saharan Africa. Rome, Italy: FAO; 2004

[10] Sarkiyayi S, Agar TM. Comparative analysis on the nutritional and anti-nutritional contents of the sweet and bitter cassava varieties. Advance Journal of Food Science and Technology. 2010;**2**(6):328-334

[11] United state Department of Agriculture (USDA). Basic Report: 11134, Cassava, raw. National nutrient database for standard reference release 28. [Internet]. 2016. Available from: https://ndb.nal.usda.gov/ndb/foods/ [Accessed 2017-08-19]

[12] Tewe OO. The Global Cassava Development Strategy. U.N. Food and Agriculture Organization; 2004

[13] Montagnac JA, Davis CR, Tanumiharjo SA. Nutritional value of cassava for use as a staple food and recent advances for improvement. Comprehensive Reviews in Food Science and Food Safety. 2009;**8**(3):181-194. DOI: 10.1111/j.1541-4337.2009. 00077.x

[14] Burns AE, Gleadow RM, Zacarias AM, Cuambe CE, Mille RE, Cavagnaro TR. Variations in the chemical composition of cassava (*Manihot esculenta* Crantz) leaves and roots as affected by genotypic and environmental variation. Journal of Agricultural and Food Chemistry. 2012;**60**(19):4946-4956. DOI: 10.1021/jf2047288

[15] El-Sharkawy MA. Cassava biology and physiology. Plant Molecular Biology. 2003;**53**: 621-641

[16] Charles AL, Chang YH, Co WC, Sriroth K, Huang TC. Influence of amylopectin structure and amylose content on the gelling properties of five cultivars of cassava starches. Journal of Agricultural and Food Chemistry. 2005;**53**(7):2717-2725

[17] Buitrago JA, Gil JL, Opina B. The use of cassava in animal feeding. In: Ospina B, Cellabos H, editors. Cassava in the Third Millennium: Modern Production, Processing, Use and Marketing Systems. Columbioa: Centro Internacional de Agricultura Tropical; 2002. pp. 526-569

[18] Emmanuel OA, Clement A, Agnes SB, Chiwona-Karltun L, Drinah BN. Chemical composition and cyanogenic potential of traditional and high yielding CMD resistant cassava (*Manihot esculenta* Crantz) varieties. International Food Research Journal. 2012;**19**(1):175-181

[19] Koehler P, Wieser H. Chemistry of cereal grains. In: Gobbetti M, Gänzle M, editors. Handbook on Sourdough Biotechnology. Boston, MA: Springer; 2013

[20] Chauynarong N, Elangovan AV, Iji PA. The potential of cassava products in diets for poultry. World's Poultry Science Journal. 2009;**65**:23-36

[21] Nuss E, Tanumihardjo SA. Maize: A paramount staple crop in the context of global nutrition. Comprehensive Reviews in Food Science and Food Safety. 2010;**9**:417-436. DOI: 10.1111/j.1541-4337.2010.00117.x

[22] Ngiki YU, Igwebuike JU, Moruppa SM. Effects of replacing maize with cassava root-leaf meal mixture on the performance of broiler chickens. International Journal of Science and Technology. 2014;**3**(6):352-362

[23] Salvador EM, Steenkamp V, McCrindle CME. Production, consumption and nutritional value of cassava (*Manihot esculenta*, Crantz) in Mozambique: An overview. Journal of Agricultural Biotechnology and Sustainable Development. 2014;**6**(3):29-38

[24] Ogunsipe MH, Adejumo JO, Agbede JO, Asaniyan EK. Effect of roxazyme G2G supplementation on cassava plant meal fed to broiler chickens. Livestock Research for Rural Development. 2015;**27**:240. Available from: http://www.lrrd.org/lrrd27/12/ogun27240.htm Accessed 2017-08-20

[25] Aro SO, Aletor VA, Tewe OO, Agbede JO. Nutritional potentials of cassava tuber wastes: A case study of a cassava starch processing factory in south-western Nigeria. Livestock Research for Rural Development. 2010;**22**:213. Available from: http://www.lrrd.org/lrrd22/11/aro22213.htm [Accessed 20-08-2017]

[26] O'Brien GM, Wheatley CC, Iglesias C, Poulter NH. Evaluation, modification and comparison of two rapid assays for cyanogens in cassava. Journal of Science of Food and Agriculture. 1994;**65**:391-399

[27] Cardoso AP, Mirione E, Ernesto M, Massaza F, Cliff J, Haque MR, Bradbury JH. Processing of cassava roots to remove cyanogens. Journal of Food Composition and Analysis. 2005;**18**:451-460

[28] Fakir MSA, Jannat M, Mostafa MG, Seal H. Starch and flour extraction and nutrient composition of tuber in seven cassava accessions. Journal of Bangladesh Agricultural University. 2012;**10**(2):217-222

[29] FAO/WHO. Joint FAO/WHO Food standards program, codex Alimentarius Commission, XII. Rome, Italy: FAO/WHO; 1991 (suppl. 4)

[30] Nambisan B. Strategies for elimination of cyanogens from cassava for reducing toxicity and improving food safety. Food Chemistry and Toxicology. 2011;**49**(3):690-693. DOI: 10.1016/j.fct.2010.10.035

[31] Adepoju OT, Adekola YG, Mustapha SO, Ogunola SI. Effect of processing methods on nutrient retention and contribution of cassava (*Manihot* spp.) to nutrient intake of Nigerian consumers. African Journal of Food, Agriculture, Nutrition and Development. 2010;**10**(2):2099-2211

[32] Ernesto M, Cardoso AP, Cliff J, Bradbury JH. Cyanogens in cassava flour and roots and urinary thiocyanate concentration in Mozambique. Journal of Food Composition and Analysis. 2000;**13**:1-12

[33] Ukachukwu SN. Studies on the nutritive value of composite cassava pellets for poultry: Chemical composition and metabolizable energy. Livestock Research for Rural Development. 2005;**17**:125. Available from http://www.lrrd.org/lrrd17/11/ukac17125.htm [Accessed 20-08-2017]

[34] Hahn SK, Reynolds L, Egbunike GN. Cassava as livestock feed in Africa. In: Proceedings of IITA/ILCA/ University of Ibadan Workshop on the Potential Utilization of Cassava as Livestock Feed in Africa; 14-18 November 1988; Ibadan, Nigeria. Ibadan, Nigeria: International Institute of Tropical Agriculture (IITA), and Addis Ababa, Ethiopia: International Livestock Centre for Africa (ILCA)

[35] Food and Agricultural Organisation (FAO). Better farming series 46: Use of cassava and sweet potatoes in animal feeding. 1995. p. 47

[36] Balagopalan C, Padmaja G, Nanda SK, Moorthy SN. Cassava in Food, Feed and Industry. Boca Raton, Florida: CRC Press; 1988

[37] dos Santos VLF, de Andrade Ferreira MA, dos Santos GT, Damasceno JC, de Oliveira K, Agostino BC, Olivo PM, Soares LFP, de Lima Silva L. Cassava peel as a replacement for corn in the diet of lactating cows. Tropical Animal Health and Production. 2015;**47**(4):779-781. DOI: 10.1007/s11250-015-0776-9

[38] Oboh G. Nutrient enrichment of cassava peel using a mixed culture of *Saccharomyces cerevisiae* and *Lactobacillus* spp. solid media fermentation. Electronic Journal of Biotechnology. 2006;**9**:46-49

[39] Panigrahi A. Review of the potential for using cassava root meal in poultry diets. In: Kurup GT, Palaniswami MS, Potty VP, Padmaja G, Kabeerathumma S, Pillai SV, editors. Tropical Tuber Crops: Problems, Prospects and Future Strategies. New Delhi, India: Oxford and IBH Pub. Co; 1996. pp. 416-428

[40] Kong S, Ogle B, Preston TR, Borin K. Feed selection and growth performance of local chickens offered different carbohydrate sources in fresh and dried form supplemented with protein-rich forages. Livestock Research for Rural Development. 2010;**22**:225 Retrieved August 22, 2017, from http://www.lrrd.org/lrrd22/12/saro22225.htm

[41] Akapo AO, Oso AO, Bamgbose AM, Sanwo KA, Jegede AV, Sobayo RA, Idowu OM, Fan J, Li L, Olorunsola RA. Effect of feeding cassava (*Manihot esculenta* Crantz) root meal on growth performance, hydrocyanide intake and haematological parameters of broiler chicks. Tropcal Animal Health and Production. 2014;**46**(7):1167-1172. DOI: 10.1007/s11250-014-0622-5

[42] Garcia M, Dale N. Cassava root meal for poultry. Journal of Applied Poultry Research. 1999;**8**:132-137

[43] Ravindran V. Preparation of cassava leaf products and their use as animal feeds. In: Roots, Tubers, Plantain and Bananas in Animal Feeding. Vol. 95. Rome, Italy: Food and Agriculture Organisation; 1991. p. 11e25

[44] Tewe OO. Detoxification of cassava products and effects of residual toxins on consuming animals. In: D. Machin, S. Nyold, (eds.), Feeding Proceedings, the FAO Expert Consultation Held in CIAT; 21-25 January, 1991; Columbia; 1991

[45] Cooke RD, Maduagwu EN. The effect of simple processing on the cyanide content of cassava chips. Journal of Food Technology. 1985;**13**:299e306

[46] Oyewole OB. Application of biotechnology to cassava processing in Africa. In: Agbor Egbe T, Brauman A, Griffon D, Trèche S, editors. Transformation Alimentaire du Manioc. Paris: ORSTOM Éditions; 1995. pp. 278-286

[47] Olomo V, Ajibola O. Processing factors affecting the yield and physico-chemical properties of starch from cassava chips and flour. Starch–Stärke. 2003;**55**:476-481

[48] Abera S, Rakshit SK. Comparison of physicochemical and functional properties of cassava starch extracted from fresh root and dry chips. Starch-Stärke. 2003;**5**:287-296

[49] Amerah AM, Ravindran V, Lentle RG, Thomas DG. Feed particle size: Implications on the digestion and performance of poultry. World's Poultry Science Journal. 2007;**63**(03):439-455. DOI: 10.1017/S0043933907001560

[50] Amerah AM, Ravindran V, Lentle RG, Thomas DG. Influence of feed particle size on the performance, energy utilization, digestive tract development, and digesta parameters of

broiler starters fed wheat- and corn-based diets. Poultry Science. 2008;**87**(11):2320-2328. DOI: 10.3382/ps.2008-00149

[51] Zhu F. Composition, structure, physicochemical properties, and modifications of cassava starch. Carbohydrate Polymers. 2015;**122**:456-480. DOI: 10.1016/j.carbpol.2014.10.063

[52] Karam LB, Grossmann MVE, Silva RSSF, Ferrero C, Zaritzky NE. Gel textural characteristics of corn, cassava and yam starch blends: A mixture surface response methodology approach. Starch–Stärke. 2005;**57**:62-70

[53] Karam LB, Ferrero C, Martino MN, Zaritzky NE, Grossmann MVE. Thermal, microstructural and textural characterisation of gelatinised corn, cassava and yam starch blends. International Journal of Food Science and Technology. 2006;**41**:805-812

[54] Yin Y, Deng ZY, Huang HL, Zhong HY, Hou ZP, Gong J, Liu Q. Nutritional and health function of carbohydrate for pigs. Journal of Animal and Feed Science. 2004;**13**(4):523-538

[55] Chung HJ, Liu Q, Huang R, Yin Y, Li A. Physicochemical properties and in vitro starch digestibility of cooked rice from commercially available cultivars in Canada. Cereal Chemistry. 2010;**87**(4):297-304

[56] Araújo DD, Amorim AB, Saleh MAD, Curcelli F, Perdigón PL, Bicudo SJ, Berto DA. Nutritional evaluation of integral cassava root silages for growing pigs. Animal Nutrition. 2016;**2**(3):149-153. DOI: 10.1016/j.aninu.2016.04.006

[57] Phuc BHN, Ogle B, Lindberg JE. Effect of replacing soybean protein with cassava leaf protein in cassava root meals based diets for growing pigs on digestibility and N retention. Animal Feed Science and Technology. 2000;**83**:223-235

[58] Hong TTT, Lindberg JE. Effect of cooking and fermentation of a piglet diet on gut environment and digestibility in growing pigs. Journal of Livestock Science. 2007;**109**:135-137

[59] Morgan NK, Choct M. Cassava: Nutrient composition and nutritive value in poultry diets. Animal Nutrition. 2016;**2**(4):253-261. DOI: 10.1016/j.aninu.2016.08.010

[60] Collares RM, Miklasevicius LVS, Bassaco MM, Salau NPG, Mazutti MA, Bisognin DA, Terra LM. Optimization of enzymatic hydrolysis of cassava to obtain fermentable sugars. Journal of Zhejiang University-Science B (Biomedicine & Biotechnology). 2012;**13**(7):579-586. DOI: 10.1631/jzus. B1100297

[61] Olanbiwoninu AA, Odunfa SA. Production of fermentable sugars from organosolv pretreated cassava peels. Advanced Microbiology. 2015;**5**:117-122

[62] Franco CML, Ciacco CF. Studies on the susceptibility of granular cassava and corn starches to enzymatic attack. 1. Study of the conditions of hydrolysis. Starch–Stärke. 1987;**39**:432-435

[63] Villena GK, Gutierrez-Correa M. Production of lignocellulolytic enzymes by *Aspergillus niger* bio-films at variable water activities. Journal of Biotechnology. 2007;**10**:124-140

[64] Adeyemo AI, Sani A, Aderibigbe TA, Abdurrasheed MO, Agbolade JO. A study of *Aspergillus niger*-hydrolyzed cassava peel meal as a carbohydrate source on the histology of broiler chickens. SpringerPlus. 2014;**3**:31. DOI: 10.1186/2193-1801-3-31

[65] Sornyotha S, Kyu KL, Ratanakhanokchai K. An efficient treatment for detoxification process of cassava starch by plant cell wall-degrading enzymes. Journal of Bioscience and Bioengineering. 2010;**109**(1):9-14. DOI: 10.1016/j.jbiosc.2009.06.021

[66] Rajan A, Abraham TE. Enzymatic modification of cassava starch by bacterial lipase. Bioprocess and Biosystems Engineering. 2006;**29**:65-71

[67] Rajan A, Sudha JD, Abraham TE. Enzymatic modification of cassava starch by fungal lipase. Industrial Crops and Products. 2008;**27**:50-59

[68] Oso AO, Li L, Zhang B, Uo R, Fan JX, Wang S, Jiang G, Liu H, Rahoo T, Tossou MC, Pirgozliev V, Oduguwa OO, Bamgbose AM. Effect of fungal fermentation with *Aspergillus niger* and enzyme supplementation on metabolizable energy values of unpeeled cassava root meal for meat-type cockerels. Animal Feed Science and Technology. 2015;**210**:281-286. DOI: 10.1016/j.anifeedsci.2015.09.015

[69] Mathivanan R, Selvaraj P, Nanjappan K. Feeding of fermented soybean meal on broiler performance. International Journal of Poultry Science. 2006;**5**:868-872

[70] Khempaka S, Thongkratok R, Okrathok S, Molee W. An evaluation of cassava pulp feedstuff fermented with *A. oryzae*, on growth performance, nutrient digestibility and carcass quality of broilers. Journal of Poultry Science. 2014;**51**:71-79

[71] Boonnop K, Wanapat M, Nontaso N, Wanapat S. Enriching nutritive value of cassava root by yeast fermentation. Scientia Agricola. 2009;**66**:616e20

[72] Poungchompu O, Wanapat M, Wachirapakorn C, Wanapat S, Cherdthong A. Manipulation of ruminal fermentation and methane production by dietary saponins and tannins from mangosteen peel and soapberry fruit. Archives of Animal Nutrition. 2009;**63**:389e400

[73] Polyorach P, Wanapat M, Wanapat S. Increasing protein content of cassava (*Manihot esculenta* Crantz) using yeast in fermentation. Khon Kaen Agricultural Journal. 2012;**40**(Suppl. 2):178e82

[74] Wanapat M, Kang S, Polyorach S. Development of feeding systems and strategies of supplementation to enhance rumen fermentation and ruminant production in the tropics. Journal of Animal Science and Biotechnology. 2013;**4**(1):32. DOI: 10.1186/2049-1891-4-32

[75] Wanapat M, Kang S. Cassava chip (*Manihot esculenta* Crantz) as an energy source for ruminant feeding. Animal Nutrition. 2015;**1**(4):266-270. DOI: 10.1016/j.aninu.2015.12.001

[76] Moorthy SN, George M. Cassava fermentation and associated changes in physicochemical and functional properties. Critical Reviews in Food Science and Nutrition. 1998;**38**:73-121

[77] Asaoka M, Blanshard JMV, Rickard JE. Seasonal effects on the physicochemical properties of starch from 4 cultivars of cassava. Starch–Stärke. 1991;**43**:455-459

[78] Kemdirim OC, Chukwu OA, Achinewhu SC. Effect of traditional processing of cassava on the cyanide content of gari and cassava flour. Plant Foods and Human Nutrition. 1995;**48**(4):335-339

[79] Padmaja G. Cyanide detoxification in cassava for food and feed uses. Critical Reviews in Food Science and Nutrition. 1995;**35**(4):299-339

[80] Siritunga D, Sayre R. Transgenic approaches for cyanogen reduction in cassava. Journal of Association of Official Analytical Chemists International. 2007;**90**(5):1450-1455

[81] Zvauya R, Muzondo MI. Reduction of cyanide levels in cassava during sequential sun drying and solid state fermentation. International Journal of Food Science and Nutrition. 1995;**46**(1):13-16

[82] Liu J, Zheng Q, Ma Q, Gadidasu KK, Zhang PJ. (2011) Cassava genetic transformation and its application in breeding. Integrated Plant Biology. 2011;**53**(7):552-569

[83] Girma G, Bultosa G, Abera S. Effect of cassava (*Manihot esculenta* Crantz) variety, drying method and blending ratio on the proximate composition and sensory properties of cassava-wheat composite bread. European. Journal of Food Science and Technology. 2015;**3**(4):41-54

[84] Kudra T. Energy aspects in drying. Drying Technology. 2004;**22**(5):917-932

[85] ITunde-Akintunde TY, Afon AA. Modelling of hot-air drying of pretreated cassava chips. Agricultural Engineering International: the CIGR Ejournal. 2009;**12**:34-41

[86] Nebiyu A, Getachew E. Soaking and drying of cassava roots reduced Cyanogenic potential of three cassava varieties at Jimma, Southwest Ethiopia. World Journal of Agricultural Sciences. 2011;**7**(4):439-443

[87] United Nation Convention. Convention on biological diversity. [Internet]. 1992. Available from https//www.cbd.int/doc/legal/cbd-en.pdf. [Accessed 2017-08-20]

[88] Zidenga T, Siritunga D, Sayre RT. Cyanogen metabolism in Cassava roots: Impact on protein synthesis and root development. Front Plant Science, 2017;**24**(8):220. DOI: 10.3389/fpls.2017.00220

[89] Siritunga D, Sayre R. Engineering cyanogen synthesis and turnover in cassava (*Manihot esculenta*). Plant Molecular Biology. 2004;**56**(4):661-669

[90] Oboh G, Akindahunsi AA. Biochemical changes in cassava products (flour and gari) subjected to *Saccharomyces cerevisiae* solid media fermentation. Food Chemistry. 2003;**82**(4):599-602

[91] Ihemere U, Arias-Garzon D, Lawrence S, Sayre R. Genetic modification of cassava for enhanced starch production. Plant Biotechnology Journal. 2006;**4**(4):453-465

[92] Leyva-Guerrero E, Narayanan NN, Ihemere U, Sayre RT. Iron and protein biofortification of cassava: Lessons learned. Current Opinion in Biotechnology. 2012;**23**(2):257-264. DOI: 10.1016/j.copbio.2011.12.009

[93] Welch RM, Graham RD. Breeding for micro nutrients bin staple food crops from a human nutrition perspective. Journal of Experimental Botany. 2004;**55**(396):354-364

[94] National Research Council (NRC). Panel on the Applications of Biotechnology to Traditional Fermented Foods. Applications of Biotechnology to Fermented Foods: Report of an Ad Hoc Panel of the Board on Science and Technology for International Development. Washington (DC): National Academies Press (US); 1992. 12, Cassava Processing in Africa. Available from: https://www.ncbi.nlm.nih.gov/books/NBK234676/. [Accessed 2017-80-19]

[95] Nwokoro O, Anya FO. Linamarase enzyme from Lactobacillus delbrueckii NRRL B-763: Purification and some properties of a β-glucosidase. Journal of the Mexican Chemical Society. 2012;**55**(4):246-250

[96] Silva HO, Da Fonseca RA, Guedes RD. Productive performance and digestibility of cassava leaf meal in broiler diets with or without addition of enzymes. Brazilian Journal of Animal Science. 2000;**29**:823-829

[97] Iyayi EA, Losel DM. Protein enrichment of cassava by-products through solid state fermentation by fungi. Journal of Feed Technology in Africa. 2001;**6**:116-118

[98] Ngiki YU, Igwebuike JU, Moruppa SM. Utilisation of cassava products for poultry feeding: A review. International Journal of Science and Technology. 2014;**2**(6):48-59

[99] Tewe OO, Egbunike GN. Utilization of cassava in non-ruminant livestock feeds. In: Hahn SK, Reynolds L, Egbunike GN editors. Cassava as livestock feed in Africa. Proceedings of IITA/ILCA/University of Ibadan Workshop on the Potential Utilisation of Cassava as Livestock Feed in Africa; 1992; Ibadan, Nigeria. 1992. pp. 28-38.

[100] Organisation for the Prohibition of Chemical Weapons. Hydrogen cyanide. 2013. http://www.opcw.org/about-chemical-weapons/types-of-chemical-agent/blood-agents/hydrogen-cyanide/. Retrieved on 22-08-2017

[101] Oladunjoye IO, Ojebiyi OO, Rafiu TA. High methionine supplementation improves the nutritional value of cassava peel meal for broiler chicken. Livestock Research for Rural Development. 2014;**26**:63. Available from http://www.lrrd.org/lrrd26/4/olad26063.htm [Accessed 22-08-2017]

[102] Olugbemi TS, Mutayoba SK, Lekule FP. Effect of Moringa (*Moringa oleifera*) inclusion in cassava based diets fed to broiler chickens. International Journal of Poultry Science. 2010;**9**:363-367

[103] Tesfaye EB, Animut GM, Urge ML, Dessie TA. Cassava root chips and *Moringa oleifera* leaf meal as alternative feed ingredients in the layer ration. The Journal of Applied Poultry Research. 2014;**23**(4):614-624. DOI: 10.3382/japr.2013-00920

[104] Moyo B, Patrick JM, Arnold H, Voster M. Nutritional characterization of moringa (*Moringa olifera* lam.) leaves. African Journal of Biotechnology. 2011;**10**:12925-12933

[105] Ukachukwu SN. Effect of composite cassava meal with or without palm oil and/or methionine supplementation on broiler performance. Livestock Research for Rural Development. 2008;**20**(53). Retrieved August 22, 2017 from http://www.lrrd.org/lrrd20/4/ukac20053.htm

[106] Kana JR, Fulefack HD, Gandjou HM, Ngouana R, Noubissie MM, Ninjo J, Teguia A. Effect of cassava meal supplemented with a combination of palm oil and cocoa husk as alternative energy source on broiler growth. Archiva Zootechnica. 2012;**15**(4):17-25

Chapter 15

Use of Co-Products from the Processing of Cassava for the Development of Adsorbent Materials Aiming Metal Removal

Daniel Schwantes, Affonso Celso Gonçalves Jr.,
Marcelo Angelo Campagnolo,
César Ricardo Teixeira Tarley,
Douglas Cardoso Dragunski, Jéssica Manfrin and
Andréia Da Paz Schiller

Additional information is available at the end of the chapter

http://dx.doi.org/10.5772/intechopen.71048

Abstract

Nowadays the contamination of water resources by the most varied pollutants have been accelerated. Technologies of decontamination of water are too costly, however, the development of low cost adsorbents, have proven to be efficient, promising and cheap alternatives for this purpose. The use of adsorbents from cassava residues has shown great potential for use as an adsorbent. The productive chain of this crop involves the production and processing of its roots, generating a large volume of solid waste. Aiming the sustainability of production systems, productive chains should optimize the production of cassava barks residues, since these do not present significant uses or benefit. In this scenario, this chapter gathers information from the literature on the use of solid waste from the cassava agroindustry and its use as adsorbents, aiming the removal of toxic metals, as well as their potential for the treatment of other contaminants. Several authors have demonstrated through studies the potentiality of the use of agroindustrial cassava residues as adsorbents. Because of a range of characteristics observed these adsorbents present viability for large-scale use, being in very similar to activated carbon. Thus, the use of these adsorbent materials represents an extremely viable and sustainable alternative.

Keywords: cassava adsorbents, cassava wastes, cassava bark, cassava bagasse, adsorption of metals, biosorbents

1. Introduction

Cassava, *Manihot esculenta* Crantz, is a perennial and shrubby plant, native to the South American continent, probably from Brazil, since the Indians already cultivated cassava, when the country was discovered.

It is a plant well adapted to variations in temperature and rainfall, being found in areas of different edaphoclimatic conditions and between latitude of 30°N and 30°S. It is growing in regions from sea level up to 2300 m of the altitude and in areas considered marginal to other crops: soils with low fertility and annual rainfall less than 600 mm in the semi-arid or above 1500 mm in the humid and sub-humid tropics [1].

The residues produced by the industrialization of cassava roots do not have many destinations; the main purpose is being animal feed [1], which is an agroindustrial problem to be faced by the cassava production segment.

Another global problem that has been aggravating in the last decades concerns the organic and inorganic pollution of the waters, being the adsorption process is one of the most promising alternatives for the removal of these contaminants. In this context, commercial activated carbon is one of the most widely used adsorbents for a large variety of pollutants in water, but its high cost is one of its main disadvantages.

In this context, the adsorption of metallic pollutants is more promising when using natural adsorbents such as agroindustrial waste. These are a promising alternative for chemical remediation because of their high adsorption capacity, low cost, and high availability.

Several authors have been researching alternative biosorbents for the removal of metal ions from contaminated solutions, such as banana and orange barks [2], cocoa barks [3], dry mass of *Eichhornia crassipes* [4], mussel shells [5], and bagasse of natural and modified sugar cane [6]. However, few studies report the adsorptive capacity of the solid fraction of residues from the industrialization of cassava roots in the context of their use as natural or modified adsorbents in contaminated water.

It is important to emphasize some advantages of the use of plants residues to the treatment of wastewater. They are operational facility, low processing, good adsorption capacity, selective adsorption of metal ions, low cost, high availability, and easy regeneration [7].

The subsequent items will describe some of the characteristics of the cassava crop, its agroindustrial residues, as well as reports on the hydrological contamination by toxic metals, and some results from studies describe the use of natural and modified adsorbents obtained from cassava agroindustrial residues as a viable possibility.

2. Production chain of cassava

Brazilian annual production of cassava has remained at 23 million tons [8], while the harvested area has declined, giving way to the cultivation of other more profitable crops.

The consumption of natural cassava tends to lose space, as it moves away from the producer poles, because it is a perishable merchandise and it requires immediate consumption after harvest. **Table 1** shows the data of the quantity produced in tons of cassava in recent years.

Cassava processing in Brazil is concentrated in flour. It is estimated that 80% of the roots are intended for this purpose. About 3% can be counted as being destined for the extraction of starch and its modifications, with the rest probably destined for animal feed [1].

The processing of cassava roots generates solid residues called bark (bark + weaves), being the main destinations are animal feed and its use as biofertilizers. It is estimated, on average, that about 11 million cassava barks are produced annually only in Brazilian territory, and normally, the barks are disposed together with the bran/bagasse, which are supplied for animal feed or disposed to the soil for decomposition and incorporation of organic matter [9]. It should be noted that a greater amount of waste compose the cassava bagasse or bran, that represents the final residue of cassava starch extraction, and its main destination is animal feed.

Region	2013	2014	2015	2016
North	7,467,943	8,037,507	7,971,127	8,620,328
Northeast	4,803,212	5,668,126	5,311,813	5,787,657
Southeast	2,491,229	2,524,993	2,477,465	2,279,168
South	5,477,417	5,583,682	5,489,019	5,163,158
Midwest	1,244,417	1,427,756	1,507,383	1,207,362
Brazil	23,242,064	23,242,065	22,756,807	23,057,673

Source: SEAB [8].

Table 1. Brazilian production of cassava in metric tons from 2013 to 2016 by regions of Brazil.

3. Water contamination by metals

Only 0.8% of the water available on the planet can be easily used for public supply. Of this small fraction, only 3% is in the form of surface water, while the other 97% corresponds to groundwater. As important as the amount of available water is the water quality, because the quality of water around the planet has deteriorated more and more, especially in the last 50 years, mainly in the regions where industrial centers and high population density exist.

In fact, one of the major problems faced in the last decades is the water pollution, coming, in large part, to the inadequate management of pesticides, low quality of the water used in irrigation, and the indiscriminate disposition of industrial or domestic waste. This contamination can cause the accumulation of substances that can be toxic to plants and, when entering the food chain, may become dangerous to animals and humans.

The industry contaminates water by eviction of disinfectants, detergents, solvents, toxic metal ions, radioactive waste, petroleum products, agrochemicals, and other compounds in rivers and lakes. Among the metals, some are strongly polluting and have several harmful effects on ecosystems, causing physical and chemical changes in water and a decrease in their quality and the mortality of flora and fauna, and consequently, human health.

The expression heavy metal applies to elements having a specific mass greater than 5 g cm^{-3} or having an atomic number greater than 20. Some metals are used in the biological metabolism, and in this way, they can be considered essential, as is the case of Cu^{2+}, Zn^{2+}, Ni^{2+}, and Cr^{3+}, and at higher levels, they can become toxic. On the other hand, Pb^{2+} and Cd^{2+} are considered toxic even at low concentrations [4].

For the treatment of contaminated water, there are several types of treatments considered conventional (physical, chemical, and biological), among which centrifugation, distillation, filtration, flocculation, sedimentation, ultrafiltration, electrodialysis, reverse osmosis, air entrainment, liquid-liquid extraction, catalysis, hydrolysis, neutralization, oxidation, reduction, precipitation, photolysis, ozonolysis, activated sludge, aerated lagoons, enzymatic treatments, and anaerobic digestion, among others can be mentioned. All these have advantages and disadvantages; however, it can be mentioned that the great majority is not always effective in the removal of small levels of metals without elevation of the costs.

In this scenario, the search for alternative adsorbents, which presents high availability and low cost of acquisition, can help and reduce the treatment of water and effluents with toxic metals. In the sequence, results from research that support this hypothesis will be presented.

4. Results from researches

4.1. Case of study: remediation of waters using solid wastes from the processing of cassava roots as biosorbents

The search for alternatives to conventional methods that have low cost and high efficiency pushed, in recent years, the research on the use of different biosorbents. Among the biosorbents that have been studied by several authors, cassava agroindustrial residues are also viable alternatives to their use as adsorbents. As presented by some authors, such as the studies, Schwantes et al. [9] applied the following solid residues of cassava: bark, bagasse, and the mixture of both, obtaining promising results for the use of these materials as natural adsorbents of Cu^{2+} and Zn^{2+}. Schwantes et al. [10] applied the biosorbents from bark, bagasse, and the mixture of both in the removal of Cr^{3+} from contaminated water. Or also, Schwantes et al. [11]

applied modified cassava barks with solutions 0.1 mol.L^{-1} of H_2O_2, H_2SO_4, and NaOH in the removal of Cd^{2+}, Pb^{2+}, and Cr^{3+}. Schwantes et al. [12] and Schwantes et al. [13] used cassava biomass in Pb^{2+} and glyphosate removal, among others.

As already commented in this chapter, the adsorption of metallic ions is extremely viable when using natural adsorbents such as, for example, industrial and agricultural waste. These adsorbents are a promising alternative for chemical remediation due to their great adsorption capacity, low cost, and high availability. In nature, there are a large number of biosorbents, which in their natural state and properly applied provide values of adsorption capacity similar or sometimes better than those presented by chemically modified materials.

In this way, it is known that the industrial processing of cassava generates significant quantities of waste that may cause serious environmental problems, and the correct treatment is rarely given to these residues. Some results of the literature will be demonstrated in the following items, which demonstrate the benefits of using cassava residues as adsorbents and their efficiency in the removal of pollutants from water.

4.1.1. Characterization of cassava residues and their possible use as adsorbents

It is usually observed that adsorbents with high capacity to remove pollutants from the environment have superficial structures varying from fibrous to spongy, with irregular structures and a large number of cavities on their surface.

Moreover, a good adsorbent is one that has a large number of active sites available for interaction with the species of interest. The sites are adsorption points constituted for one or more chemical functional groups distributed on the surface of the biosorbent. Clusters such as carboxyl, phenols, nitrogen groups, alkanes, and others may generate active sites that are favorable to adsorption of metallic species in addition to other pollutants.

Another outstanding characteristic in good adsorbents refers to textured parameters, such as high volume and varied pore diameters, aiming for adsorption of compounds with different ranges of hydration rays and perhaps the most outstanding characteristic and high-specific surface area.

According to the published studies [9–13], residues of the cassava crop have potential in the use of adsorption processes in function of physical, chemical, and morphological characteristics.

When studying the morphology of the barks and bagasse of cassava, as well as the mixture of both, Schwantes et al. [9] and Schwantes et al. [10] observed that the surfaces of the biosorbents are endowed with fibrous and spongy aspect, with irregular and heterogeneous structure. Many of these cavities can be evidenced, demonstrating that the material presents characteristics that suggest favorable conditions for adsorption (**Figure 1**).

The infrared (IV) characterization contributes to the understanding of the dynamics or the adsorption mechanism, since it provides information about the functional groups present

Figure 1. SEM of the adsorbents originating of cassava residues: barks (3000×) bagasse (3000×), and mixture of both fractions (5000×). Source: Refs. [9, 10].

in the structure of the adsorbents [7]. The biosorption is the result of electrostatic interaction and formation of complexes between ions (pollutants) and functional groups of biomass [6].

In the case of cassava agroindustrial residues, barks, bagasse, and their mixture, all in their natural form, Schwantes et al. [9] when applied them in the removal of Cu^{2+} and Zn^{2+} and Schwantes et al. [13] when used in the removal of glyphosate from water found the following characteristic bands for the infrared spectra: 3440, 2920, 1730, 1650, 1420, and 1030 cm^{-1}, according to **Figure 2**.

According to **Figure 2**, the presence of a big and strong band at 3440–3330 cm^{-1} is observed, which can be attributed to the vibrational elongation of the O-H bond, possibly characterized by the vibrational stretching of the hydroxyl groups present in carbohydrates, fatty acids, proteins, lignin units, cellulose, and absorbed water [14], while the band at 2920 cm^{-1} can be attributed to a vibrational elongation of the C-H bond in alkanes.

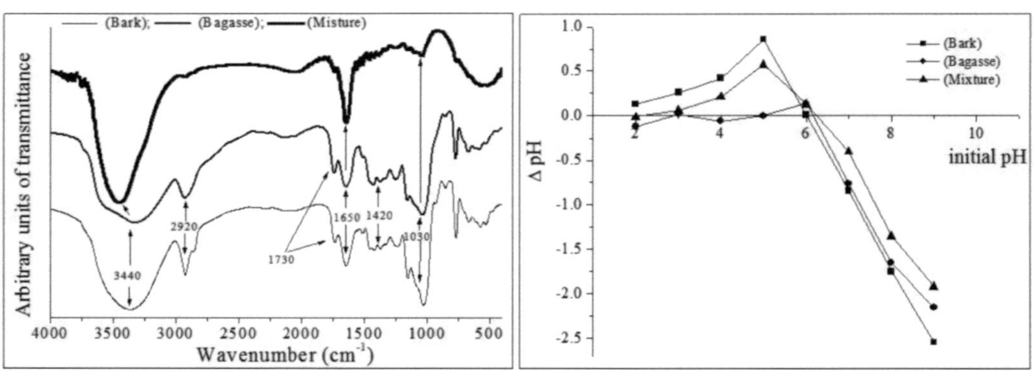

Figure 2. In the left: FT-IR to bark, bagasse, and mixture; in the right: pH_{PZC} in KCl 0.5 mol.L^{-1} to the adsorbents bark, bagasse, and mixture of both. Source: Refs. [9, 10, 13].

The bands in 1730 cm^{-1} are indicative of the presence of the starch, or the aldehyde and ketone functions present in lignin and holocellulose. Bands at 1420–1650 cm^{-1} can be attributed to the vibrational stretching of the C-O link of amides and carboxylic groups. Peaks at 1030 cm^{-1} can be attributed to the C-O stretch [14].

The adsorption of metals by residues of plants, natural materials, and agroindustrial residues can be attributed to the presence of some functional groups such as lignin, alcohols, carboxylic groups, proteins, and carbohydrates [15].

pH_{PZC}, or point of zero charge, is defined as the pH at which the surface of the solid has a neutral charge [16]. When the pH > pH_{PZC}, the surface of the adsorbent is electronegative, favoring the adsorption of cations; in case, if the pH < pH_{PZC}, the surface of the adsorbent is electropositive, and in this state, H^+ ions compete with the metallic cations, repelling them of the surface decreasing the adsorption. The characterization of the adsorbent in relation to its point of zero charge is very important, since it will be possible to predict the behavior of surface charges that vary with the pH of the medium and that, in one way or another, may influence the adsorption process of pollutants, favoring their removal from the environment.

According to authors Schwantes et al. [9], Schwantes et al. [10], and Schwantes et al. [13], the adsorbents originating of the cassava residues, bark, bagasse, and mixture of both, indicate that the pH corresponding to the equivalence point between positive and negative charges for the adsorbent bark is 6.00, for the bagasse material is 6.17, and for the mixture of both is 6.24 (**Figure 2**).

In this way, according to the cited authors, the adsorption of cations such as Cd^{2+}, Pb^{2+}, Cr^{3+}, Cu^{2+}, Zn^{2+}, and others should be favored by pH values higher than the values found for pH_{PZC} (6.00, 6.17, and 6.24 for bark, bagasse, and mixture, respectively).

It is important to remember that what is usually called the "cassava bark" in industry is the result of the cleaning of the roots in the moment of their reception in industry; in this time, the cassava barks are removed, as well as part what is denominated weaves. Both these residues constitute the first agroindustrial co-product. The average composition of the bark and bark + weaves (mixture), which configure the first agroindustrial residues related have the following composition (**Table 2**).

It is important to emphasize that the chemical composition of the adsorbent materials provides a basis for the verification of which chemical elements may possibly return the solution during the exposure of the adsorbent to the medium, since there is always the possibility of the occurrence of the inverse process called desorption.

Another very important aspect relative to the chemical composition of the materials is about the presence or absence of the component to be removed in the structure of the adsorbent. For example, a natural adsorbent whose composition is rich in Zn is possibly not efficient for the removal of this metal from solution and can increase the Zn concentration in the solution because of the diffusion of this from adsorbent to the medium.

Parameters	Bark	Mixture	Parameters	Bark	Mixture
	g.100 g^{-1} dry mass			mg.100g^{-1} dry mass	
Volatile solids		26.23	Total CN	0	23.9
Ashes	4	1.45	Free CN	60	120
Soluble carbohydrates		7.86	Phosphor	110	60
Starch	0	32	Sulfur	18	320
Lipids	3	0.63	Boron		18
Nitrogen	0.64	2.1			
Fiber	41				
Lignin		6.46			

Source: Alves [1].

Table 2. Mean values of several determinations carried out on cassava residues.

When evaluating the natural adsorbents from the residues of cassava roots, Schwantes et al. [9] and Schwantes et al. [10] found higher concentrations for K, Ca, Mg, Cu, Fe, Mn, and Zn in the cassava barks, when compared to bagasse and the mixture of both. This result is possibly because of the fact that the bagasse is the result of an industrial chemical and physical processing, where there is isolation and extraction of the starch, resulting in the removal of part of these chemical elements.

The authors also emphasize the presence of small concentrations of Pb in the adsorbent materials evaluated (**Table 3**), which according to the researchers may be an indication of the presence of this toxic metal in the soil, which was absorbed by the roots of cassava during growth and development of culture in the countryside.

It should be noted that soil contamination by toxic metals, such as Pb, could have different origin, for example, from inadequate disposal of Pb batteries, contaminated effluents, and disposal of contaminated waste in soil, atmospheric pollution originated of the burning fossil fuels containing Pb, pollution and atmospheric deposition from industries that work with Pb, and the use of fertilizers and agricultural inputs [17].

4.1.2. Experimental results involving natural adsorbents of cassava and remediation of toxic heavy METAS in waters

According to researchers in the literature that relate the use of adsorbent materials based on residual biomass of cassava and removal of metals from water, cassava biosorbents are efficient in the removal of metals such as Cd^{2+} [18], Pb^{2+} [12], Cr^{3+} [10], Cu^{2+}, and Zn^{2+} [9]. The literature mention results where the influence of the pH of the medium is evaluated, the proportion between adsorbent and the volume of the adsorbate, adsorption kinetics,

Adsorbent	K	Ca	Mg	Cu	Fe	Mn	Zn	Cd	Pb	Cr
	g.kg^{-1}			mg.kg^{-11}						
Bark	24.10	35.03	6.83	14.33	35.67	123.33	32.00	<0.005	11.00	<0.01
Bagasse	5.77	23.23	4.58	5.67	24.50	27.67	18.67	<0.005	14.67	<0.01
Mixture	7.77	22.58	5.12	6.00	26.00	34.00	17.00	<0.005	3.33	<0.01

ND: not detected by EAA/flame method; LQ (quantification limits): K = 0.01; Ca = 0.005; Mg = 0.005; Cu = 0.005; Fe = 0.01; Mn = 0.01; Zn = 0.005; Cd = 0.005; Pb = 0.01; Cr = 0.01. Source: Refs. [9, 10].

Table 3. Chemical characteristics of adsorbent materials.

equilibrium studies, as well as the potential reuse of these materials in other sorption processes. Some of these results will be described further in the next items in this chapter.

4.1.2.1. Influence of the pH of the contaminant solution and the proportion between cassava biosorbents and adsorbate

The pH is one of the most important parameters in the adsorption process, because its interference occurs in the solid-solution interface, influencing the loads of the active sites of the biomass and in the behavior of the adsorbates. The pH controls the surface properties of the adsorbents, functional groups, and ionic state of the metallic species, affecting a lot the adsorption of metallic ions [15]. This occurs because in acidic pH, the H^+ ions strongly compete between them and with the metals in solution by the active sites in the adsorbent. At basic pH, the connection sites may not be activated as a function of pH, or precipitation of these metals can occur, and it does not occur as contact between adsorbent/adsorbate.

According to Mimura et al. [16], the adsorption of metallic species in the positive form will be favored at the pH in which negative species predominate on the adsorbent surface. As verified in **Figure 3**, proportions between adsorbent/adsorbate higher than 4 g.L^{-1} do not result in high removal of ions. Also can be observed that there is a little variation of the % removal in the studied pH range; however, it should be considered that metals such as Cd and Pb present an easy precipitation in pH ranges close to neutrality, making the adsorption process impracticable under these chemical conditions.

The authors of **Figure 3** mention that these tests were executed in low concentrations (10 mg.L^{-1}), being, therefore, the adsorbent materials bark, bagasse, and mixture capable of almost completely removing the contents of Cd^{2+} and Pb^{2+} from contaminated solutions. It can be observed that in the studied pH range, the removal reached values higher than 90% in all materials, especially the adsorbent consisting of "bark + bagasse," which almost completely removed the metals contained in the contaminant solution. According to the authors of this study, 4 g of adsorbent per liter of solution contaminated with Cd^{2+} or Pb^{2+} is the ideal ratio between adsorbent and adsorbate.

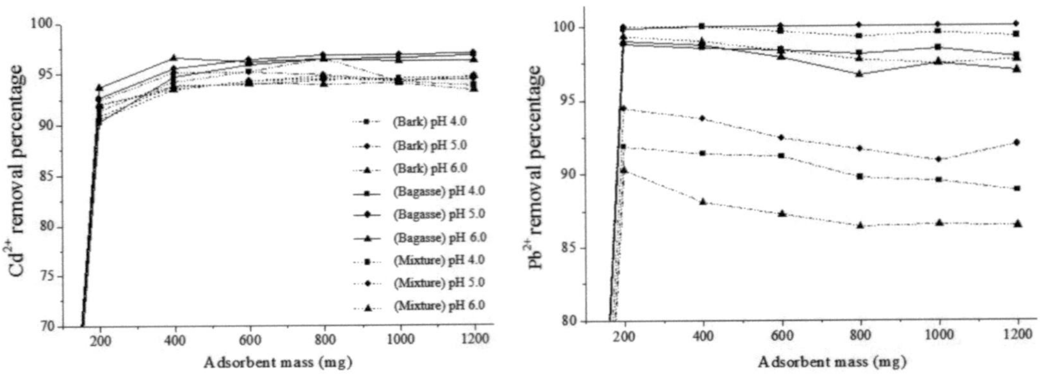

Figure 3. Adsorbent mass and pH of solution on the % of removal of Cd^{2+} [18] and Pb^{2+} [12] for the absorbents barks, bagasse, and mixture of cassava. The values in the x-axis corresponding to the proportion adsorbent/adsorbate: 200 mg (4 $g.L^{-1}$), 400 mg (8 $g.L^{-1}$), 600 mg (1.2 $g.L^{-1}$), 800 mg (1.6 $g.L^{-1}$), 1000 mg (2.0 $g.L^{-1}$), and 1200 mg (2.4 $g.L^{-1}$).

Other authors also verified removal percentages of varied metals similar to those found for cassava adsorbents, and the pH range studied, because it was slightly acidic, did not cause great influences in the adsorption process, for example, in the case of the adsorbent materials using barks of *Pinus elliottii* [19], rice barks [10], peanut barks [20], among others.

4.1.2.2. Contact time influences between cassava biosorbents and metals

Another relation that must be observed because of its importance is the influence of the contact time between adsorbent and adsorbate also called adsorption kinetics. The kinetics of this process is dependent on the relative velocity between four successive stages: transport within the solution, transport by diffusion, transport through the pores, and adsorption.

In the **Figure 4**, it is possible to observe that, independent of the metal ion, Cd^{2+} or Pb^{2+}, or adsorbent material studied, occurs an increase in the adsorption process with the passage of time, and that in general, after 20 min of agitation, the system enters into dynamic equilibrium, indicating a quick adsorption. It is also possible to observe that the material denominated mixture presents adsorption superior for Cd^{2+} and Pb^{2+}, demonstrating that even in low concentrations, the mixture of the two materials (barks + bagasse) produces superior results than acting in isolation.

By linearizing these results according to the Pseudo-second order [21] and intraparticle diffusion models, as cited by Yang and Al-Duri [22], we have the following results.

According Ho and McKay [21], the pseudo-second-order model presents a good fit for many adsorbent and polluting materials, which are not different for cassava biosorbents studied by Schwantes et al. [12], Schwantes et al. [13], and Schwantes [18]. These authors suggest

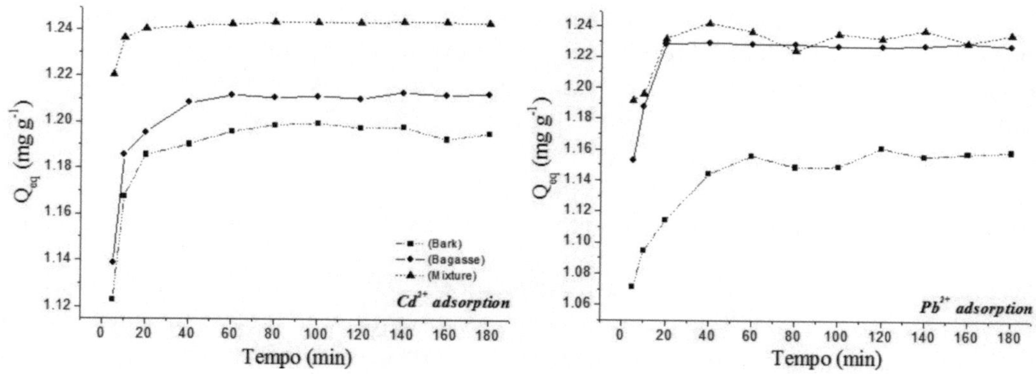

Figure 4. Effect of time on the amount of adsorbed ions ($mg.g^{-1}$) of the ions Cd^{2+} [18] and Pb^{2+} [12] for the adsorbents bark, bagasse, and mixture.

predominance of chemical adsorption of Pb^{2+}, that is, with the occurrence of chemical bonds between the polluting metal and the adsorbent biomass.

Figure 5 can be observed that the author Schwantes [18] fragmented the contact time between adsorbent and adsorbate and their close relationship with the adsorption capacity of Cd^{2+} in several lines, searching for multilinearity. According Neta et al. [23], when performing the fragmentation for the data, there is the possibility of graphically observing the phenomenon of movement of adsorbate into adsorbent particle.

In the case of adsorbents based on cassava residues, for Cd^{2+} adsorption, we can observe two to three phases in the retention process of this metal, the first phase being represented by the boundary layer effect, with external mass transfer, in that Cd^{2+} ions are rapidly adsorbed by

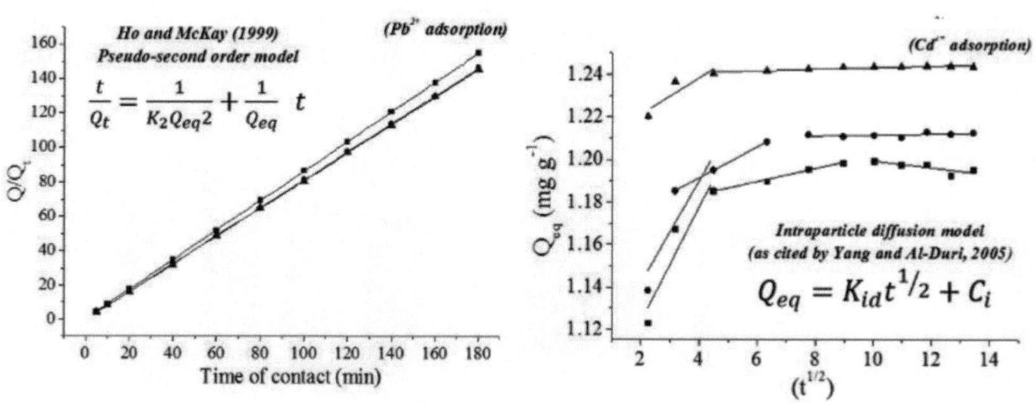

Figure 5. Linearization by the pseudo-second order model for adsorbed amount of Pb^{2+} by natural cassava adsorbents [12] and linearization by intraparticle diffusion model for Cd^{2+} quantity adsorbed [18] by natural adsorbents of cassava solid wastes.

the cassava adsorbents. After a certain time, the adsorption velocity decreases, resulting in the second phase that refers to the diffusion of the molecules to the adsorption sites inside the adsorbent [23]. Finally, in the third and last phase, equilibrium is observed, that is, the intraparticle diffusion begins to decay because of the low concentration of solute in the solution as well as a lower availability of sites for adsorption.

4.1.2.3. Adsorption equilibrium studies in cassava biosorbents

According to published studies [12, 18], the adsorbent materials based on natural cassava, under some specific conditions, may present efficiency close to that of activated carbon (**Figure 6**). It is important to emphasize that commercial activated carbon is the result of physical and chemical modifications, and the production costs of this material are high, while the natural adsorbents present low cost and high availability, since they rarely have a final destination that adds value to the product; in most cases, they are only agroindustry waste.

As presented in **Figure 6**, in relation to the adsorption of Cd^{2+}, it is observed that the adsorption capacity of the biosorbents decreases significantly with the increase in the initial concentration (C_0), whereas the adsorbent activated carbon (AC) also decreases, however, in a smaller proportion.

According to authors Schwantes et al. [12], the adsorbents based in cassava (barks, bagasse, and mixture) showed high values of Q_{eq} for the metal ion Pb^{2+} (**Figure 6**), comparable to the adsorbent activated carbon, which removed Pb^{2+} to below the limit of quantification (LQ = 0.01 mg.L^{-1}) of the used method (FAAS—Flame atomic absorption spectrometry). The results presented in **Figure 6** also can be graphically studied through the construction of adsorption isotherms, which are presented in the **Figure 7**.

According to Giles [24], the isotherms in the **Figure 7** fit in the "L group" (of Langmuir) and in the subgroup "1," this subgroup indicates slow saturation of the surface and characteristics of adsorbent materials of high adsorption capacity. Langmuir, Freundlich, and

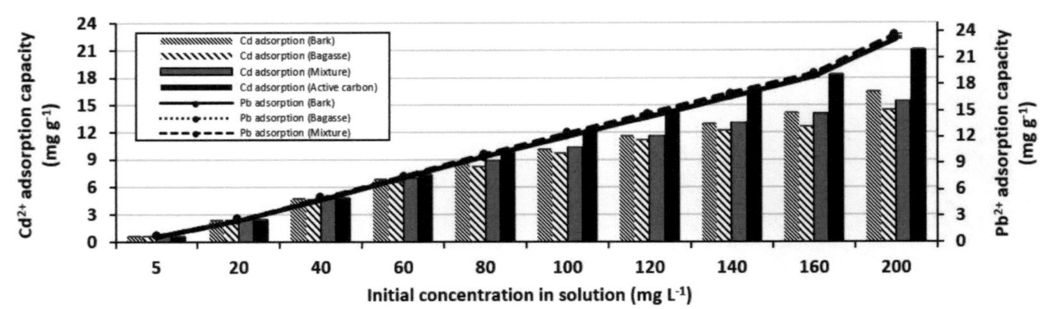

Figure 6. Adsorbed quantity (Q_{eq}) of the ions Cd^{2+} and Pb^{2+} by adsorbents barks, bagasse, mixture, and activated carbon, in increasing initial concentrations (C_0), varying from 5 to 200 mg.L^{-1}, n = 3. Limit of quantification (LQ): Cd = 0.005 mg.L^{-1}; Pb = 0.01 mg.L^{-1}. Source: Refs. [12, 18].

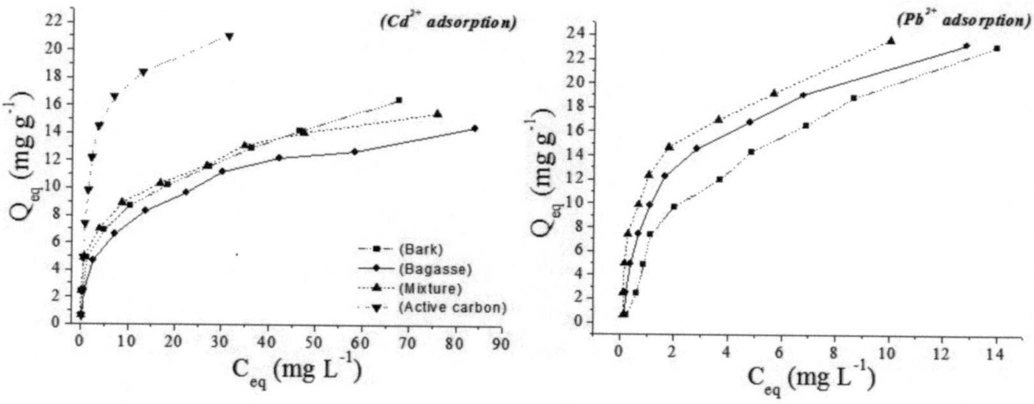

Figure 7. Adsorption isotherms of Cd^{2+} and Pb^{2+} from adsorbents (C_0: 5–200 mg.L^{-1}; 400 mg; pH 5.5, 60 min; 200 rpm; 25°C). Source: Refs. [12, 18].

Dubinin-Radushkevich (D-R) models linearized the results presented in the isotherms in **Figure 7**, for Schwantes [18] and Schwantes et al. [12], presented in **Table 4**.

It can be observed that cassava bark shows adsorption of Cd^{2+} in monolayers, as adjusted by Langmuir. However, the materials bagasse and mixture presented better adjustment for the Freundlich model, suggesting, in this case, adsorption in Cd^{2+} predominantly in multilayers.

Some results presented by Schwantes [18] are remarkable, as the parameter "Q_m" of Langmuir, which measures the maximum amount of adsorption, was observed values of the order of 16.66, 14.88, and 15.45 mg of Cd^{2+} per liter of contaminated solution, for the adsorbents bark, bagasse, and mixture. These values are close from the values obtained for activated carbon (22.69 mg.L^{-1}), suggesting that the natural adsorbents of cassava, even if they constitute residual biomass, present great potential in the removal of this metal toxic.

The model of Dubinin-Radushkevich (D-R), according Schwantes [18] and Schwantes et al. [12], explains satisfactorily the adsorption of Cd^{2+} and Pb^{2+} by the cassava adsorbents, presented a good fit (R^2) (**Table 4**). According to **Table 4**, the values of "E" in the majority of the cases assume values superior to 8 KJ.mol^{-1}, suggesting predominance of chemical adsorption [7] to cassava biosorbents. It indicates that the quantity of the toxic metal adsorbed by cassava biomass, in normal conditions, rarely returns to the solution, being this result, according to the interpretation of the authors of this chapter, a great advantage to decontamination of the waters containing toxic metals.

4.1.2.4. Possibility of reuse of the cassava biosorbents

The desorption phenomenon, according to Mimura et al. [16], corresponds to the removal of the metal from the connecting site of the surface of the adsorbent; for this, usually, it used acids in solution containing the adsorbents after sorption, so that H^+ ions can replace the

Linear models		Bark	Bagasse	Mixture	Actived carbon	Bark	Bagasse	Mixture
		Adsorption of Cd^{2+}				Adsorption of Pb^{2+}		
Langmuir [25]	Q_m	16.66	14.88	15.45	22.69	29.26	25.16	24.81
$\frac{C_{eq}}{q_{eq}} = \frac{1}{q_m b} + \frac{C_{eq}}{q_m}$	K_L	0.025	0.034	0.017	0.005	0.052	0.003	0.002
	R_L	0.168	0.128	0.226	0.506	0.489	0.628	0.746
	R^2	0.983	0.982	0.979	0.984	0.961	0.989	0.983
Freundlich [26]	K_f	9.949	4.863	3.479	4.400	3.814	6.180	8.643
$\log q_{eq} = \log K_f + \left(\frac{1}{n}\right) \log C_{eq}$	n	4.246	3.684	3.049	3.331	1.224	1.449	1.704
	R_2	0.920	0.995	0.993	0.991	0.870	0.772	0.712
D-R [27] $\ln Q_{eq} = \ln Q_d - Bd\varepsilon^2$	Q_d	0.0004	0.0003	0.0003	0.0004	0.0015	0.0006	0.0004
	E	12.004	12.461	13.814	16.807	10.733	13.245	14.712
	R^2	0.984	0.992	0.994	0.979	0.997	0.984	0.991

Q_m (mg.g^{-1}): maximum adsorption capacity; K_L or b (L.mg^{-1}): constant related to the adsorbent/adsorbate interaction forces; R_L: Langmuir constant; R^2: determination coefficient; K_f (L.mg^{-1}): related to the adsorption capacity; *n:* related to the heterogeneity of the solid; Q_d (mol.g^{-1}): maximum adsorption capacity; E (Kj.mol^{-1}): mean sorption energy. Source: Refs. [12, 18].

Table 4. Linearization of the data by Langmuir, Freundlich, and Dubinin-Radushkevich (D-R) to the biosorption process of Cd^{2+} and Pb^{2+} by bark, bagasse, and mixture and activated carbon.

adsorbed cations by the ion exchange mechanism. For this practice, it is fundamental knowledge of the interaction characteristics between the adsorbate and the adsorbent for their resistance for reuse purposes in new adsorption processes.

Figure 8 shows that the recuperation of the material in acid solution (HCl) is high, varying by 60% to near of 90% to cassava adsorbents used in the removal of Cd^{2+} [18], Pb^{2+} [12], Cu^{2+}, and Zn^{2+} [9]. However, it is observed that, when using the adsorbents of cassava to

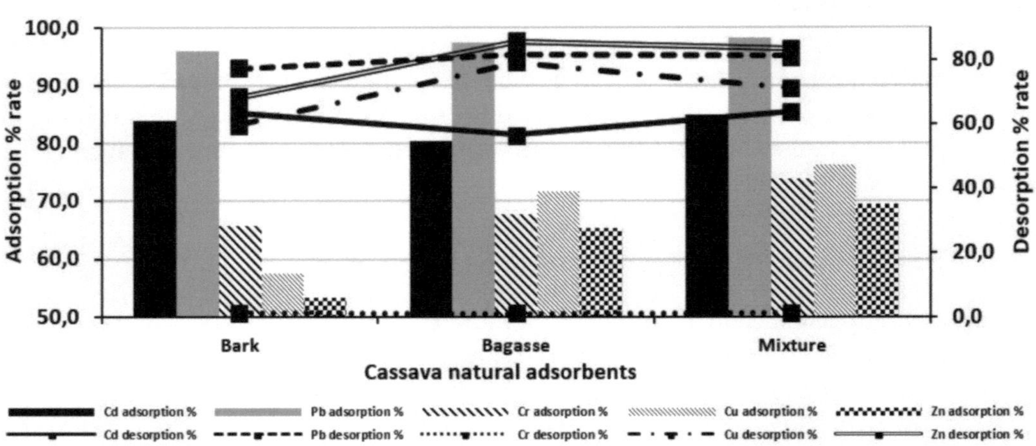

Figure 8. Possibility of cassava adsorbents reuse to new cycles of metallic ions sorption. Source: Refs. [9, 10, 12, 18].

remove Cr^{3+}, desorption rates are 1%, that is, recovery of cassava biosorbents is compromised, which according to Schwantes et al. [10] is a strong indicative of chemical adsorption of this metal with referred natural adsorbents.

4.2. Case studies: chemical modifications to cassava biosorbents and their use in the removal of toxic metals in waters

As already mentioned in this chapter, a good adsorbent is one that has a large number of active sites available for interaction with the polluting species of interest. The sites are adsorption points consisted of one or more chemical functional groups distributed on the surface of the biosorbent. It is important to emphasize that some biosorbents may have their surfaces chemically modified in order to increase the amount of active sites and, consequently, to increase the adsorption capacity of metallic ions.

As mentioned above, it is not novelty to use modifying agents in improving the characteristics of the adsorbents. The literature reports the use of chemically modified adsorbents and their use in the removal of metals, agrochemicals and other pollutants from liquid solution, such as modified adsorbents with HCl [28], jute fiber modified with H_2O_2 in the removal of Cu^{2+}, Ni^{2+}, and Zn^{2+} [29], wheat barks treated with H_2SO_4 in the removal of Cu^{2+} [30], and bagasse ashes modified with H_2O_2 in the removal of Pb^{2+} [31], among other examples.

The principal modifications included delignification, esterification of carboxyl and phosphate groups, methylation of amine groups and hydrolysis of carboxylate groups, acid treatments, basic treatments, and peroxide treatments, among others cited in the literature.

The aforementioned modifications aim to eliminate the coloration, turbidity, and other unfavorable characteristics of the treated water (after adsorption), as well as increase the adsorptive capacity of metals and other pollutants [7].

In contrast, when using acid treatments to plant biomass, such as sulfuric acid, hydrochloric, or nitric acid, all in their diluted form, the hydrolysis of cellulose is accelerated, resulting in a higher capacity of adsorption of the resulting adsorbents.

It is important to mention that this whole modification process generally does not overtax the final product, that is, this process results in adsorbents of greater adsorption capacity without generating great costs to the final adsorbent.

Related to the use of modified cassava adsorbents, the literature is scarce. Schwantes et al. [11] realized chemical modifications to the cassava barks through H_2O_2, H_2SO_4, and NaOH solution of 0.1 mol.L^{-1} in the proportion of 7 g of biosorbent per liter of modified solution. The authors kept the biosorbents in contact with the modifying solution for 6 h in a Dubnoff system, with shaking at 150 rpm at 60°C. Subsequently, successive washes were carried out on the material with distilled and deionized H_2O to remove eventual remnants of the modifying agent, resulting in three modified adsorbents, nominated by the authors according to the modification applied in M. H_2O_2, M. H_2SO_4, and M. NaOH.

Without these citations, the literature presents failures (with the exception of the methods to production of activated carbon), while other chemical and/or physical modifications applied to biosorbents originated from cassava agroindustry, being necessary for more studies about this topic. In the subsequent items, the physical, chemical, and morphological characteristics obtained by the authors aforesaid to the adsorbents M. H_2O_2, M. H_2SO_4, and M. NaOH, as well as the adsorption capacity of various polluting ions will be described.

4.2.1. *Characterization of the cassava-modified adsorbents and their possible utilization in the removal of toxic metals*

According to Schwantes et al. [9] and Schwantes et al. [10], the micrographs observed in **Figure 9a**, to the adsorbent M. *in natura*, showed a surface with spongy and fibrous aspect, with irregular and heterogeneous structure. Some spheres present in the micrograph can be noted, possibly indicating the presence of residual starch granules from the originating material. It is also observed that the modified adsorbents (**Figure 9b–d**) have a surface that remind the originating material (**Figure 9a**), however, with some differential aspects depending on each treatment applied.

As can be seen in **Figure 9b–d**, corresponding to the chemical modifications of the chemical reagents H_2O_2, H_2SO_4, and NaOH, respectively, the modified adsorbents retain the characteristics of their precursor (M. *in natura*, **Figure 9a**), however, with some marked modifications. In these cases, structures that remember the cells of the original plant are observed.

Hydrogen peroxide, a powerful oxidizing agent, caused some changes in the surface of the adsorbent M. H_2O_2. It is observed in **Figure 9b** that M. H_2O_2 presents with heterogeneous surface, remembering the form of scales. In **Figure 9c**, referring to M. H_2SO_4, unlike the others, a relatively homogeneous surface is observed, with a flat structure, apparently little porous, reminding the shape of scales, also presenting some prominent cracks, possibly because of the dehydration from sulphuric acid, which is a powerful dehydrator.

Caustic soda is a strong, highly soluble, and recognizably corrosive base that, upon contact with the cassava barks, caused changes in its surface structure, as shown in **Figure 9d**, which illustrates an irregular, heterogeneous surface with cracks (breaks) of the adsorbent, with a surface that reminds scales.

As previously mentioned in this chapter, one of the main objectives of the application of chemical modifications in biosorbents is the promotion of new functional groups, aiming to increase the adsorption capacity by the modified biomass. According to the observation in

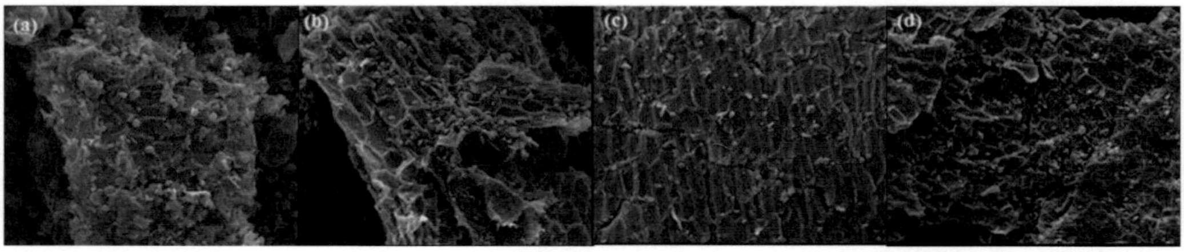

Figure 9. Micrographs of the adsorbents M. *in natura* (a) [9, 10], M. H_2O_2 (b), M. H_2SO_4 (c), and M. NaOH (d) in 800, 400, 400 and 400×, respectively. Source: The authors.

the studies of Schwantes et al. [9] and Schwantes et al. [10], the infrared spectra of the natural adsorbent, when superimposed on the modified cassava adsorbents, contrast with the appearance of some peaks, such as in 2855, 1161, and 579 cm^{-1}, as well as the peaks obtained in 1019 cm^{-1}, specifically for the M. H_2SO_4 adsorbent.

The peaks obtained in 1161 cm^{-1} possibly indicate the presence of the vibrational stretches of C-O connections present in carboxylic acids, one of the main responsible for the formation of active sites on the adsorbent surface. Peaks at 2855 cm^{-1} may be indicative of the vibrational elongation of C-H connections present in aldehydes [32]. Peaks at 579 cm^{-1} suggest groupings with S-S connections, suggesting the presence of amino acids like cysteine "[$SCH_2CH(NH_2)CO_2H)_2$]", lipoic acids "($C_8H_{14}O_2S_2$)", and others.

In general, the authors corroborate that the results obtained in the infrared spectra suggest that solutions of H_2O_2, H_2SO_4, and NaOH caused modifications of cassava barks, forming functional groups that may provide adsorbents with good sorption characteristics.

In the treatment of biomass of cassava barks with modifying solutions, changes in the point of zero charge occur because the material M. *in natura* presented pH_{PZC} of 6.02 [9, 10], and after modifications, this value was changed to 3.98 in the modification with H_2O_2, 2.05 for modification with H_2SO_4, and about 7.07 for modification with NaOH [11] (**Figure 10**).

The adsorption of metal cations is favored when they are in solutions in which the pH values are higher than pH_{PZC}, because in these cases, the surface of the adsorbent presents predominance of the negative charges. These results suggest that the simple washing of the adsorbent materials of cassava can change the point of zero charge of the adsorbent; in other words, the point of zero charge of an adsorbent can be handled, according to the need for values higher or lower than the biosorbent of origin.

Another curious fact is the simple washing of the cassava biosorbent with modifying solutions caused extraction of part of the constituent elements of the biomass, as evidenced in **Table 5**.

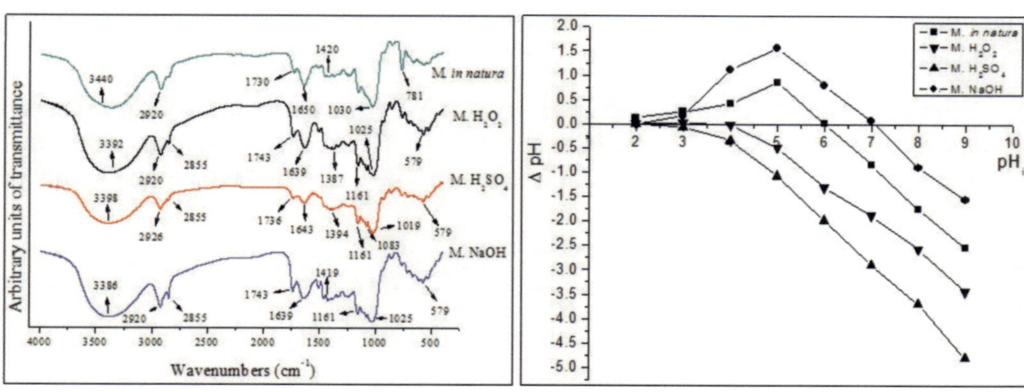

Figure 10. In the left: infrared spectra for adsorbents based in natural cassava barks [9, 10] and modified with H_2O_2, H_2SO_4, and NaOH. Source: The authors. In the right: pH_{PZC} of adsorbents M. *in natura* [9, 10], M. H_2O_2, M. H_2SO_4, and M. NaOH [11].

Adsorbents	K	Ca	Mg	Cu	Zn	Mn	Fe	Cd	Pb	Cr
	g.kg^{-1}			mg.kg^{-1}						
M. *in natura* [9, 10]	24.10	35.03	6.83	14.33	32.00	123.33	335.66	<LQ	13.00	<LQ
M. H_2O_2 [11]	7.84	5.68	1.27	10.60	32.20	121.50	333.70	<LQ	10.40	<LQ
M. H_2SO_4 [11]	5.78	3.41	0.43	4.30	20.40	115.70	330.90	<LQ	5.10	<LQ
M. NaOH [11]	11.22	6.52	1.49	4.80	32.60	122.00	331.60	<LQ	11.50	<LQ

LQ (limit of quantification): K = 0.01; Ca = 0.005; Mg = 0.005; Cu = 0.005; Fe = 0.01; Mn = 0.01; Zn = 0.005; Cd = 0.005; Pb = 0.01; Cr = 0.01 (mg kg^{-1}).

Table 5. Average values of element concentrations in the studied adsorbents.

According to Schwantes et al. [11], the modifying agents act as extractors of the metallic elements in the biosorbent, with reduction of K (67%), Ca (84%), Mg (81%), Cu (70%), Fe (1%), and Pb (20%) to the cassava adsorbent modified with H_2O_2. As for the adsorbent modified with H_2SO_4, it can evidence the reduction of K (76%), Ca (90%), Mg (94%), Cu (70%), Zn (36%), Mn (6%), Fe (1%), and Pb (61%), while for the adsorbent modified with NaOH, it resulted in reduction of K (53%), Ca (81%), Mg (78%), Cu (67%), Mn (1%), Fe (1%), and Pb (12%).

This possibly occurs in function of the actions of modifying solutions (H_2O_2, H_2SO_4, and NaOH), in addition to the loss of these metals caused by the postmodification washing, eliminating a good part of these elements. It is important to emphasize the high reductions of the metal contents when using the modifying solution of H_2SO_4, a strong acid, recognized as a potent dehydrator [11].

As previously quoted in this chapter, one of the many characteristics of a good adsorbent is the high-specific surface area, as well as considerable values for the pore volume parameters and a good distribution to the average diameters of these values.

In the **Table 6**, it can be seen that the specific surface area (SSA) of the cassava adsorbents is generally small when compared to the activated carbon. Among the values obtained, it is important to highlight the adsorbent M. H_2O_2, which presents the highest values for SSA (0.91 m^2.g^{-1}), followed by M. NaOH (0.70 m^2.g^{-1}) and M. H_2SO_4 (0.46 m^2.g^{-1}).

It can be observed that SSA of the modified cassava is small when compared to values obtained for activated carbon originated from coffee barks (130 a 391 m^2.g^{-1}), but similar to the other residues of low economic value, such as Oliveira et al. [33], which obtained 0.46 m^2.g^{-1} for

Parameters	M. *in natura*	M. H_2O_2	M. H_2SO_4	M. NaOH
Superficial area (m^2.g^{-1})	0.5583	0.9156	0.4637	0.7017
Pore volume (cm^3.g^{-1})	0.001137	0.00307	0.00179	0.00146
Pore diameter (nm)	1.922	1.734	3.295	1.924

Source: The authors.

Table 6. Texture of the cassava adsorbents.

rice bran. Also, Penha et al. [34], that even using chemical treatment of the rice barks, do not obtain higher values than 1.13 $m^2.g^{-1}$, with mean pores volume equal to 1.94 $cm^3.g^{-1}$ and mean diameter of the pores equal to 6.9 nm, that is, predominance of mesoporous.

The volumes of the pores obtained in this research (0.0031 and 0.0018 cm^3 g^{-1} to M. H_2O_2 and M. NaOH) are similar to those obtained for Penha et al. [34] in modified rice barks (0.0019 $cm^3.g^{-1}$). Pores with diameter situated between 2.0 and 50.0 nm are considered mesoporous, being the cassava adsorbents are predominantly microporous and, to a lesser degree, mesoporous (**Table 6**).

Although low surface area and low pore volume, which are initially disadvantages the adsorption process, this is not the only determining condition in a high-pollutant remediation capacity of the adsorbent.

4.2.2. Experimental results involving modified adsorvents of cassava and remediation of toxic metals in waters

4.2.2.1. Influences of the pH of the contaminant solution and the proportion between the cassava-modified adsorbents and metallic adsorbents

Schwantes et al. [11], when evaluating the influence of different pH ranges and modified adsorbent mass on the removal of Cd^{2+}, Pb^{2+}, and Cr^{3+}, observed significant difference in relation to the amount of mass of the adsorbents studied.

These results demonstrate that the adsorbents of cassava in their natural form and modified with H_2O_2, H_2SO_4, and NaOH depend closely on the amount of adsorbent used but did not show this relationship with the pH variable in the range studied by the authors, which was 3.60–7.00. The removal of metals from the solutions was higher for the lower masses tested by the authors (close to 4 $g.L^{-1}$), according to **Figure 11**.

In a similar study, in which activated carbon with $ZnCl_2$ was produced from residues of cassava bark, with the purpose of using it to remove Pb^{2+} from contaminated water, there was also an increase in the removal efficiency of Pb^{2+} as the doses of mass of the adsorbents studied by the authors increased, reaching efficiency of 74% [35].

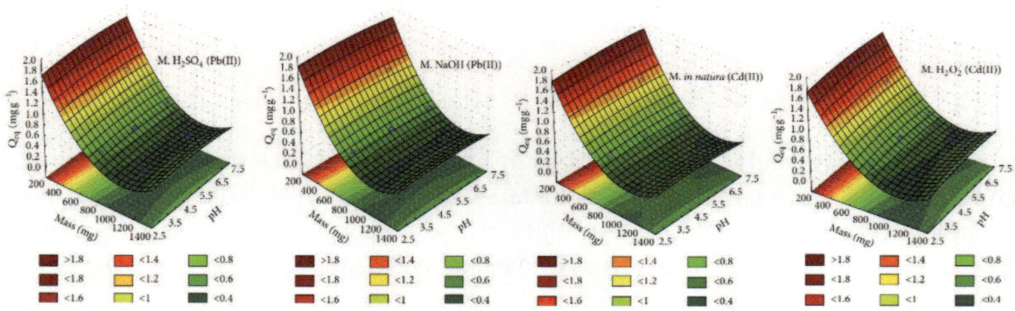

Figure 11. Response surfaces for interaction between the mass of the natural cassava adsorbent and modified with H_2O_2, H_2SO_4, and NaOH, in the removal of Cd^{2+} and Pb^{2+}. Source: Schwantes et al. [11].

These authors point out that the higher dosage of adsorbents increases adsorption, and this is mainly because the greater number of surfaces and functional groups of the adsorbents, with which the metal can interact, as well as to an improvement in the dissolved oxygen value and a reduction in the electrical conductivity of the wastewater, that may occur as a function of higher doses of the adsorbent used.

According to Ilaboya et al. [35] that tested pHs between 2 and 12, observed that the metallic ions removal of the samples they evaluated were related to the pH of the samples, observing that the amount of Pb^{2+} ions removed raised to pH 8 and then began to decrease, they considered that the pH together with the temperature were the most effective variables for adsorption of ions of this metal.

The elevation of Pb^{2+} removal as the pH is high may be related to the lower occurrence of hydrogen ions and the decrease of positive charges on the surface of the adsorbent, resulting in lower electrostatic repulsive forces between the surface and the ions of the metal.

According to Schwantes et al. [11], the fact that in their studies, the pH range studied did not influence the adsorption process is an excellent result, because the adsorbents proposed by the authors can be used in waters and effluents with different pH values, without affecting the removal efficiency of these pollutants.

The fact that the results observed by Ilaboya et al. [35] and Schwantes et al. [11] differ in the behavior of removal as a function of pH may have relation with the form of activation of the material used (cassava barks). In addition, Ilaboya et al. [35] used physical activation through pyrolysis, which changes the characteristics of the adsorbent, demonstrating that adsorbent materials from cassava barks have several possibilities of use. In addition, it should be remembered that the activation reagent in these cases differed.

Although the differences occurred in both works, it is possible to observe that the use of adsorbents of cassava barks is efficient in cases when it is activated only chemically and in cases where physical and chemical activation occurs, that is, with addition of pyrolysis.

It is important to emphasize that in all the researches mentioned above, the proportion between the adsorbent mass and the volume of contaminated water is one of the great differentials of this technology, since the higher adsorption efficiencies of metals occur with values around 4–5 g of adsorbent per liter of adsorbate.

4.2.2.2. Contact time influences between modified cassava adsorbents and toxic metals

According to study conducted by published by Schwantes et al. [11], showed in **Figure 12**, it can be observed that after 20–40 min of contact time with the modified cassava adsorbents, in solutions containing 10 $mg.L^{-1}$ of Cd^{2+} and Pb^{2+}, the system enters in dynamic equilibrium. The process of adsorption of these metals by the modified cassava adsorbents is a fast process, easily applicable in large scale.

Schwantes et al. (2016) applied the results obtained in **Figure 12** the linear model of Pseudo-second order, proposed by Ho and McKay [21], being observed good mathematical adjustments (R^2), suggesting in this way the occurrence of chemical adsorption of Cd^{2+} and Pb^{2+} by the adsorbents of cassava modified with H_2O_2, H_2SO_4, and NaOH.

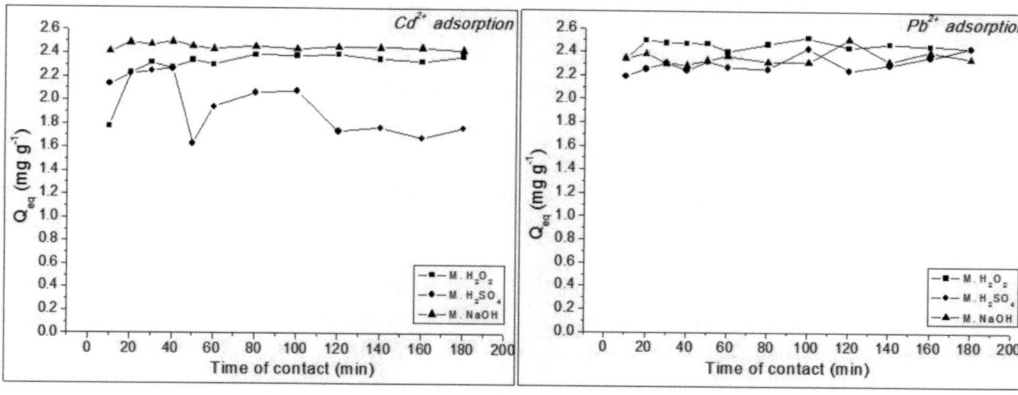

Figure 12. Adsorbed amount of metals Cd^{2+} and Pb^{2+} in function of contact time between modified cassava adsorbents. Source: Schwantes et al. [11].

4.2.2.3. Equilibrium studies of adsorption in modified cassava adsorbents

In the studies of Schwantes et al. [11], the adsorption isotherms of Cd^{2+}, Pb^{2+}, and Cr^{2+} were linearized by the mathematical models of Langmuir and Freundlich. These authors observed a predominance of good adjustments for the Langmuir models, suggesting the occurrence of adsorption in monolayers for cassava bark modified with H_2O_2, H_2SO_4, and NaOH in the adsorption of Cd^{2+}, H_2O_2, and NaOH in the adsorption of Pb^{2+} (**Table 7**).

Comparing the results published by Schwantes et al. [12] and Schwantes et al. [11], increases in the maximum capacity of adsorption of Cd^{2+} and Pb^{2+} by cassava barks about 17 and 45%, respectively, can be observed, when compared to the unmodified adsorbent.

Models and parameters		**M. *in natura* [18]**	**M. H_2O_2 [11]**	**M. H_2SO_4 [11]**	**M. NaOH [11]**	**Activated carbon [18]**	**M. *in nature* [12]**	**M. H_2O_2 [11]**	**M. H_2SO_4 [11]**	**M. NaOH [11]**
		Adsorption of Cd^{2+}					**Adsorption of Pb^{2+}**			
Langmuir [25]	Q_m	16.66	13.420	7.058	19.539	22.696	29.265	21.678	24.004	42.463
$\frac{C_{eq}}{q_{eq}} = \frac{1}{q_m b} + \frac{C_{eq}}{q_m}$	K_L	0.025	0.026	0.017	0.006	0.0049	0.005	0.022	0.019	0.002
	R_L	0.165	0.163	0.224	0.467	0.506	0.489	0.188	0.207	0.667
	R^2	0.967	0.993	0.980	0.996	0.984	0.961	0.994	0.938	0.996
Freundlich [26]	K_f	9.949	1.967	1.565	3.634	4.400	3.814	2.510	1.692	3.393
$\log q_{eq} = \log K_f + \left(\frac{1}{n}\right) \log C_{eq}$	n	4.246	2.287	1.020	1.678	3.331	1.224	1.764	1.488	0.762
	R^2	0.920	0.934	0.977	0.902	0.991	0.870	0.980	0.977	0.931

Q_m (mg.g^{-1}): maximum capacity of adsorption; K_L ou b (L.mg^{-1}): constant related to the adsorbent/adsorbate interaction forces; R_L: Langmuir constant; R^2: coefficient of determination; K_f (L.mg^{-1}): related to the adsorption capacity; n: related to the heterogeneity of the solid.

Table 7. Parameters referring to the Langmuir and Freundlich linear models for adsorption of Cd^{2+} and Pb^{2+} by cassava adsorbents.

Moreover, it is observed that the modification of the cassava bark with NaOH generated an adsorbent with adsorption capacity of Cd^{2+} similar to the commercial activated carbon tested by Schwantes et al. [12]. This result is an unprecedented and very interesting result from the point of economic view, since the cost of acquiring commercial activated carbon is several times higher than that of the modified adsorbent of cassava barks. In addition, these authors observed good adjustments for the Freundlich model, which suggests multilayer adsorption of Cd^{2+} for M. H_2SO_4 and M. H_2O_2 and H_2SO_4 in the adsorption of Pb^{2+}.

The models of Langmuir and Freundlich also were studied by Omotosho and Sangodoyin [36] to linearize the adsorption data obtained by investigating the efficiency of the cassava bark carbon activated with zinc chloride at the activation levels of 1:3, 2:3, and 1:1 in the removal of NO_3^- wastewater from cassava processing.

The results linearized by Langmuir for the four levels of activation of that author show that all had coefficients values (R^2) between 0.972 and 0.994. These results demonstrate, as verified by Schwantes et al. [11], which the material followed the assumptions of Langmuir theory, that is, the carbon surface was adsorbing at specific monolayer sites, the authors concluded that this characteristic implies that the activated carbon of cassava barks without and with chemical activation are effective adsorbents.

4.2.2.4. Possibility of reuse of the cassava-modified adsorbents

When evaluated the possibility of reuse of the cassava-modified adsorbents, Schwantes et al. [11], when proceeding the acid elution with HCl 0.1 M, obtained the follow rates of desorption to Cd^{2+}: M. H_2O_2 (60%); M. H_2SO_4 (62%) and M. NaOH (74%); to desorption of Pb^{2+}: M. H_2O_2 (65%); M. H_2SO_4 (53%) and M. NaOH (56%); and to desorption of Cr^{3+}: M. H_2O_2 (1%); M. H_2SO_4 (3%); and M. NaOH (1%).

According to the authors, the results about the desorption of the metallic ions confirm the possibility of reuse of the adsorbents through the action of a strong acid solution, which promotes the extraction of these elements from the active site, as observed for the use of adsorbents of cassava for Cd^{2+} and Pb^{2+}.

However, it is observed that when the modified adsorbents are used in Cr^{3+} adsorption cycles, low desorption values are observed, which, according to Schwantes et al. (2016), is a strong indication that this ion is adsorbed by chemical bonds (chemisorption), which makes it impossible to reuse them in new sorting cycles. This effect was observed for other authors noted by the literature, such as jatropha seed pie [37], cassava barks in their natural form [11], and others.

5. Final considerations

The use of these adsorbent materials from cassava residual biomass is a viable alternative in the removal of water pollutants, since they represent an important increase in the cassava agroindustrial chain. The reuse of these materials is in according to the sustainability, since such residues can be used in remediation of environmental compartments contaminated with toxic metals such as Cd, Pb, Cr, Cu, Zn, and others.

The results of the researches presented in this chapter demonstrate that cassava agroindustrial residues have potential for use in the form of adsorbents, and they can be used in their natural or modified form, presenting high efficiencies and comparable to conventional commercial adsorbents, however, with a lower cost of production.

The authors of this chapter also reiterate that although the barks, bagasse, and other solid residues from the processing of the cassava roots have proven potential for adsorption of several contaminants (especially metals), further research on these materials is still needed, especially related to pesticides, POP, and other pollutants.

Author details

Daniel Schwantes[1], Affonso Celso Gonçalves Jr.[2*], Marcelo Angelo Campagnolo[3], César Ricardo Teixeira Tarley[4], Douglas Cardoso Dragunski[5], Jéssica Manfrin[2] and Andréia Da Paz Schiller[2]

*Address all correspondence to: affonso133@hotmail.com

1 Department of Engineering and Exact Sciences, Federal University of Paraná, Palotina, Paraná, Brazil

2 Program of Post-Graduation in Agronomy, State University of Western Paraná, Marechal Cândido Rondon, Paraná, Brazil

3 Department of Environmental Engineering, Pontifical Catholic University of Paraná, Toledo, Paraná, Brazil

4 Department of Chemistry, Londrina State University, Londrina, Paraná, Brazil

5 Program of Environmental Sciences, State University of Western Paraná, Toledo, Paraná, Brazil

References

[1] Alves AAC. Fisiologia da mandioca. In: Aspectos socioeconômicos e agronômicos da mandioca. 1st ed. Cruz das Almas – BA: EMBRAPA Mandioca e Fruticultura Tropical, Cap. 7; 2006. pp. 139-169

[2] Annadurai G, Juang RS, Lee DJ. Adsorption of heavy metals from water using banana and orange peels. Water Science and Technology. 2002;**47**(1):185-190

[3] Meunier N, Laroulandie J, Blais JF, Tyagi RD. Cocoa shells for heavy metal removal from acidic solutions. Bioresource Technology. 2003;**90**(3):255-263

[4] Gonçalves AC Jr, Selzlein C, Nacke H. Uso de biomassa seca de aguapé (*Eichornia crassipes*) visando à remoção de metais pesados de soluções contaminadas. Acta Scientiarum Technology. 2009;**31**(1):103-108

[5] Peña-Rodríguez S, Fernández-Calviño D, Nóvoa-Muñoz JC, Arias-Estévez M, Núñez-Delgado A, Fernández-Sanjurjo MJ, Álvarez-Rodríguez E. Kinetics of hg (II) adsorption and desorption in calcined mussel shells. Journal of Hazardous Materials. 2010;**180**:622-627

[6] Dos Santos VCG, Souza JVTM, Tarley CRT, Caetano J, Dragunski DC. Copper ions adsorption from aqueous medium using the biosorbent sugarcane bagasse in natura and chemically modified. Water, Air, and Soil Pollution. 2011;**216**:351-359

[7] Wan Ngah WSW, Hanafiah MAKM. Removal of heavy metal ions from wastewater by chemically modified plant wastes as adsorbents: A review. Bioresource Technology. 2008;**99**(10):3935-3948

[8] SEAB. Secretaria de Estado da Agricultura e do Abastecimento. Available from: http://www.agricultura.pr.gov.br/arquivos/File/deral/Prognosticos/mandiocultura_2013_14.pdf [Accessed: May 2017]

[9] Schwantes D, Gonçalves AC Jr, Miola AJ, Coelho GF, Dos Santos MG, Leismann EAV. Removal of cu (II) and Zn (II) from water with natural adsorbents from cassava agroindustry residues. Acta Scientiarum Technology. 2015;**37**(3):409-417

[10] Schwantes D, Gonçalves Jr AC, Casarin J, Pinheiro A, Pinheiro IG, Coelho GF. Removal of Cr(III) from contaminated water using industrial waste of the cassava as natural adsorbents. African Journal of Agriculture Research. 2015;**10**(46):4241-4251

[11] Schwantes D, Gonçalves AC Jr, Coelho GF, Campagnolo MA, Dragunski DC, Tarley CRT, Miola AJ, Leismann EA. Chemical modifications of cassava peel as adsorbent material for metals ions from wastewater. Journal of Chemistry. 2016;**2016**:1-15

[12] Schwantes D, Gonçalves AC Jr, Strey L, Schwantes V, Nacke H. Reuse and recycling techniques: Kinetics, equilibrium and thermodynamics of the adsorption process of lead using cassava industry wastes. In: Bártolo H, Duarte JP, editors. Green Design, Materials and Manufacturing Processes. 1st ed. Vol. 1. Boca Raton: CRC Press Taylor & Francis Group; 2013. pp. 417-422

[13] Schwantes D, Gonçalves AC Jr, Coelho GF, Casarin J, Stangarlin JR, Pinheiro A. Reuse and recycling techniques: Equilibrium of the adsorption process of glyphosate using wastes from the cassava industry. In: Bártolo H, Duarte JP, editors. Green Design, Materials and Manufacturing Processes. 1st ed. Vol. 1. Boca Raton: CRC Press Taylor & Francis Group; 2013. pp. 423-428

[14] Han R, Zhang L, Song C, Zhang M, Zhu H, Zhang L. Characterization of modified wheat straw, kinetic and equilibrium study about copper ion and methylene blue adsorption in batch mode. Carbohydrate Polymers. 2010;**79**:1140-1149

[15] Pehlivan E, Altun T, Cetin S, Bhanger MI. Lead sorption by waste biomass of hazelnut and almond shell. Journal of Hazardous Materials. 2009;**167**(1-3):1203-1208

[16] Mimura AMS, Vieira TVA, Martelli PB, Gorgulho HF. Utilization of rice husk to remove Cu2+, Al3+, Ni2+ and Zn2+ from wastewater. Química Nova. 2010;**33**(6):1279-1284

[17] Gonçalves AC Jr, Nacke H, Schwantes D, Coelho GF. Heavy metal contamination in Brazilian agricultural soils due to application of fertilizers. In: Hernández-Soriano MC, editor. Environmental Risk Assessment of Soil Contamination. 1st ed. Vol. 4. Rijeka: InTech; 2014. pp. 105-135

[18] Schwantes D. Kinetics, Equilibrium and Thermodynamic of the Adsorption Process of Metallic Ions Using Wastes from the Cassava Industry as Biosorbents. 18-02-2013. 103p. Universidade Estadual do Oeste do Paraná. Campus de Marechal Cândido Rondon, PR: Dissertação de Mestrado; 2013

[19] Gonçalves AC Jr, Strey L, Lindino CA, Nacke H, Schwantes D, Seidel EP. Applicability of the Pinus bark (Pinus Elliottii) for the adsorption of toxic heavy metals from aqueous solutions. Acta Scientiarum Technology. 2012;**34**(1):79-87

[20] Runping H, Han P, Cai Z, Zhao Z, Tang M. Kinetics and isotherms of neutral red adsorption on peanut husk. Journal of Environmental Sciences. 2008;**20**:1035-1041

[21] Ho YS, McKay G. Pseudo-second order model for sorption processes. Process Biochemistry. 1999;**34**:451-465

[22] Yang X, Al-Duri B. Kinetic modeling of liquid-phase adsorption of reactive dyes on activated carbon. Journal of Colloid and Interface Science. 2005;**287**(1):25-34

[23] Neta JJS, Silva CJ, Moreira GM, Reis C, Reis EL. Removal of the reactive blue 21 and direct red 80 dyes using seed residue of Mabea fistulifera Mart. As biosorbent. Revista Ambiente & Água. 2012;**1**:7

[24] Giles CH, Macewan TH, Nakhwa SN, Smith D. Studies in adsorption. Part XI. A system of classification of solution adsorption isotherms, and its use in diagnosis of adsorption mechanisms and in measurement of specific surface areas of solids. Journal of the Chemical Society. 1960. p. 3973-3993

[25] Langmuir I. The adsorption of gases on plane surfaces of glass, mica and platinum. Journal of the American Chemical Society. 1918;**40**:1361-1403

[26] Freundlich H. Adsorption in solution. Physical Chemistry Society. 1906;**40**:1361-1368

[27] Dubinin MM, Radushkevich LV. Equation of the characteristic curve of activated charcoal. Chemisches Zentralblatt. 1947;**55**:331-337

[28] Argun ME, Dursun S. Removal of heavy metal ions using chemically modified adsorbents. Journal of International Environmental Application & Science. 2006;**1**(1-2):27-40

[29] Shukla SR, Pai RS. Adsorption of cu(II), Ni(II) and Zn(II) on modified jute fibres. Bioresoure Technology. 2005;**96**:1430-1438

[30] Ozer A, Ozer D, Ozer A. The adsorption of copper(II) ions on to dehydrated wheat bran (DWB): Determination of the equilibrium and thermodynamic parameters. Process Biochemistry. 2004;**39**:2183-2191

[31] Gupta VK, Ali I. Removal of lead and chromium from wastewater using bagasse fly ash – A sugar industry waste. Journal of Colloid and Interface Science. 2004:271-321

[32] Stuart BH. Infrared Spectroscopy: Fundamentals and Applications. Hoboken, New Jersey: John Wiley and Sons; 2004;**1**:224

[33] Oliveira EA, Montanhera SF, Andrade AD, Nóbrega JA, Rollemberg MC. Equilibrium studies for the sorption of chromium and nickel from aqueous solutions using raw rice bran. Process Biochemistry. 2005;**40**:3485

[34] Penha RS, Santos CC, Cardoso JJF, Silva HAS, Santana SAA, Bezerra CWB. Chemically treated Rice husk as low-cost adsorbent for metal ions uptake (Co^{2+} and Ni^{2+}). Revista Virtual Química. 2016;**8**(3):588-604

[35] Ilaboya IR, Oti EO, Ekoh GO, Umukoro LO. Performance of activated carbon from cassava peels for the treatment of effluent wastewater. Iranica Journal of Energy & Environment. 2013;**4**(4):361-375

[36] Omotosho OA, Sangodoyin AY. Production and utilization of cassava peel activated carbon in treatment of effluent from cassava processing industry. Water Practice Technology. 2016;**8**:215-224. DOI: 10.2166/wpt.2013.023

[37] Gonçalves AC Jr, Nacke H, Daniel S, Campagnolo MA, Miola AJ, Tarley CRT, Dragunski DC. Adsorption mechanism of chromium(III) using biosorbents of *Jatropha curcas* L. Environmental Science and Pollution Research. 2017;**1**:1-13

Chapter 16

Review of Various Harvesting Options for Cassava

Shadrack Kwadwo Amponsah, Ahmad Addo and
Byju Gangadharan

Additional information is available at the end of the chapter

http://dx.doi.org/10.5772/intechopen.71350

Abstract

Harvesting plays a critical role in the cassava production value chain. A review of some existing cassava harvesting options is necessary to facilitate the proper adaption and uptake of improved harvesting methods applicable to farmers from different parts of the globe. In terms of capacity, manual, semi-manual and fully mechanised harvesting options respectively require about 22–51 man-hha-1, 16-45 man-hha-1 and 1–4 man-hha-1. An added advantage with mechanised options is that the field is left ploughed after harvesting with savings on fuel, time and cost. Mechanised harvesters work best on ridged fields with minimal trash or weeds and relatively dry soils (12–16% d.b. moisture content). Earlier attempts at mechanised harvesting have been affected by constraints such as soil characteristics, nature and size of tubers, depth and width of cluster and bond between tubers and the soil, leading to high tuber damage. Though less research attention is given to cassava harvesting mechanisation, that aspect of the global cassava transformation agenda has always been the problem. There is still room for improvement in the provision of appropriate harvesting options for cassava worldwide and a more concerted effort from both the government and private sector is vital.

Keywords: cassava, harvesting, mechanised, manual, adoption, improved

1. Introduction

Cassava has become an important food security and the world's third most important crop. The crop is an essential source of food and income throughout the tropics providing livelihood for countless farmers, processors and traders worldwide. Almost 60 percent of world production is concentrated in five countries Nigeria, Brazil, Thailand, Indonesia and the Congo Democratic Republic [1]. In Africa, cassava is the single most important source of dietary energy for a large proportion of the population living in the tropical areas [2]. According to Tufan [3], no other continent depends on cassava to feed as many people as does Africa, where over 500 million people consume it daily.

Harvesting is one of the serious bottlenecks in the cassava production value chain. Manual harvesting is slow and associated with drudgery and high root damage, especially under arid conditions [4]. This situation tends to increase the total cost of production because more farm hands are usually required to harvest in order to meet industrial and local demands coupled with an increase in cassava prices on the market.

Over the years, various mechanised harvesting options have been developed for use in different parts of the world to overcome these challenges. Earlier attempts at mechanising cassava harvesting have been challenged mainly by inappropriate method of planting, field topography and scale of cultivation. A review of various harvesting options for cassava is crucial to ensure proper adaption and adoption of improved harvesting methods applicable to farmers from different parts of the globe.

2. Cassava harvesting

The most difficult operation in cassava production is harvesting [5]. This is so because cassava is a highly perishable crop and begins to deteriorate as early as 1–3 days after harvest. It is therefore important to harvest cassava at the right time and in the proper manner. Harvesting too early results in low yield and poor eating quality; on the other hand, when the roots are left too long in the soil, the central portion becomes woody and inedible. It also ties the land unnecessarily to one crop whilst exposing the roots to pests. Cassava is ready for harvest as soon as there are storage roots large enough to meet the requirements of the consumer, starting from 6 to 7 months after planting (MAP), especially for most of the new cassava cultivars [6]. Matured roots are clustered around the base of the plant and extend about 60 cm on all sides. It is for these roots, which contain from 15–40% starch that the crop is cultivated.

Under the most favourable conditions, yields of fresh roots can reach 90 t/ha while average world yields from mostly subsistence agricultural systems are 10 t/ha [7]. Cassava is traditionally harvested by hand lifting the lower part of stem and pulling the roots out of the ground, then removing them from the base of the plant by hand. The upper parts of the stems with the leaves are usually removed before harvest. Levers and ropes can be used to assist harvesting. A mechanical harvester can also be used. Mechanical harvesters, like those developed in Brazil would grab onto the stem and lift the roots from the ground [8]. Harvesting cassava during relatively dry weather is the best since the soil does not stick to the harvesting implement or roots easily [9].

2.1. Methods of cassava harvesting

Mechanisation in terms of harvesting, like most of the other root crops, is still in the development stage with very few commercial technologies in existence. Development of labour-saving technology for cassava harvesting has become the most critical challenge in the cassava transformation worldwide. Earlier attempts at mechanised harvesting have been affected by constraints such as soil characteristics, nature and size of tubers, depth and width of cluster, and bond between tubers and the soil, leading to high tuber damage. Amponsah et al. [10] stated that farm size and level of root tuber breakage are critical factors that are considered in the selection and adoption of any type of cassava harvesting method. There are basically three cassava harvesting options available to farmers across the globe; manual, semi-manual and mechanised.

2.2. Manual harvesting

This is the traditional method of harvesting cassava using the bare hands with or without the use of indigenous tools such as hoe, cutlass, mattock, earth chisel etc. Usually, these tools are used to dig round the standing stem to facilitate the pulling of the roots from the soil before detaching the uprooted roots from the base of the plant. **Figure 1** shows various manual cassava harvesting options.

Harvesting cassava manually is laborious especially during the dry season when soil moisture is at lower levels. According to Nweke et al. [11], manual harvesting requires about 22–62 man days per hectare.

Manual lifting of cassava with the bare hands requires about 23–47 man h/ha as compared to the use of a hoe which requires between 42 to 51 man h/ha [4]. The use of manual harvesting tools is preferable on relatively dryer (hard) soils, whereas manual uprooting technique is best suited for soils with relatively higher moisture content. However, best efficiency of manual harvesting is achieved when the upper cassava plant biomass is removed or coppiced before harvesting.

2.3. Semi-manual harvesters

Semi-manual harvesters are harvesting aids that usually adopt the lever principle to ensure that little human effort is used in uprooting the cassava. Various harvesting aids can be found in different cassava growing regions across the globe.

The CRI harvester (**Figure 2**) was developed at the CSIR-Crops Research Institute (CRI), Kumasi with the intention of decreasing the toil farmers go through as a result of excessive waist bending when using existing manual harvesting tools. The original design, adopted from the International Institute of Tropical Agriculture (IITA) in Nigeria, has undergone several design modifications to ensure best efficiency is achieved using the implement [12].

The CRI harvester operates according to the "grip and lift" principle and is made up of a frame with a steel plate to which an immovable griping jaw is fixed. A chisel tip serves as a base which allows for lifting of cassava roots from the soil when using the gripping jaw. It also facilitates the uprooting of cassava especially in hard and dry soils by employing the "dig and lift" principle. This comes in handy where the "grip and lift" principle fails. The harvester has

Figure 1. Different manual harvesting options using a hoe (a), bare hands (b), mattock (c), machete (d) and earth chisel (e).

a mechanical advantage of 4.5 when operating under the second class lever principle. With a total weight of 5 kg, even women and children can easily operate and use the tool for harvesting cassava.

Field assessment of the performance of the CRI harvester showed that it is faster harvesting vertically planted cassava though cassava planted slanted offered the least root tuber breakage and drudgery, regardless of cassava variety. **Table 1** presents some performance evaluation results of the CRI harvester according to Amponsah et al. [10].

Figure 2. The CRI harvester in use.

The National Centre for Agricultural Mechanisation (NCAM) in Nigeria also developed and commercialised a semi-mechanised cassava lifter/harvester [13]. The NCAM harvester (**Figure 3**), consists of a frame to which a footboard and immovable griping jaws are attached and a lever (handle) which is hinged to the frame. Both implements have been tested to harvest up to 200 plants per man-hour and can be classified under semi-manual types of cassava harvesters since they require some degree of human effort to be able to use them effectively for harvesting compared to the mechanised types.

The CTCRI cassava harvester (**Figure 4**) was developed at the Central Tuber Crops Research Institute (CTCRI), Kerala, India with the aim of reducing the level of drudgery associated with the use of other manual cassava harvesting tools. The tool, with a mechanical advantage of 3.4 and total weight is 8 kg, operates on the second class lever principle and has a self-tightening mechanism used to grip the cassava stem. The height of the fulcrum at the far end

Parameter	Value
Field capacity (man h/ha)	49.9–156
Root tuber breakage (%)	4.3–19.6
Energy expenditure (W)	470.3–773.7

Table 1. Performance evaluation results of the CRI harvester.

Figure 3. The NCAM harvester.

Figure 4. The CTCRI semi-manual harvester.

of the lever can be adjusted to facilitate uprooting of cassava plants raised on different land preparation methods (flat, mounds or ridges). The CTCRI harvester requires about 16–40 man h/ha and uses 547–639 W of physical energy during cassava harvesting [4].

2.4. Mechanised harvesters

Harvesting cassava mechanically involves the use of a harvesting implement integrally hitched to a tractor to dig out the cassava roots. Manual effort may be needed after cassava uprooting to collect and detach the cassava root tubers. The following field requirements/conditions are also necessary to allow for an optimum mechanical cassava harvesting operation: a field free from hidden obstructions (rocks, roots, stumps etc. down to 40 cm deep) of sizes that can interfere with lifting the tubers; good weed control as weeds block the lifters; Cutting down (coppicing) the cassava plant to a stalk level of about 30 cm prior to harvesting to allow the tractor operator to work in a regular manner. Ridge cultivation of cassava in rows is preferred to facilitate better orientation of stems for tractor operation during harvest.

Mechanised harvesters can be classified into semi-mechanised and fully mechanised. Whereas all processes from digging of roots, lifting of uprooted roots onto soil surface to transport are mechanically done in fully mechanised harvesters, only the root digging process is mechanised in the case of semi-mechanised harvesters.

The digging, lifting and transport of cassava root cluster into a windrow have been demonstrated under Ghanaian condition using a prototype fully mechanised cassava harvester developed at the Leipzig University, Germany [14]. The harvester reduces the heavy physical work involved in manual cassava harvesting using the hoe and cutlass, especially in the dry season. Design goals for the Leipzig mechanical harvester prototype were, cutting of soil, digging of soil, raising of soil containing the cassava root cluster, transporting the cassava root cluster into windrow behind the tractor to ease manual tuber detachment from stem, reducing the number of moving parts, improvement in the flow of soil and residue to prevent blockade and fuel conservation during seedbed preparation for next cropping. The structural arrangement of the harvester consists of a digging share rising into a conical shaped mouldboard between two legs, a frame of digging tool, a stem guiding device, a frame for stem pulling device and hydraulically operated belt pulling elements. The 1 m wide harvester which is a fully mounted implement operates according to the "dig and pull" principle. It cuts and loosens the growth area of the root cluster by two vertical beams, and a share attached to the base plate.

Figure 5 shows the Leipzig mechanical harvester prototype. The cassava root cluster is loosened carefully, lifted to about 20 cm and delivered to the transport unit made of two belts and a set of steel/plastic press rollers. The windrowed root clusters are then detached with hand or cutlass and finally collected. The harvesting process produces a well pulverised field, thus effectively eliminating the tedious and energy intensive conventional primary tillage operation. Additional advantages for using the harvester include, lowering of the total production

Figure 5. The Leipzig mechanical cassava harvester.

cost, increase in labour productivity and considerable decrease in harvesting losses and root damage.

The harvester was introduced into Ghana in 1991. However, field testing only started in 1993. As a result, it could not be evaluated extensively and further investigation on the performance of the harvester was expected to be conducted in other agro-ecological zones of the country. **Table 2** shows the summarised performance evaluation results after testing the Leipzig mechanical cassava harvester prototype on the TMS 30572 cassava variety for some agro-ecological zones in Ghana according to Bobobee et al. [14].

The Latin American and Caribbean Consortium to Support Cassava Research and Development (CLAYUCA) conducted some research on the adaptation and evaluation of semi-mechanised harvesting systems for cassava in Columbia. This evaluation process became important due to the excessive cost of manual harvesting. A semi-mechanised cassava harvester prototype developed in Brazil was imported and its performance was evaluated under specific conditions in the main cassava growing regions of Columbia [15].

The prototype harvester has a front cutting disk that facilitated the harvesting process and was able to work even on dry soils where manual harvesting was not possible [15]. For a smooth operation, however, it required the cutting of cassava stems prior to harvesting to a height of 20–40 cm. **Figure 6** shows the CLAYUCA mechanised harvester model P600.

http://dx.doi.org/10.5772/intechopen.71350

Parameter	Value
Draft requirement (kN)	11.94–16.2
Working depth (cm)	25
Soil moisture content (% d.b.)	3.5–5.8
Soil bulk density (g/cm^3)	1.82
Cone Index (MPa)	0.88–2.5
Average fuel consumption (l/ha)	40.3
Working speed (km/h)	2.4–4.1
Field capacity (ha/h)	0.25–0.38
Tractor power requirement (kW)	55–80

Table 2. Performance evaluation results for the Leipzig mechanical cassava harvester.

Figure 6. The CLAYUCA mechanised harvester model P600.

The technical and performance characteristics of the CLAYUCA harvester prototype is presented in **Table 3**.

The main effect of the use of the harvester is the improvement in the efficiency of labour. Under the traditional system, in which the cassava roots are harvested by hand, a good performance for a worker is around 500 kg roots/day [15]. With the use of the harvester Model P600, CLAYUCA has been able to measure the harvest of around 1100 kg roots/day. In more developed cassava producing systems, such as those found in South Brazil, a good performance using mechanical harvesters is around 1500 kg roots harvested/day. The economic importance of the use of mechanical harvesters is in the reduction in the number of workers that are needed to harvest a cassava field. Ospina et al. [16] reiterated that the introduction of the CLAYUCA harvester prototype allows a reduction of 53% in labour cost for harvesting resulting in a reduction of 43% of the cost of harvest, and a further reduction of 12% of the total production costs.

According to Oni [17], the National Centre for Agricultural Mechanisation (NCAM) in Nigeria developed a mechanised cassava harvester which was adapted for use in most farming communities in Nigeria. The harvester consists of a combination of a standard chisel plough preceding a serrated disc plough, both mounted on a tractor-drawn toolbar. The harvester has a field capacity of 0.8–1.2 ha/h. **Figure 7** shows the NCAM tractor-drawn cassava harvester.

Odigboh and Moreira [18] reported that mechanisation of cassava harvesting has attracted a great deal of research attention but with very modest successes achieved. Catalogues of agricultural machines produced by Brazilian manufacturers contain no cassava harvesters. What exists in Brazil, as elsewhere in the world, are few models of cassava harvesting aids in limited production and on trial use by a few farmers. Also, there are many problems associated with cassava harvesting. Some of these problems are as a result of the serious difficulties created by the random growth patterns of the roots and the equally random branching of the stems. In addition, cassava does not have a specific harvesting season. According to Odigboh and Moreira [18], an effective harvester must therefore be able to operate in the parched hard soils of the dry season, the drenched muddy soils of the tropical rainy season,

Parameter	Value
Working width (m)	2.4
Working depth (cm)	30–40
Harvester weight (kg)	200
Average working speed (km/h)	7
Field capacity (ha/h)	0.63–1.1
Tractor power requirement (kW)	67

Table 3. Performance evaluation results for the CLAYUCA cassava harvester.

Figure 7. NCAM semi-mechanised cassava harvester [15].

as well as in soils the consistencies of which vary between those two extremes. Agbetoye et al. (2000) reported that most of the experimental cassava harvesters in literature are based on the elevator digger principle whereby the share cuts through the soil 0.3–0.4 m deep and 0.7–0.8 m wide and handling about 0.23 m^3 or about 500 kg of soil to harvest a plant. All these unique characteristics must be appropriately considered to design an effective harvester for cassava.

The TEK mechanical cassava harvester was developed and manufactured at the Department of Agricultural and Biosystems Engineering, Kwame Nkrumah University Science and Technology, Kumasi. This harvester was developed after the Leipzig to suit local prevailing field conditions. However, unlike the Leipzig which was fully mounted with a hydraulic transport system, the TEK harvester did not have that. One thing that was evident during the field evaluation of the Leipzig was that most tractors found on farmer's fields were not able to support the hydraulic system of the harvester. This necessitated the disabling of the hydraulic transport system in the design of the TEK mechanised harvester. The TEK cassava harvester (**Figure 8**) basically has the following parts; digger, shakers consisting of a slatted mould conical mouldboard, the linkage points and the vertical support.

The TEK mechanical harvester, though semi-mechanised, is a fully mounted implement which operates according to the 'dig and pull' principle. Having met the necessary field conditions prior to harvest, the implement hitched to the tractor is gently lowered to set the required depth of penetration (depending on root depth of the cassava variety to be harvested). As the digger goes through the soil, the roots are brought onto the surface for collection and detachment facilitated by the inclination of the slatted conical mouldboard (B). Due to the large quantity of soil and trash that is dug out together with the roots, there is often an increase in the resistance behind the tractor leading to increased fuel consumption. When the soil is moist and sticky, the slatted conical mouldboard serves as shakers to sieve the soil clods and reduce adhesion. This helps to accelerate the harvesting process resulting in an increase

A – Beam to which digging unit is attached
B – Conical mouldboard
C – Top link hitching point
D – Digger
E - Vertical support
F – Lower link hitching points
G – Slatted rods for shaking off soil

Figure 8. The TEK mechanical cassava harvester.

in the efficiency of the tractor and harvesting implement. **Table 4** presents the field evaluation results of the TEK mechanised harvester.

An added advantage after mechanical harvesting of cassava is that the land is ploughed for subsequent crop establishment. Only harrowing and ridging may be needed, thus total cost of production for the subsequent season is reduced. Careless use of machinery for harvesting however, can damage tubers, resulting in rapid deterioration that will lower the value of the end product.

Parameter	Value
Working width (m)	1
Working depth (cm)	23–29
Harvester weight (kg)	300
Average working speed (km/h)	5
Field capacity (ha/h)	0.4–0.52
Draft power requirement (kN)	10.33

Table 4. Performance evaluation results for the TEK cassava harvester.

Author details

Shadrack Kwadwo Amponsah[1]*, Ahmad Addo[2] and Byju Gangadharan[3]

*Address all correspondence to: skamponsah@hotmail.com

1 CSIR - Crops Research Institute, Kumasi, Ghana

2 Department of Agricultural and Biosystems Engineering, KNUST, Kumasi, Ghana

3 ICAR – Central Tuber Crops Research Institute, Sreekariyam, Trivandrum, India

References

[1] FAOSTAT. FAO Statistical Yearbook. Rome, Italy: Food and Agricultural Organisation of the United Nations; 2016

[2] Cock JH. Cassava: A basic energy source in the tropics. In: Howeler RH, editor. The Cassava Handbook: A Reference Manual Based on the Asian Regional Cassava Training Course. Thailand: Centro Internacional de Agricultura Tropical (CIAT); 2011. p. 23

[3] Tufan HA. Next Generation Cassava Breeding Project [WWW Document]. 2013. http://www.nextgencassava.org/about.html (accessed 3-31-15)

[4] Amponsah SK, Sheriff JT, Byju G. Comparative evaluation of manual cassava harvesting techniques in Kerala. India. Agric. Eng. Int. CIGR J. 2014;**16**:41-52

[5] Agbetoye LAS. Engineering challenges in developing indigenous machinery for cassava production and processing. In: Proc. Ann. Conf.Of the Nigerian Soc. Eng. (Lagelu 2003). Ibadan, Nigeria. 2003. pp. 80-86

[6] Ekanayake IJ, Osiru DSO, Porto MCM. Agronomy of Cassava. IITA. 1997

[7] USDA, NRCS. Plant guide – cassava: *Manihot esculenta* Crantz. Baton Rouge, Louisiana and Pacific Islands, Mongmong, Guam. Available at: http://plants.usda.gov/plantguide/pdf/cs_maes.pdf: National Plant Data Centre; 2003. (accessed 24-06-2011)

[8] Kuiper L, Ekmekci B, Hamelinck C, Hettinga W, Meyer S, Koop K. Bio-ethanol from Cassava. Ecofys Netherlands BV, 1-38. 2007. Available at: http://www.probos.net/biomassa-upstream/pdf/FinalmeetingEcofys.pdf. (accessed 24-06-2011)

[9] Kouakou J, Nanga SN, Plagne-Ismail C, Pali AM, Ognakossan KE. Cassava Production and Processing. The Netherlands: CTA publications; 2016

[10] Amponsah SK, Berchie JN, Manu-aduening J, Danquah OE, Adu JO, Agyeman A, Bessah E. Performance of an improved manual cassava harvesting tool as influenced by planting position and cassava variety. African J. Agric. Res. 2017;**12**:309-319

[11] Nweke F, Nweke FI, Lynam JK, Spencer DS. The Cassava Transformation. Michigan State University Press; 2002

[12] Amponsah SK. Performance evaluation of the Tek mechanical cassava harvester in three selected locations of Ghana. M.Sc thesis, Agricultural Engineering Dept., Kwame Nkrumah University of Science and Technology, Kumasi. 2011. Available at: http://dspace.knust.edu.gh:8080/jspui/bitstream/123456789/3960/1/Final.pdf. (Accessed: October 14, 2014)

[13] Oni FC, Eneh MCC. Strategies for reducing the cost of production and processing of cassava. Facilitation and Sensitization Workshop for Cassava Producing States under the Presidential Initiatives on Cassava Production and Export. 8th–12th November, 2004. Federal Ministry of Agriculture and Rural Development, Abuja

[14] Bobobee EYH, Okyere JB, Twum A, Neumann RH, Knechtges H. Performance Evaluation of a Mechanical Cassava Harvester. Proceedings of the International Commission of Agricultural Engineering (CIGR). XII World Congress. Ag. Eng' 94. Milano 94. 1994. Report No. 94-D-066

[15] Ospina B, Cadavid LF, García M, Alcalde C. Mechanization of Cassava Production in Colombia, Cassava Research and Development in. Centro Internactional de Agricultura Tropical (CIAT): Asia; 2002

[16] Ospina B, Cadavid LF, Garcia M, Alcalde C. Mechanisation of cassava production in Columbia. Latin American and Caribbean Consortium to Support Cassava Research and Development (CLAYUCA), CIAT. Cali, Columbia: Apartado Aereo. 2007. pp. 67-13

[17] Oni KC. Cassava processing as a tool for employment generation and rural development. In: Workshop on Cassava Research and Development. Abuja, Nigeria: Raw Materials Research and Development Council (RMRDC); 2005

[18] Odigboh EU, Moreira CA. Development of a Complete Cassava Harvester 1 – Conceptualisation. Journal of Agricultural Mechanisation in Asia, Africa and Latin America. 2002;**4**(33):1-7